AF477135

Advances in Laser Materials Processing

Advances in Laser Materials Processing

Editors

Sergey N. Grigoriev
Marina A. Volosova
Anna A. Okunkova

MDPI • Basel • Beijing • Wuhan • Barcelona • Belgrade • Manchester • Tokyo • Cluj • Tianjin

Editors

Sergey N. Grigoriev
Department of High-Efficiency Machining Technologies, Moscow State University of Technology STANKIN, 127055 Moscow, Russia

Marina A. Volosova
Department of High-Efficiency Machining Technologies, Moscow State University of Technology STANKIN, 127055 Moscow, Russia

Anna A. Okunkova
Department of High-Efficiency Machining Technologies, Moscow State University of Technology STANKIN, 127055 Moscow, Russia

Editorial Office
MDPI
St. Alban-Anlage 66
4052 Basel, Switzerland

This is a reprint of articles from the Special Issue published online in the open access journal *Metals* (ISSN 2075-4701) (available at: https://www.mdpi.com/journal/metals/special_issues/advances_laser_materials_processing).

For citation purposes, cite each article independently as indicated on the article page online and as indicated below:

LastName, A.A.; LastName, B.B.; LastName, C.C. Article Title. *Journal Name* **Year**, *Volume Number*, Page Range.

ISBN 978-3-0365-4887-6 (Hbk)
ISBN 978-3-0365-4888-3 (PDF)

Cover image courtesy of Moscow State University of Technology STANKIN, Moscow, Russian Federation.

Contents

About the Editors

Sergey N. Grigoriev

Doctor of Engineering Science, Professor, Head of the Department of High-Efficiency Processing Technologies of Moscow State University of Technology STANKIN, completed his higher education in 1982. The scientific team at the department has received awards from the most prestigious international scientific competitions and exhibitions under the supervision of Dr. of Eng. Sci., Prof. S. N. Grigoriev. The most notable achievements correspond to advances in laser, plasma, and sintering technologies. The researchers and academic staff of the department are multiple winners of grants from the President of the Russian Federation for the leading scientific schools in the field of Engineering and Technical Sciences. Research and academic staff at the department have developed projects in the fields of cutting-tool efficiency, the efficiency of the most advanced and new technologies, adaptive processes, multi-component and nanocomposite structures, and electrophysical and electrochemical processes. Prof. S.N. Grigoriev has initiated and participated in various invited lecturers at conferences. He serves as editor of the most authoritative international scientific journals all over the world and has authored over 180 patents and more than 400 international scientific articles published in journals indexed in the Web of Science and Scopus.

Marina A. Volosova

Cand. of Engineering Science (Ph.D.), Docent, a leading researcher in the Department of High-Efficiency Processing Technologies of Moscow State University of Technology STANKIN, completed her higher education in 2000 at the Moscow State University of Technology STANKIN as a Chartered Engineer of Mechanical Engineering. She was awarded a Ph.D. in Engineering Science with a specialization on 05.02.07—Technology and Equipment for Mechanical and Physical-Technical Processing in 2003 and a Docent degree in 2010. She is a Laureate of the Government of the Russian Federation in Science and Engineering for Young Scientists, winner of many international and national competitions and awards for works that are mostly related to laser, plasma, and sintering processing in the context of ceramic cutting tool coating improvements, and the author of approximately 50 patents and more than 200 international scientific articles in journals indexed in the Web of Science and Scopus.

Anna A. Okunkova

Cand. of Engineering Science (Ph.D.), a senior researcher in the department of High-Efficiency Processing Technologies of Moscow State University of Technology STANKIN, completed her higher education in 2005 at the Nizhniy Novgorod State Technical University as a Chartered Engineer in the Design of Technical and Technological Complexes. In 2009, she was awarded a Ph.D. in Engineering Science with a specialization in Automation and Control Technological Processes and Productions. In 2015, she was honored as a Laureate of the Government of the Russian Federation in Science and Engineering for Young Scientists (N10904) for the work on the mobile laser complexes for the reparation of specialized machines. She has developed 25 patents and published more than 70 international scientific articles in the journals indexed in the Web of Science and Scopus.

Preface to "Advances in Laser Materials Processing"

Innovative laser processing technologies developed in recent years make it possible to solve any engineering problems—from repair technologies and technologies for volumetric growth and improvements in the surface and structure of products from a broad class of materials to micromachining and laser polishing. The traditional scientific approach to improving laser technology and the methods for post-processing formed surfaces and the structure of functional products concerning a wide range of materials has proven its effectiveness in transitioning from laboratory conditions to real production. The high efficiency of this approach can be explained by the fact that, under various operating conditions of a functional product, in all cases, its surface layer is the most loaded, and the service life of the entire machine-building product or mechanism under specific operating conditions depends on its quality and structure. The second important parameter of the viability of a product is its volumetric properties associated with the structure—primarily the volumetric strength and thermal stability in specific conditions.

Many years of experience of the scientific team of the Moscow State Technological University STANKIN in the field of electrophysical processes, as well as world-class scientists in the field of laser materials processing and highly qualified specialists, made it possible to present the results of the latest scientific research in the field of improving laser technologies and post-processing for relevant machinery products as a collection of scientific articles. In addition, the presented achievements contribute to the development of innovative products in the field of laser technologies in machinery and other related industries and present the results of developments in the form of patented methods, techniques, and equipment, ready for complete implementation in enterprises in the context of the novelty for transition to the sixth technology paradigm.

The Special Issue is dedicated to the most recent achievements in the laser processing of various materials, such as cast irons, tool steels, high-entropy alloys, hard-to-remelt materials, cement mortars, and innovative manufacturing and post-processing methods based on a laser.

Sergey N. Grigoriev, Marina A. Volosova, and Anna A. Okunkova
Editors

metals

MDPI

Editorial

Advances in Laser Materials Processing

Sergey N. Grigoriev, Marina A. Volosova and Anna A. Okunkova *

Department of High-Efficiency Machining Technologies, Moscow State University of Technology STANKIN, 127055 Moscow, Russia; s.grigoriev@stankin.ru (S.N.G.); dr.volosova@gmail.com (M.A.V.)
* Correspondence: annaokunkova@rambler.ru; Tel.: +7-909-913-1207

1. Introduction and Scope

Today, laser processing is becoming more and more relevant due to its fast adaptation to the most critical technological tasks, its ability to provide processing in the most rarefied and aggressive mediums (e.g., vacuum conditions), a wide field of potential applications, and green aspects related to the absence of industrial cutting chips and dust. With the development of 3D technologies in production, laser processing has received a new round of interest associated with its abilities in selective high-precision powder melting or sintering [1–4] that are now available even for oxide ceramics without a binder for remelting [5,6]. New technologies and equipment, which improve and modify laser optic parameters [7], contribute to the better absorption of laser energy by metals or powder surface [8] and allow an increase in laser power up to a few kilowatts. This can positively influence the industrial spread of the laser in mass production and advance existing manufacturing methods.

The latest achievements in laser processing have become a relevant topic in the most authoritative scientific journals and conferences over the last half-century. Advances in laser processing have received multiple awards in the most prestigious competitions and exhibitions worldwide and at international scientific events.

This Special Issue is devoted to the most recent achievements in the field of the laser processing of metals and innovative manufacturing methods based on lasers.

2. Contributions

Eleven research articles and reviews are published in the presented Special Issue. The touched upon subjects cover the most critical topics in laser materials processing. Among them are questions regarding:

1. Distributing/redistributing energy in the laser beam spot (round and quadratic) in laser additive manufacturing (experimentally and numerically) and surface treatment for various types of metals alloys, such as cast iron and steel [9–12];
2. Laser surface treatment to repair surface defects and cracks in irons, steels, and composites [13–16];
3. The post-processing (mechanical, heat and plasma treatment) of laser welds and parts produced by laser additive manufacturing [17–19].

    In particular,

- Beam profiling in the laser-additive manufacturing of metals and alloys [9],
- The influence of specific energy on the microstructure and properties of laser cladded high entropy alloy [10];
- Spacing distribution on the surface of gray cast iron by laser remelting using a biomimetic design model [11];
- A square, top-hat (flat-top) shaped intensity distribution in the laser micro polishing tool steels [12];
- Laser scabbling to remove the defect surfaces of cement mortar composites [13];
- Laser melting to bionically repair thermal fatigue cracks in ductile iron [14];
- Biomimetic laser treatment on five types of low and medium carbon steels [15];

Citation: Grigoriev, S.N.; Volosova, M.A.; Okunkova, A.A. Advances in Laser Materials Processing. *Metals* **2022**, *12*, 917. https://doi.org/10.3390/met12060917

Received: 19 May 2022
Accepted: 23 May 2022
Published: 27 May 2022

- The heat treatment and ultrasonic peening of the laser-welds on metals and alloys [17];
- The heat treatment, vibratory tumbling, and ultrasonic cavitation finishing of the anticorrosion steels of the austenite and martensitic class [18].

These subjects can be highlighted as the main, crucial research directions in laser materials processing. The topics of the laser surface treatment of cast irons [16] and defect minimization in the laser powder bed fusion of metals and alloys [19] are reviewed.

Beam profiling or shaping in the laser additive manufacturing of metals and alloys [9] provides the redistribution of the laser energy flux in a laser beam spot. It leads to a decrease in material losses through evaporation by more than 2.5 times when switching from the classical Gaussian mode (TEM_{00}, laser beam spot of 109 µm) to the inverse Gaussian (donut) distribution (airy distribution of the first harmonic, $TEM_{01*} = TEM_{01} + TEM_{10}$, laser beam spot of 310 µm) during the dynamic simulation of a CoCr type of alloy. The calculation of the Péclet number (a similarity criterion characterizing the relationship between convective and molecular processes of heat transfer (convection to diffusion) in a material flow in the liquid phase) shows that the cylindrical (top-hat or flat-top) distribution ($TEM_{FT} = TEM_{01*} + TEM_{00}$ mode, laser beam spot of 210 µm) is effective in a narrow temperature range (oxide ceramics).

Experimental research of a critical laser cladding parameter, such as the specific energy in processing high entropy alloys (HEA, 24.44% of Fe, 26.18% of Co, 22.43% of Cr, Ni in balance) [10], showed significant differences in the microstructure and properties of the coatings. The increase of specific energy plays a positive role in

- The microstructure evolution that is mainly composed of the face-centered and body-centered cubic phases, precipitating a small amount in the Fe-Cr phase and Laves phase;
- Promoting the diffusion of Ti from the substrate of Ti6Al4V to the HEA coating, mainly composed of columnar crystal and shrinkage cavities;
- Subsequently affecting the microhardness of samples up to 1098HV, which is ~200% higher than for the substrate.

In order to improve the wear resistance and thermal fatigue resistance of gray cast iron (3.41% of C, 1.61% of Si, 0.96% of Mn, Fe in balance) surfaces, the original uniform distribution laser melting strengthening model was designed as a nonuniform distribution model [11]. The adjacent melting zones affect each other, resulting in heat preservation and a tempering effect; simultaneously, the area of the melting zone and grain size increase, and the hardness decreases from 765–820 HV to 570–620 HV. The phase transformation law of the microstructure in the melting zone was determined at the later stage of thermal fatigue. The maximum residual tensile stress was 204 MPa in the melting zone and 103.4 MPa in the phase transformation zone. The surface wear morphology of different unit combinations after 600 thermal fatigue cycles showed that different degrees of wear appeared on the material surface, caused by microcrack peeling with an increase in fatigue times.

The influence of three different quadratic top-hat (flat-top) laser beam sizes (100 µm, 200 µm, 400 µm side length) and fluences from 3.0 up to 12.0 J/cm^2 on the resulting surface topography and roughness of the 1.2379 tool steel product (AISI D2, 1.56% C, 0.4% Si, 11.86% Cr, 0.83% Mo, Fe in balance) in laser micro polishing showed [12] that chromium carbides are the source of undesired surface and dimples. The laser process parameters were: the frequency of 20 kHz, scanning speed of 200, 400, 800 mm/s, track offset of 10, 20, 40 µm, and laser power of 8–24, 32–96, 128–384 W, correspondingly. Particularly for high laser fluences, a noticeable stripe structure was observed, which is typically a continuous remelting characteristic. The micro-roughness was significantly reduced from 32 to 3 nm when the macro-roughness was increased from 0.002 to 0.200 µm. The results show that smaller laser polishing fluences are required for larger laser beam dimensions. Additionally, the same or even a lower surface roughness and less undesired surface features were created for larger laser beam dimensions. Therefore, a potential path for the industrial applications of laser micro polishing, where area rates of up to several m^2/min might be achievable with commercially available laser beam sources.

In laser scabbling with a pulsed laser with a power density of 1.45×10^5 W/mm^2, the effects of silica sand proportion in color changing and penetration depth samples were studied for the five types of cement mortar samples, mainly composed of CaO and SiO$_2$ [13]. Increasing the silica sand proportion resulted in a decrease in scabbling penetration depth and fewer surface cracks on the top and bottom surfaces. The evaporation of material was the dominant mechanism of this scabbling due to the high-power density laser. The chemical changes in the cement mortar were an increase in Si and a decrease in Ca. The difference is explained by the boiling point of silicon dioxide (3220 K) being higher than for calcium oxide (3120 K).

The thermal fatigue cracking of ductile iron (3.65% C, 2.42% Si, 0.60% Mn, Fe in balance) machine parts can be a severe problem in abrasive wear conditions when the presence of graphite in the material complicates the repair of crack defects [14]. A novel method for remanufacturing ductile iron brake discs based on coupled bionics to repair thermal fatigue cracks discontinuously, using bioinspired crack blocking units fabricated by laser remelting, showed that the crack could be fully closed at a laser energy of 165.6^{+19}_{-15} J/mm^2 with a pulse duration of 8 ms, an electric current of 155 A, laser beam spot of 1 mm, and laser power of 347.2 W to a sufficient depth of 0.59 mm. The microhardness was 680HV0.2, comparing 298HV0.2 of the untreated sample. The sample treated at the mentioned energy level exhibited the highest tensile force of 40.68 kN, which was 37.11% higher than the unrepaired one. After 2000 thermal fatigue cycles, the crack width of the unrepaired specimen increased by 499.21 μm, while the crack width of the repaired specimen increased between 118.31 and 412.34 μm. The result of 118.31 μm was shown at the laser energy level of 165.6^{+19}_{-15} J/mm^2, which was 23.70% of the unrepaired sample. Research has shown the beneficial effects of reducing the spacing of units from 7 to 3 mm by inhibiting thermal fatigue crack propagation from 150.62 to 57.68 μm, which was 61.70% smaller than that of the unrepaired sample.

Five kinds of low and medium carbon steels with different carbon element contents (Steel 1—0.15%, Steel 2—0.25%, Steel 3—0.37%, Steel 4—0.45%, Steel 5—0.58%) were studied by laser remelting [15]. Laser process parameters were: an energy level of 16.88 J, pulse a duration of 8 ms, a frequency of 5 Hz, a scanning speed of 1 mm/s, and a beam diameter of 1.59 mm. Compared with the untreated samples, when the carbon content is 0.15–0.45%, the tensile strength of the laser biomimetic samples of Steel 2 is 532 MPa, which is higher than for the untreated samples (440 MPa) by 20.91%. For other groups of steels, the difference ranged from 3.14 to 1.26%, or even to 5.21% for Steel 5. With an increase in carbon content, the tensile strength increases first and then decreases, while the plasticity of the biomimetic samples decreases continuously from 19 to 2% compared with untreated samples (from 39 to 15%). The difference in plasticity degree between untreated and treated samples for low carbon samples (Steel 1) was 34.48% and for medium carbon samples (Steel 5) was 86.67%. The bionic samples have better wear resistance than that of the untreated samples. For bionic specimens with different carbon elements, wear resistance increases with an increase of carbon element content. The difference in weight loss reduction percentage was 31.01% for Steel 1, 31.72% for Steel 2, 36.08% for Steel 3, 48% for Steel 4, and 67.76% for Steel 5.

The review discusses the main experimental aspects of the laser surface treatment of four cast-iron groups: gray (lamellar) cast, pearlitic ductile (nodular), austempered ductile, and ferritic ductile iron [16]. The review summarizes the typical laser types used in surface treatments, i.e., CO$_2$, Nd:YAG, or fiber (diode); the coating is graphite or manganese/zinc phosphate; shielding gas is argon, nitrogen, or helium; the power distribution is Gaussian, uniform; and the spot geometry is circular, elliptical, or rectangular. The use of diode and Nd:YAG lasers without coatings allows for greater energy absorption than when using conventional CO$_2$ lasers, which is responsible for a cooling rate that ensures the formation of a more resistant martensitic structure. Lasers with circular geometry and Gaussian energy distribution generate a parabolic heat-affected zone in the transverse section, where properties are not uniform within a constant layer depth. The heat-affected zone's hardness is 700–800 HV, which decreases as the fraction of retained austenite increases in favor of the martensitic phase. The compressive residual stresses in the hardened zone are explained by the volumetric increase associated with the transformation from austenite to harder martensite. Single-pass with rectangular-shaped and uniform energy distribution lasers and adjacent discrete laser spots can be used to avoid a tempering effect of overlapping that reduces the austempered ductile iron wear resistance by 10–100 times. Ductile irons and austempered ductile irons present lower wear damage under comparable conditions than gray cast irons since graphite flakes act as stress raisers, favoring crack nucleation and growth during friction, as in dry sliding tests with a SiC platform (frequency of 2.5 Hz, a load of 5 kg). Regardless of the initial microstructure of the cast iron, linear energy is the critical parameter since it considers the combined effect of experimental parameters, such as laser power, absorption layer thickness, and scanning velocity. It is suggested to apply surface hardening without melting on ductile irons and austempered ductile irons to achieve higher wear resistance because of the residual stresses created during phase transformations (compressive in laser surface hardening and tensile in laser surface melting).

The effect of laser offset and defocusing on microstructure, geometry, and mechanical property responses of 2 mm-thick dissimilar AA6061/Ti-6Al-4V laser welds were studied in [17]. In order to reduce residual stresses, the joints were both stress relaxation heat-treated (at 530 °C for 2 h followed by air cooling) and mechanically treated by ultrasonic peening (for 1 min/side, the diameter of the peening needle was 3 mm, and the vibration frequency was 20 kHz). The welds microstructure was martensitic in the Ti-6Al-4V fusion zone, columnar dendritic in the AA6061 fusion zone, and partially martensitic in the Ti-6Al-4V heat-affected zone. Intermetallic compounds of the Al–Ti system were detected at the AA6061/Ti-6Al-4V interface and in the aluminum fusion zone. Both negative defocusing and a higher laser offset of 0.3 mm compared to 0.1 mm improved the tensile strength of the welds (from 110.2 ± 6 to 158.1 ± 8 MPa or by 43.6% for top weld and from 100.0 ± 3 to 172.7 ± 8 or by 72% for bottom weld), mainly by reducing the amount of brittle intermetallic compounds. The heat treatment, leading to the aging of the martensite and the increasing the intermetallic compound size, reduced the tensile strength and ductility of the joints. The average growth of the intermetallic layer was 2.60 ± 0.6 μm and 1.30 ± 0.3 μm for 0.3 mm-offset top and bottom joints correspondingly. On the contrary, for dissimilar Al–Ti welds, the mechanical treatment effectively increased joint ductility and corrosion resistance in the 3.5% NaCl solution (corrosion current density for 0.3 mm-offset bottom joint was reduced from 16.12 to 9.28 $\mu A \chi \mu^{-2}$).

The samples produced by laser additive manufacturing of two types of anticorrosion steels—20kH13 (X20Cr13, 0.16–0.25% of C, 12–14% of Cr, powder fraction of ~40 μm) and 12kH18N9T (X10CrNiTi18-10, $\leq$0.12% of C, 17–19% of Cr, 8–9.5% of Ni, $\leq$2.0% of Mn, powder fraction of 20–63 μm) steels—of the martensitic and austenitic class were subjected to cavitation abrasive finishing and vibration tumbling to research the various effects on surface quality and physical and mechanical properties [18]. The laser process parameters were a laser power of 80 and 100 W and a scanning speed of 390 and 100 mm/s, for X20Cr13 and X10CrNiTi18-10 steels, respectively, and a layer thickness of 20 μm. The roughness parameter Ra was reduced by 4.2 times for the X20Cr13 sample after cavitation-abrasive finishing when the roughness parameter Ra for the X10CrNiTi18-10 sample was reduced by 2.8 times after vibratory tumbling. The tensile strength was increased by 15% for X20Cr13 steel (1584.09 ± 3.58 MPa for additively manufactured sample comparing 1485 MPa of cast sample after quenching and low tempering) and 20% for X10CrNiTi18-10 steel (657.03 ± 3.32 MPa for additively manufactured sample comparing 540 MPa of cast sample after quenching at 1050–1100 °C with cooling in water) than for cast and traditionally heat-treated cast samples. The wear resistance of 20kH13 (X20Cr13) steel correlated with measured hardness (44.2 HRC_z/46.2 HRC_{xy} for tempering at 240 °C in air, 38.7 HRC_z/39.1 HRC_{xy} for tempering at 680 °C in oil, 33.4 HRC_z/35.8 HRC_{xy} for annealing at 760 °C in air compared with 43.0 HRC and 22.3 HRC of cast samples after quenching at

1030 °C and tempering at 240 °C in air, 680 °C in oil, correspondingly) and decreased with an increase in tempering temperatures.

The compressive review [19] observes the defects in the additively manufactured machinery products of the anticorrosion steels of the martensite and austenite classes; difficult to process materials such as pure titanium, nickel, and their alloys; super and high entropy alloys; and triple fusions. Studies were conducted on the structural defects observed in such products to improve their quality by eliminating residual stress, reducing porosity, and improving surface roughness. Electrophysical and electrochemical treatment methods (ultrasound, plasma, laser, spark treatment, induction cleaning, redox annealing, gas flame, plasma beam, and plasma spark treatment) for removing oxide phase formed during the melting and remelting of deposed tracks in layers are considered. Surface plasma cleaning methods demonstrate their ability to remove it, smooth the surface, and recrystallize (grain size decrease) by heat treatment in metal products. The proposed type of cleaning can be used for laser powder bed fused products to provide conditions for reliable product growing that can be even more durable in extreme conditions. Special attention is focused on the atmospheric plasma sources based on a dielectric barrier and other discharges as a part of a production setup that presents the critical value of the conducted review in the context of the novelty for transition to the sixth technology paradigm associated with the Kondratieff's waves [20–22].

3. Conclusions and Outlook

Topics such as various types of laser processing, including additive manufacturing, ablation, polishing, and micromachining, are covered by this Special Issue, presenting recent developments and achievements in the laser processing of traditional and the most-advanced metal-based materials. However, it should be noted that there are still many issues related to transferring some of the most outstanding ideas to applications in real production conditions. The Guest Editors of the Special Issue hope that the most outstanding results will contribute to and accelerate the switch to the sixth technological paradigm in the near future.

Funding: The work was supported by the state assignment of the Ministry of Science and Higher Education of the Russian Federation, Project No. 0707-2020-0025.

Acknowledgments: The Guest Editors highly appreciate the high requirements for the quality of presentation, the scientifically valuable content of the presented papers, and the kind efforts of the reviewers, editors, and assistants in contributing to this Special Issue. Many thanks to the Metals Editorial Office and personally to Toliver Guo, Assistant Editor, for his kind help and assistance in publishing the most valuable research results.

Conflicts of Interest: The authors declare no conflict of interest.

References

1. Doubenskaia, M.; Pavlov, M.; Grigoriev, S.; Tikhonova, E.; Smurov, I. Comprehensive Optical Monitoring of Selective Laser Melting. *J. Laser Micro Nanoeng.* **2012**, *7*, 236–243. [CrossRef]
2. Yadroitsev, I.; Bertrand, P.H.; Antonenkova, G.; Grigoriev, S.; Smurov, I. Use of track/layer morphology to develop functional parts by selectivelaser melting. *J. Laser Appl.* **2013**, *25*, 5–052003. [CrossRef]
3. Smurov, I.; Doubenskaia, M.; Grigoriev, S.; Nazarov, A. Optical Monitoring in Laser Cladding of Ti6Al4V. *J. Spray Tech.* **2012**, *21*, 1357–1362. [CrossRef]
4. Kotoban, D.; Grigoriev, S.; Okunkova, A.; Sova, A. Influence of a shape of single track on deposition efficiency of 316L stainless steel powder in cold spray. *Surf. Coat. Technol.* **2017**, *309*, 951–958. [CrossRef]
5. Khmyrov, R.S.; Grigoriev, S.N.; Okunkova, A.A.; Gusarov, A.V. On the possibility of selective laser melting of quartz glass. *Phys. Procedia* **2014**, *56*, 345–356. [CrossRef]
6. Khmyrov, R.S.; Protasov, C.E.; Grigoriev, S.N.; Gusarov, A.V. Crack-free selective laser melting of silica glass: Single beads and monolayers on the substrate of the same material. *Int. J. Adv. Manuf. Technol.* **2016**, *85*, 1461–1469. [CrossRef]
7. Gusarov, A.V.; Grigoriev, S.N.; Volosova, M.A.; Melnik, Y.A.; Laskin, A.; Kotoban, D.V.; Okunkova, A.A. On productivity of laser additive manufacturing. *J. Mater. Process. Technol.* **2018**, *261*, 213–232. [CrossRef]

8. Grigoriev, S.; Peretyagin, P.; Smirnov, A.; Solis, W.; Diaz, L.A.; Fernandez, A.; Torrecillas, R. Effect of graphene addition on the mechanical and electrical properties of Al2O3 -SiCw ceramics. *J. Eur. Ceram. Soc.* **2017**, *37*, 2473–2479. [CrossRef]
9. Grigoriev, S.N.; Gusarov, A.V.; Metel, A.S.; Tarasova, T.V.; Volosova, M.A.; Okunkova, A.A.; Gusev, A.S. Beam Shaping in Laser Powder Bed Fusion: Péclet Number and Dynamic Simulation. *Metals* **2022**, *12*, 722. [CrossRef]
10. Wang, L.; Gao, Z.; Wu, M.; Weng, F.; Liu, T.; Zhan, X. Influence of Specific Energy on Microstructure and Properties of Laser Cladded FeCoCrNi High Entropy Alloy. *Metals* **2020**, *10*, 1464. [CrossRef]
11. Yang, H.; Zhou, T.; Wang, Q.; Zhou, H. Effects of Laser Melting Distribution on Wear Resistance and Fatigue Resistance of Gray Cast Iron. *Metals* **2020**, *10*, 1257. [CrossRef]
12. Temmler, A.; Cortina, M.; Ross, I.; Küpper, M.E.; Rittinghaus, S.-K. Laser Micro Polishing of Tool Steel 1.2379 (AISI D2): Influence of Intensity Distribution, Laser Beam Size, and Fluence on Surface Roughness and Area Rate. *Metals* **2021**, *11*, 1445. [CrossRef]
13. Huynh, T.-V.; Seo, Y.; Lee, D. The Effect of Silica Sand Proportion in Laser Scabbling Process on Cement Mortar. *Metals* **2021**, *11*, 1914. [CrossRef]
14. Ma, S.; Zhou, T.; Zhou, H.; Chang, G.; Zhi, B.; Wang, S. Bionic Repair of Thermal Fatigue Cracks in Ductile Iron by Laser Melting with Different Laser Parameters. *Metals* **2020**, *10*, 101. [CrossRef]
15. Chang, G.; Zhou, T.; Zhou, H.; Zhang, P.; Ma, S.; Zhi, B.; Wang, S. Effect of Composition on the Mechanical Properties and Wear Resistance of Low and Medium Carbon Steels with a Biomimetic Non-Smooth Surface Processed by Laser Remelting. *Metals* **2020**, *10*, 37. [CrossRef]
16. Catalán, N.; Ramos-Moore, E.; Boccardo, A.; Celentano, D. Surface Laser Treatment of Cast Irons: A Review. *Metals* **2022**, *12*, 562. [CrossRef]
17. Leo, P.; D'Ostuni, S.; Nobile, R.; Mele, C.; Tarantino, A.; Casalino, G. Analysis of the Process Parameters, Post-Weld Heat Treatment and Peening Effects on Microstructure and Mechanical Performance of Ti–Al Dissimilar Laser Weldings. *Metals* **2021**, *11*, 1257. [CrossRef]
18. Grigoriev, S.N.; Metel, A.S.; Tarasova, T.V.; Filatova, A.A.; Sundukov, S.K.; Volosova, M.A.; Okunkova, A.A.; Melnik, Y.A.; Podrabinnik, P.A. Effect of Cavitation Erosion Wear, Vibration Tumbling, and Heat Treatment on Additively Manufactured Surface Quality and Properties. *Metals* **2020**, *10*, 1540. [CrossRef]
19. Okunkova, A.A.; Shekhtman, S.R.; Metel, A.S.; Suhova, N.A.; Fedorov, S.V.; Volosova, M.A.; Grigoriev, S.N. On Defect Minimization Caused by Oxide Phase Formation in Laser Powder Bed Fusion. *Metals* **2022**, *12*, 760. [CrossRef]
20. Glaziev, S.Y. The discovery of regularities of change of technological orders in the central economics and mathematics institute of the soviet academy of sciences. *Econ. Math. Methods* **2018**, *54*, 17–30. [CrossRef]
21. Korotayev, A.V.; Tsirel, S.V. A spectral analysis of world GDP dynamics: Kondratiev waves, Kuznets swings, Juglar and Kitchin cycles in global economic development, and the 2008–2009 economic crisis. *Struct. Dyn.* **2010**, *4*, 3–57. [CrossRef]
22. Perez, C. Technological revolutions and techno-economic paradigms. *Camb. J. Econ.* **2010**, *34*, 185–202. [CrossRef]

 *metals*

Review

On Defect Minimization Caused by Oxide Phase Formation in Laser Powder Bed Fusion

Anna A. Okunkova *, Semen R. Shekhtman, Alexander S. Metel, Nadegda A. Suhova, Sergey V. Fedorov, Marina A. Volosova and Sergey N. Grigoriev

Department of High-Efficiency Processing Technologies, Moscow State University of Technology STANKIN, 127055 Moscow, Russia; s.shekhtman@stankin.ru (S.R.S.); a.metel@stankin.ru (A.S.M.); n.suhova@stankin.ru (N.A.S.); sv.fedorov@stankin.ru (S.V.F.); m.volosova@stankin.ru (M.A.V.); s.grigoriev@stankin.ru (S.N.G.)
* Correspondence: a.okunkova@stankin.ru; Tel.: +7-909-913-12-07

Abstract: The article is devoted to the compressive review of the defects observed in the products of the machinery usage made mainly of anti-corrosion steels of the martensite-austenite group, difficult to process materials such as pure titanium, nickel, and their alloys, super and high entropy alloys and triple fusions produced by laser additive manufacturing, particularly the laser powder bed fusion. Studies were conducted on the structural defects observed in such products to improve their quality in the context of residual stress elimination, porosity reduction, and surface roughness improvement. Electrophysical and electrochemical treatment methods of removing oxide phase formation during melting and remelting of deposed tracks in layers are considered (such as ultrasound, plasma, laser, spark treatment, induction cleaning, redox annealing, gas–flame, plasma–beam, plasma–spark treatment). Types of pollution (physical and chemical) and cleaning methods, particularly plasma-based methods for oxide phase removing, are classified. A compressive comparison of low- and high-pressure plasma sources is provided. Special attention is focused on the atmospheric plasma sources based on a dielectric barrier and other discharges as a part of a production setup that presents the critical value of the conducted review in the context of the novelty for transition to the sixth technology paradigm associated with the Kondratieff's waves.

Keywords: surface cleaning; laser powder bed fusion; selective laser melting; atmospheric plasma sources; dielectric barrier discharge; nickel alloy; titanium alloy; anticorrosion steel

Citation: Okunkova, A.A.; Shekhtman, S.R.; Metel, A.S.; Suhova, N.A.; Fedorov, S.V.; Volosova, M.A.; Grigoriev, S.N. On Defect Minimization Caused by Oxide Phase Formation in Laser Powder Bed Fusion. *Metals* **2022**, *12*, 760. https://doi.org/10.3390/met12050760

Academic Editors: Atef Saad Hamada and Pasquale Cavaliere

Received: 22 February 2022
Accepted: 19 April 2022
Published: 28 April 2022

Publisher's Note: MDPI stays neutral with regard to jurisdictional claims in published maps and institutional affiliations.

1. Introduction

Additive technologies for the production of products are the primary trend of recent years in power engineering, automotive, mechanical engineering, biomedical engineering, aerospace, and defense industries [1–4]. Additive technologies fundamentally change the production process by layer-by-layer growing solid according to a digital 3D model [5–9]. The introduction of additive technologies provides ample opportunities to manufacture complex profile products with high accuracy, up to ±0.05 mm, depending on the used modes [10–12]. The technology also allows working with non-standard materials up to oxide ceramics and glasses [13–15] and using the technology as thermal post-processing of three-dimensional complex-shaped products made by other methods [16].

Increasing additive production's mobility and flexibility have modified the digital system of technological production preparation. It has moved it into a virtual environment allowing a quick transition to the production of new products without developing the specific technological documentation and tooling for each specific product that usually hampers rapid technical preparation for the production of new products in a short time (work preparation process can be reduced from traditional 6–12 to 1–2 months for engineering products) [17,18]. Digitalizing the production causes traditionally as well the ability to reduce the impact of the human factor on the quality of the resulting products [19] and reduce the duration of design and technological preparation on extra tooling of the first and second order (for example, stamps and dies and profile cutters for their production) [20,21].

The possibility of manufacturing parts that are not inferior and sometimes even exceed the physical and mechanical properties of parts obtained by traditional methods can be named as one of the known advantages that extends the product's life in the conditions of extreme exploitation (cyclic thermal or mechanical loads) [22,23]. An ability to manufacture parts of complex configurations with internal structures from bimetals makes additive manufacturing outstanding from traditional shaping [24,25] and allows reducing the grown object weight by internal cavities forming. High material utilization rate due to the production of workpieces with a minimum allowance for subsequent machining and the recovery of up to 99% of uncured material for reuse in the production cycle increases production efficiency [26]. However, it should be noted that implementing traditional machining methods leads to the loss of raw materials that can reach up to 80–85% in the production of dies for injection molds that require the massive volume milling of the internal cavity of workpieces [27].

Among the most striking shortcomings, scientists identify the high costs for equipment and electric power, powder material of specific parameters, insufficient rates of newly developed material introduction, and the inability of quick technology adoption for these materials in place [4,28,29]. Among others, it can be named high post-processing costs for small production volumes and relatively low productivity for growing large-sized solids (with overall dimensions of more than 100 mm can take up to a few days) [30]. Growing large-sized parts is also associated with a deviation in accuracy that grows from the centrum of the part profile in plane to its periphery. The shape of the laser beam spot is an axisymmetric circle closer to the center when it looks like an ellipse-shaped with a more blurred edge closer to the periphery that is especially noticeable on the parts with an overall size of more than 300 mm in-plane [31,32]. Insufficient quality of single-track formation influences the quality of the part dramatically and leads to the deviations of the grown solid [33,34].

Another known advantage is that powder material reusing leads to deterioration of powder properties during re-exploitation due to the energy excess in the melt pool, ejecting partly fused particles out of the processing area [35,36]. The absence of scientifically based recommendations on the choice of LPBF factors for producing the parts from known but still not adopted for LPBF specifics steels and alloys, non-standard alloys, new metal-based materials, superalloys, high-entropy alloys, triple fusions, etc. can hampers development of the technology and transition to the next technological paradigm [37–40].

Despite all mentioned shortcomings, LPBF is considered economically feasible when shaping is impossible by traditional methods and when time on tooling production slows down the production of a prototype, and also in the case of the low utilization rate of high-cost processed materials and the production of high-tech, small-scale or personalized products [41–43] made of a broad range of engineering materials (Table 1).

Table 1. The most spread engineering materials for LPBF.

Material	References
Technical ceramics	[44]
Tool steels	[45]
Anti-corrosion steels	[46,47]
Gradient materials (anti-corrosion steel/carbides of the group 4, 5 and 6 transition metals (with the exception of chromium)	[48]
Al and Al-based alloys	[1,23,49]
Ni and Ti alloys	[50,51]
CoCr alloys	[52,53]
High reflective and thermo conductive metals and alloys of the copper subgroup	[54,55]

LPBF is used in industry to produce complex-profile parts of assembly units and assemblies, non-separable multi-element assemblies [56], dies and mold parts [57], forming inserts for chill casting [58], a wide range of engineering scaffolds [59,60].

Formed solidification defects include discontinuities and microstructural imperfections in the fabricated components that are still inevitable [61,62]. The study of the grown material defects showed the formation of such defects as intertrack lack of fusions, localized brittle zones, and grain coarsening in the heat-affected zones, leading to anisotropic ductility behavior along with layer deposition and growing solid directions. This review paper provides the associated defects of grown material structures and summarizes technologies of surface cleaning being deployed to mitigate the defects such as residual stresses, pores, oxide phase formation, and improving surface roughness.

The work is mainly devoted to the production of engineering products produced if a wide range of metallic materials to reduce their labor intensity and, in prospect, enlarge their exploitation properties for working in cyclic loads, in particularly:

- a quarter-turn lock mechanism of the aircraft that includes a pin, washer, and sleeve made of X20Cr13 (AISI 420) steel with a diameter of 11 mm, a height of 7 mm, complex-shaped, where the traditional technology is rather laborious and complicated;
- an air intake grille module made of X10CrNiTi18-10 (AISI 321) steel, which is an element for protecting the air intake duct from the objects entering it and is an obstacle to the air intake to the engine with overall dimensions of 180 mm $\times$ 100 mm $\times$ 30 mm with the minimum thickness of the inclined by 0.5–1.0° walls of 0.3 mm; traditionally the module is characterized by high labor intensity, including following operational steps such as cutting, bending, manual assembly of almost seventy parts, welding, and soldering.

The material of the washer should be wear-resistant, the material of which should differ in strength from the lock pin material by 20%, where the strength is not less than 1300 MPa, hardness is not less than 42 HRC, the density is not less than 7.7 g·cm^{-3}, arithmetic mean deviation R_a is less than 3.2 μm. Another airplane part should be produced with tensile strength not less than a standard semi-finished product with a density of less than 7.9 g·cm^{-3}, and arithmetic mean deviation R_a of less than 6.3 μm. LPBF of these parts takes the production to a new level [63–65].

Therefore, the task of studying the relationship between the LPBF factors and their influence on fused material structural defects, understanding the mechanism of defect formation, design and development of the specific atmospheric plasma source based on a dielectric barrier discharge for LPBF product surface cleaning is relevant.

The following tasks are defined to achieve minimizing structural defects of LPBF products:

- Study the formation mechanism of an undesirable oxide phase (defects);
- Conduct analysis of methods for removing the oxide phase by atmospheric plasma sources from the surface after exposure to a laser beam;
- Highlight and systematize an atmospheric plasma source principles based on the dielectric barrier and other discharges as part of a technological installation.

2. Study of Formation Mechanism of Oxide Phase (Defects)

2.1. Study of the Effect of Process Parameters on Defects in LPBF

Currently, additive technologies are implemented based on light-beam and electron-beam energy sources. However, today's most popular additive technologies based on laser radiation are related to melting, sintering, and direct metal sintering principles. When implementing the technology of selective laser sintering, the compaction of the layer of powder material occurs due to solid-phase sintering. In the basis of direct laser sintering of metals, densification occurs according to the mechanism of liquid-phase sintering due to the melting of a fusible component in a powder mixture. In the basis of the laser powder bed fusion method, the compaction of the powder material is carried out due to the complete melting and spreading of the melt. The structure of the product obtained by selective laser melting is shown in Figure 1 based on some findings in [33,34].

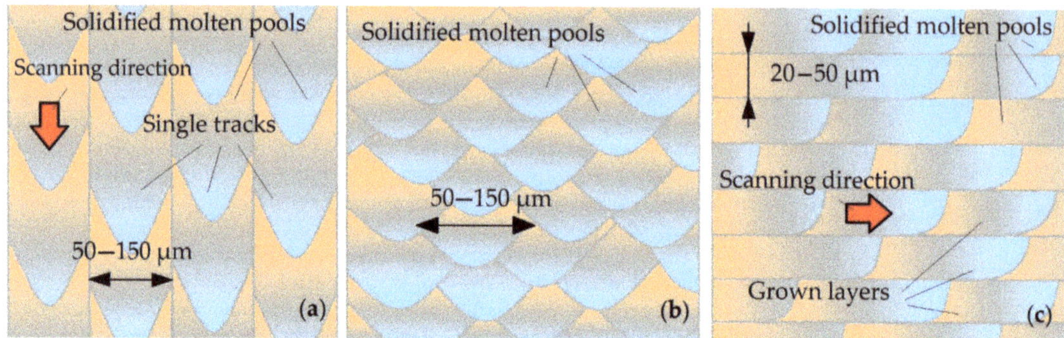

Figure 1. The schematic structure of the metal product obtained by LPBF: (**a**) Top view; (**b**) Cross section; (**c**) Longitudinal section.

Obtaining a surface of the required quality is one of the most critical issues of LPBF. However, the lack of data on the modes of laser powder melting and the powder materials themselves necessitates multiple research projects in this area [66–72]. The LPBF process is multifactorial, and a whole set of factors influences the formation of defects [33,34,71].

Further improvement of additive installations is associated with using a more powerful laser generator, a smaller diameter of the focusing spot, and a thinner layer of powder, which made it possible to use LPBF to obtain products from various metals and alloys of the required quality level.

With LPBF materials such as aluminum and copper subgroup, an important issue is their high reflectivity [1,23,49,54,55], which determines the need for a powerful laser system. However, increasing the power of the laser beam can adversely affect the dimensional accuracy of the product because if heated excessively, the powder material will melt and sinter outside the laser spot due to heat transfer. In addition, the high power of the laser generator can lead to a change in the chemical composition because of metal evaporation, which is typical for alloys containing low-melting components and having high vapor pressure.

One of the important conditions in the implementation of the LPBF is creating a protective environment that prevents the oxidation of the powder. The composition of the gaseous medium in the working chamber is an important technological parameter of the LPBF. It is known that the absorption of atmospheric gases during metal melting negatively affects its physical and mechanical properties. Therefore, a protective environment of inert gases (N_2 or Ar) is created and maintained in the working chamber. However, inert nitrogen gas as a shielding gas is limited because nitrides may form (e.g., AlN, TiN in the manufacture of aluminum and titanium alloy products), which leads to a decrease in the material's ductility. Furthermore, it has been established that the flow rate of the working gas and the direction of its flows affect the degree of porosity in titanium alloys synthesized using the LPBF technology due to the possible transfer of oxide inclusions formed during powder melting to the scanning zone [1,23,51,73,74].

Practice shows that working with technologically difficult materials such as titanium and nickel alloys leads to the formation of residual stresses, leading to warping of parts and even cracks [50,51,75,76]. Compared with stainless steel, aluminum powders have a higher powder reflectivity (over 91% for aluminum) for laser radiation and higher thermal conductivity, making the process more difficult. Oxide formation on particles when using aluminum powders can lead to defects due to the ingress of oxide films into the alloy since aluminum oxide has a higher melting point (2072 °C) than pure metal (660.3 °C). The upper oxide film from the melt pool evaporates under the action of the laser beam, whereas the oxide films located deeper in the melt are untouched and remain inside, representing defective zones. Finally, the hydrogen content of the powder can cause pores in the part if the melt pool crystallizes faster than the gas evaporates.

The quality of powder materials is characterized by granulomorphometric properties (sphericity, dispersion), geometric dimensions, and physicochemical characteristics. The study of the effect of particle size on the properties of finished products showed that in order to obtain high-quality products without porosity, the presence of small particles is necessary. During laser exposure, small particles are first melted, thereby creating favorable conditions for melting larger particles. In addition, the spherical particles improve the fluidity of the powder and can be packed more densely. A wide particle size distribution allows for a denser arrangement of particles. A uniform layer is formed with sufficient fluidity of the powder. The size distribution of particles affects the mode factors that differ for large and small particles.

An important characteristic of powders is also the structure of their constituent particles, which can be both compact and porous. In this case, the pores can access the surface (open porosity) or be closed inside the particle (closed porosity).

Used metal powders must meet the following requirements [77,78]:

- homogeneous chemical composition;
- sphericity of powder particles with a shape factor from 1.0 to 2.0 (sphericity guarantees high fluidity and packing density, which leads to rapid and reproducible distribution of powder layers);
- a narrow and uniform range of particle size distribution with an average value of 40 to 75 μm (the content of particles whose size is larger than the allowable or irregularly shaped particles can cause defects in the finished part).

Powders that meet these requirements have:

- Low coefficient of friction between particles;
- Satisfactory fluidity;
- Increased bulk density;
- Satisfactory density after shaking.

Such requirements allow uninterrupted powder supply and its deposition in thin layers in the LPBF. Depending on the required level of quality, appropriate post-processing is required [79–81].

The multifactorial nature of the LPBF, the need to determine the optimal modes of formation of products of the established quality level, and when using new alloys and materials hinder the widespread use of this method. Quality management of products obtained by LPBF technology is based on understanding the mechanism of defect formation.

2.2. Study of the Mechanism of Formation of Structural and Surface Defects

An analysis of the literature sources has shown that, to date, studies of the quantitative correlation of the formation of defects with process parameters are of a disparate nature. The main defects of LPBF technology include cracks, delamination of parts/cracking and warping, porosity, the density of the part, reduced plasticity due to the presence of residual stresses, surface roughness (Table 2). The table is suitable for nickel and titanium alloys (including high-entropy alloys), anti-corrosion steels of the martensite-austenite class, and some oxide ceramics suitable for LPBF.

Table 2. Defects in laser powder bed fusion.

Defects	Possible Reasons	Solutions	References
Cracks (surface, internal, through)	<ul><li>Too high mode than required (especcially it is actual for newly introduced metals and alloys or difficult-to-process materials and alloys of nickel-titanium group, and high-entropy alloys),</li><li>Powder quality,</li><li>Residual stresses</li></ul>	<ul><li>Use of supporting structures for fast and efficient heat dissipation;</li><li>Optimal orientation of parts in the build chamber to reduce the area of welded sections;</li><li>Heated build platform, which reduces the temperature gradient and reduces residual stresses;</li><li>Chessboard scanning strategy;</li><li>To avoid further cracking during operation, heat treatment to relieve internal stresses is recommended, etc.</li></ul>	[14,24,75,76,82,83]
Porosity	<ul><li>Too high modes;</li><li>Granulomorphometry of the powder (size, sphericity, dispersion of the powder);</li><li>The oxide film present on the surface of the powder particles;</li><li>Layer deposition density;</li><li>Shrinkage processes, capture of molecules (nitrogen, argon) during synthesis</li></ul>	<ul><li>Selection of the optimal LPBF modes (speed, laser radiation power, scanning strategy);</li><li>Selection of optimal granulomorphometry (shape, size, dispersion);</li><li>Input quality control of powders;</li><li>Consistency of the quality level of the used powder;</li><li>Carrying out subsequent processing methods, for example, hot isostatic pressing, etc.</li></ul>	[14,30,47,62,84]
Increased degree of stress-strain state	<ul><li>Residual stresses;</li><li>Product geometry;</li><li>Location of supporting structures;</li><li>Heating and cooling rates;</li><li>Technological heritage;</li><li>Choosen material physical properties;</li><li>Process parameters (laser power and speed, powder layer thickness, scanning strategy, preheating, etc.)</li></ul>	<ul><li>Selection of the optimal LPBF modes;</li><li>Carrying out heat treatment (annealing);</li><li>Rescan;</li><li>Platform heating</li></ul>	[8,45,72,82]
Increased surface roughness	<ul><li>LPBF process modes;</li><li>Peculiarities of LPBF (layer-by-layer deposition of the powder material and the formation of steps, which are more pronounced with a significant inclination of the surfaces;</li><li>Partial melting of granules from an array of powder outside the shaded section due to partial penetration of the previous deposited layer)</li></ul>	Selection of the optimal LPBF modes (reducing the scanning interval leads to an improvement in surface roughness; reducing the scanning speed reduces coagulation and improves surface roughness)	[8,18,47,82,85]

Cracks are among the most dangerous mechanical defects in the LPBF [14,86], in which low thermal conductivity and high coefficients of thermal expansion create sufficiently high internal stresses to break bonds within the material, especially along grain boundaries where dislocations are present.

Several physical phenomena are important to the LPBF, such as the absorption of laser radiation by the powder material, melt droplet coagulation phenomena that disrupt the formation of continuous layers, and thermal fluctuations experienced by the material during the process, which can lead to cracking and damage to the product being grown [8,86].

Cracks can occur due to not using a laser source with sufficient power or too fast scanning of the powder surface, which leads to insufficient metal melting and preventing a strong bonding medium for solidification [87]. Moreover, residual stresses can lead to cracking (a mechanical defect in which low thermal conductivity and high coefficients of thermal expansion create sufficiently high internal stresses to break bonds within the material, especially along grain boundaries where dislocations are present) of parts and their deformation [86]. The value of residual stresses is affected by the product's geometry, the rate of heating and cooling, and the coefficient of thermal expansion, phase and structural changes in the metal. Studies were carried out to assess the residual stresses and strains for various scanning strategies [88]. The maximum stresses along the X and Y axes were observed for samples with a scanning strategy along the contour when shifted to the center. Checkerboard scanning, also known as island scanning, has been proven to reduce residual stresses compared with other scanning strategies using a random sequence of islands.

Secondary effects due to the laser beam can unintentionally affect the properties of the structure. One such example is the formation of secondary phase precipitates in the bulk structure due to repeated heating in the solidified underlying layers as the laser beam passes through the powder layer. Depending on the composition of the precipitates, this effect can remove important elements from the bulk material. In addition, the laser power and convection currents generated in powder layers containing oxides can evaporate and "splatter" the oxides elsewhere. It is typical for the LPBF that the deposited powder is remelted several times during the construction of the product, ensuring reliable adhesion of the formed layers (Figure 2) [8,89].

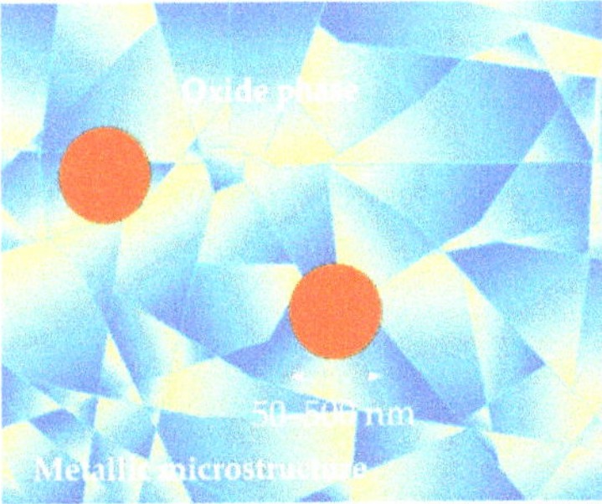

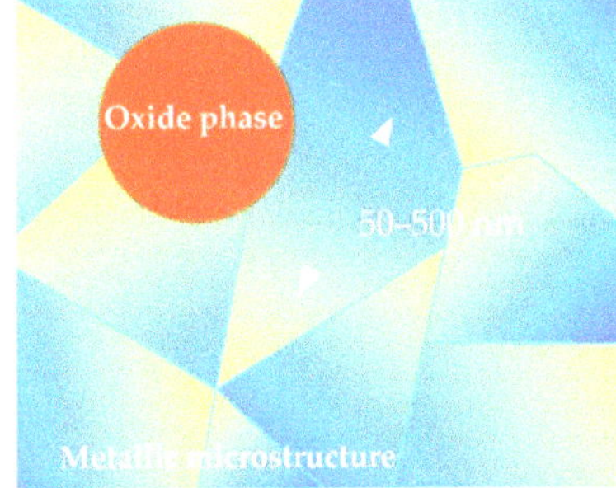

Figure 2. Oxide nanoparticles correspond to a small extent to steel obtained by LPBF.

In this case, the material is heated above the melting point, and a thin oxide film inevitably appears on the deposited surface, despite the fact that the oxygen content in the laser chamber is set to no higher than 0.15%. If this film is not removed, then during the formation of the next layer it will be walled up in the product and during the remelting of the layer formed above the film, it will remain in the structure of the product in the form of spherical nanoparticles of the amorphous oxide phase, evenly distributed over the volume of the product (Figure 2).

These oxides accumulate and do not wet, thereby forming slag, which not only removes the useful nature of the oxide in the composition, but also provides a mechanically favorable microenvironment for material cracking [8,90].

Pore formation is a very important defect in LPBF. Pores are found to form during changes in the laser scanning speed due to the rapid formation and then collapse of deep pits on the surface, which trap the inert shielding gas in the solidifying metal.

The main causes of porosity are incorrect selection of LPBF parameters and powder quality. An increase in the energy density of laser radiation leads to excessive evaporation of the material and the formation of splashes, which leads to an increase in the pore volume [85]. In addition, the high energy density of the laser radiation leads to the formation of a large melt pool, which becomes less stable and leads to the formation of separate melt droplets (balling effect) rather than a continuous melt path. Melt drops, under the action of surface tension, draw in nearby powder particles, which leads to the formation of pits around the drops and, as a result, an increase in porosity [86].

At the same time, when the energy density of laser radiation is insufficient, complete melting of the powder is not ensured, and large pores appear along the scanning lines. The size of the melt pool is too small to be in contact with adjacent scan tracks [34,35,87,91].

The speed of laser scanning affects the type of pores formed during the LPBF [30,33,71,84,85]. The porosity and pore size increase significantly with an increase in the scanning speed.

Insufficient distance between the tracks leads to an increase in porosity in the synthesized products, which is associated with the formation of a non-continuous melt track [34].

The deterioration of the process of spreading and wettability by the melt of the underlying layer is also facilitated by the presence of oxygen, which, dissolving in the metal, increases the viscosity of the melt.

The authors of [82] suggested that the oxide film present on the surface of the powder particles remains in the melt bath, thereby creating residual porosity in the crystallized volumes. The authors of [92,93] showed that the oxide film on the previous layer prevents interlayer bonding and leads to melt droplet coagulation, since metal melts usually do not chemically react with oxide films.

In addition to the LPBF parameters, the appearance of pores is influenced by the quality of the used powder materials, the size and sphericity of the particles, and the granulometric composition's uniformity (dispersity).

The more the shape of the particles approaches a spherical shape, the greater the density of the backfill. Less spherical particles form structures of the second order, which are characterized by large extended pores.

The study of the effect of particle size on the properties of finished products showed that in order to obtain high-quality products without porosity, the presence of small particles is necessary [94]. During laser exposure, small particles are first melted, thereby creating favorable conditions for the melting of larger particles. The presence of particles in the powder, the size of which is larger than the allowable or particles of irregular shape, negatively affects the porosity of the product.

Methods for reducing porosity in LPBF [95,96] include mainly powder quality control (granulomorphology and granulomorphometry, particularly degree of reflection/absorption of laser radiation by the powder), search for the optimal range of process factors, additional methods of leveling/compacting the powder in the layer, pre-drying, leveling the layer, pressing, wetting, pre-heating or heating, pre-melting, subsequent post-processing. The input control of the powder is carried out by assessing the presence of voids and their volume, particle size, shape and linear dimensions, which is important for assessing whether the particle size falls within the desired range.

In addition, subsequent processing methods such as hot isostatic pressing (HIP) can reduce porosity [97]. The use of HIP makes it possible to eliminate residual porosity and improve the physical and mechanical properties of the material.

After additive manufacturing by the LPBF, materials are characterized by anisotropy of properties, increased strength and reduced ductility due to the presence of residual stresses [8,86,97]. Annealing is carried out to remove residual stresses, obtain a more balanced structure, and increase the viscosity and plasticity of the material [98]. Sometimes remelting or premelting can be used to increase density, reduce residual stresses, etc. [99]. This strategy requires double scanning of the same layer with possibly different parameters without applying powder, although this may increase production time. Platform heating allows reducing the gradient and reducing residual stresses [100].

A checkerboard scanning strategy can also be used to reduce residual stresses in metal parts [94,101]. This strategy works by dividing the scanned area into squares of a given size, which reduces the size of the scanned sections and reduces the length of the scan vector.

However, a team consisting of scientists from the National Institute of Standards and Technology (NIST), Livermore National Laboratory Lawrence (LLNL) and other institutions found that the island scanning method actually increases residual stresses when printing certain bridge-like geometries [102]. Those, the use of the staggered scanning strategy makes it possible to reduce residual stresses, but has restrictions on the geometry of the products. A subsequent heat treatment can be used to relieve internal stresses. The grown parts, fixed by supporting structures on the building platform, are placed in an oven and aged according to the modes [30,47].

The main reason for the increased surface roughness of products obtained by the LPBF method is the way the technology is implemented: layer-by-layer melting of powder material by means of high-power laser radiation. According to the results of studies [103], the roughness of the surface layer is influenced by the parameters of the technological process such as the power of laser radiation, the scanning speed, the layer thickness and the scanning strategy. Surface roughness is directly dependent on technological modes and correlates with the nature of the change in porosity.

It was found that coagulation has a great influence on the roughness of the surface layer. Coagulation is the fusion of small powder particles into larger ones under the influence of laser radiation. A decrease in coagulation was observed with a decrease in the scanning speed, which is due to an increase in the melting time of the powder and a decrease in the viscosity of the melt. Increasing the scanning step leads to an increase in surface roughness. The surface roughness has the smallest value of 15 μm at a laser power of 200 W and a scanning speed of 3000 mm/s. The authors proposed to improve the roughness of the inclined surface due to the contour scanning of each layer with increased energy density [85].

The study of the influence of the quality of the powder material on the roughness showed that the smaller the powder particles, the more accurately it is possible to build products by reducing the scanning step and to obtain a smoother surface by reducing the thickness of the powder layer. However, in the process of product synthesis, a very rapid melting process occurs in the area of the laser spot, accompanied by boiling of the metal with the splashing of the melt and "carrying out" of fine light powder particles from the construction zone, which leads to the formation of an increased roughness of the product [52,53]. This effect can be reduced by using low-power lasers, but this significantly reduces productivity. Subsequently, processing (sandblasting, grinding, and polishing) is used to improve the roughness [30,47].

3. Analysis of Methods for Removing Oxide Phase by Atmospheric Plasma Sources after Laser Treatment

3.1. Investigation of Methods for Removing Oxide Phase

One of the main problems in the selective laser melting of metal powders is the oxide phase. It complicates the LPBF of metal powders and prevents the normal formation of layers during melting.

It is necessary to clean the formed surfaces for the further LPBF carefully to reduce the negative effect of the oxide phase. Mechanical, chemical, electrochemical, physical, thermal, and combined cleaning methods are used to remove the unwanted oxide phase.

Ultrasonic, plasma, laser, spark, induction, gas thermal, and combined methods can be used to remove the unwanted oxide phase in the LPBF operating chamber. A comparative analysis of the methods for oxide phase removal in LPBF is presented in Table 3.

Table 3. Methods for removing the oxide phase.

Nature	Method	Ability to Clean Complex Shapes	Remarks	Sources
Physical	Ultrasound	The need to use special solutions, the need for complex ultrasonic transducers	Exclusion of the use of flammable and toxic solvents	[104,105]
	Plasma	Allows you to clean products without the use of heat and the use of special liquid media, allows you to process parts of complex shape	Ability to clean geometrically complex objects	[106–108]
	Laser	does not require Additional resources, cleaning the surface retains its original appearance	Possibility of 100% localization of cleaning products	[109]
	Spark	There are restrictions on the thickness of the processed material	The complexity of the equipment, causes surface modification	[110]
Thermal	Induction cleaning, redox annealing	The technology is effective when using long bars	Possibility of surface overheating, sophisticated equipment and qualified personnel	[111]
	Gas-flame	Slight heating of parts, used to clean parts with a thickness of more than 5 mm	The need for subsequent cleaning by mechanical action, combustion products remain on the surface	[112]
Combined	Plasma-beam, plasma-spark	Allows processing of complex parts	The complexity of the equipment, the complexity of the implementation process	[113]

The performed analysis showed that one of the methods that allows to efficiently and effectively remove the oxide phase that occurs during the LPBF, providing conditions for reliable and durable formation of the product, is the plasma method.

3.2. Analysis of Methods for Removinge Oxide Phase by Sources of Atmospheric Plasma

Atmospheric pressure plasma is generated at atmospheric pressure without the use of complicated and expensive vacuum equipment. In this regard, atmospheric plasma is one of the promising processing tools in various fields of science and technology. Atmospheric low-temperature plasma and devices based on it, realizing a corona discharge, glow discharge, discharge with a dielectric barrier, and others, are widely used in various technologies [114,115].

Atmospheric plasma can be used to local clean surfaces of various materials such as metals, various types of ceramics, and other temperature-resistant metal- or ceramic-based materials.

The purpose of atmospheric pressure plasma cleaning is to remove contaminants, any substrate, even molecular traces of contaminants that form on the surface. Surface cleaning is the initial step in the surface preparation process. There are two types of pollutants—"natural" and "technological". Natural effects of the surrounding atmosphere appear, mainly containing oxygen-carbon or carbon-hydrogen bonds of organic substances, oxides, and adsorbed water particles. A technological type of pollution may occur during the part's manufacturing process.

There is another classification of pollution based on [8,47,82,116–122]. All pollution is divided into physical and chemical ones. Physical pollutions are held on the treated surface due to Van der Waals forces (electrostatically); chemical ones are due to chemical bonds. Different types of contaminants are removed by different methods (Table 4).

Table 4. Pollution classification.

Type of Pollution	Cleaning Method	References
Physical	Washing	[116]
	Ultrasonic cleaning	[82]
	Heat treatment	[47,82]
	Plasma cleaning	[8,117,118]
Chemical	Processing in solutions	[119]
	Acid and alkali etching	[120]
	Gas etching	[47]
	Laser chemfical cleaning	[121,122]

Surface treatment with plasma flows creates a low-temperature environment by using electrical energy rather than thermal energy to stimulate chemical reactions. The speed of plasma cleaning is not high (~0.4 mm/hour) [123,124]; therefore, plasma cleaning is often used as a finish cleaning, immediately before further processing (for example, coating deposition).

Plasma treatment satisfies the progressiveness criteria (the degree of progress in technology, technological equipment and organizational forms of processes, which includes qualitative and quantitative indicators [125]) such as:

–   Low duration of the process;
–   High degree of cleanliness of the treated surface;
–   Homogeneity of the cleaned surface;
–   No changes in the structure of the substrate material;
–   Environmental safety;
–   The possibility of quick disposal of cleaning products.

During the implementation of plasma surface treatment, electrons, ions, and radicals are generated. The atmospheric pressure plasma flow, interacting with the treated surface, causes the following phenomena leading to effective cleaning: heating of the surface to be cleaned, spraying, and etching. The advantages and disadvantages of these processes are shown in Figure 3.

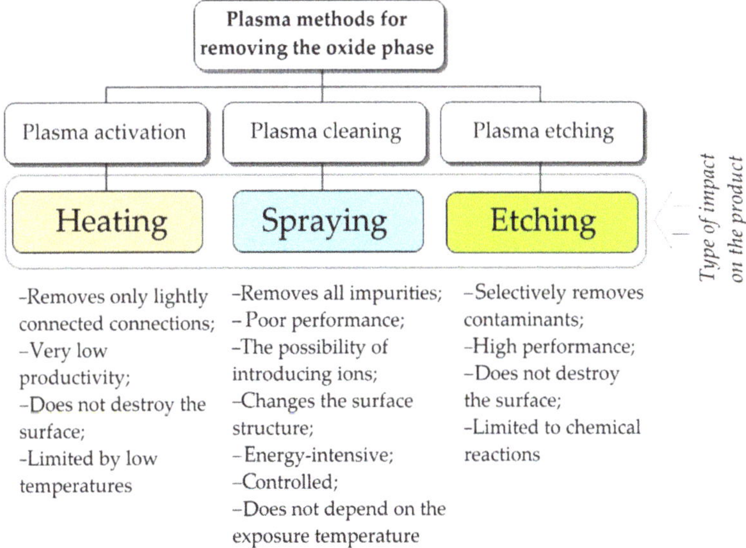

Figure 3. Plasma methods for removing the oxide phase.

Plasma treatment with atmospheric plasma reduces surface contamination by 5–6 times than chemical cleaning methods. Atmospheric plasma cleaning prepares the surface without any harmful waste.

Depending on the design solution and practical implementation of plasma surface treatment in order to clean it from the oxide phase and ensure strength properties, plasma cleaning, plasma activation, and plasma etching are used [126,127] (Figure 3).

When implementing plasma heating to clean the surface, the substrate to be cleaned is heated mainly by bombarding the surface with the electron and ion components of the plasma and by plasma radiation. It is possible to increase the plasma energy flux by applying a bias potential. Surface heating is usually limited to a low temperature (~600 °C) due to the prevention of the formation of metal carbides. At the same time, such a process can remove only lightly bound unwanted particles of a substance.

When implementing the plasma sputtering (physical vapor deposition (PVD) method of thin film deposition by sputtering that involves ejecting material from a target (source) onto a substrate (part)), a necessary condition is the presence of an additional voltage. A voltage is applied between the plasma and the substrate. Typically, the "floating potential" is about 10–20 V. Plasma sputtering is a more productive process. For example, a contaminated $h = 100$ nm layer can be removed in 500 s [128,129].

Plasma spray cleaning can be accompanied by removing the main part of the substrate material. In addition, the formation of surface defects is also possible [130].

Plasma spray cleaning can be accompanied by removing the main part of the substrate material. In addition, the formation of surface defects is also possible. Atoms or radicals chemically interact with the substrate during the plasma etching process. The first stage of plasma etching is adsorption, which is determined by the chemical composition of the substrate and temperature. Then, two ways of interaction are possible [131,132]: either adsorbed atoms and molecules interact with the surface or desorbed without chemical interaction. When volatile compounds are formed, atoms and radicals are desorbed into the gas phase and removed by pumping. Too high a temperature enhances desorption from the plasma. There is an optimum temperature when the etching process has the highest speed and slight heating of the treated surface. A combination of the above processes can give the highest cleaning efficiency.

The main parameters of surface treatment with atmospheric pressure plasma flows are the type of used inert working gas, the time of exposure to the substrate during cleaning, and the power of the plasma generator source.

These parameters affect the degree of plasma flow ionization, which determines the quality of cleaning [133].

In addition, the quality of plasma cleaning is affected by the duration of the process and the power of the plasma generator used, which implements a particular type of discharge, which must be taken into account to prevent overheating the surface of the parts during cleaning. The efficiency of plasma cleaning depends on many factors, starting with the substrate material, the duration of plasma exposure, the discharge used, the gas, the features of the physicochemical processes occurring during plasma treatment, the design features of the equipment used, and the parameters of the surrounding working environment.

The conducted analysis showed that the oxide phase can be removed effectively by one of the described plasma methods, and even more, oxide film formation can be minimized by plasma cleaning using additional ion bombardment.

4. Research of Atmospheric Plasma Sources Based on a Dielectric Barrier and Other Discharges as a Part of a Production Setup

4.1. Characteristics of Atmospheric Plasma Sources

Atmospheric plasma sources are designed to activate and clean metal and non-metal surfaces, apply thin films, air purification, surface preparation before coating deposition, gluing, painting, etc. In addition, atmospheric plasma can be used to clean the surface of organic and biological contaminants, dust particles, and oxides.

Various plasma sources are used to clean the surface from the oxide phase and other contaminants (Table 5). Special attention is paid to high-frequency (HF) and super-high-frequency (SHF) plasma sources that are carefully considered below.

Table 5. Comparison of low- and high-pressure plasma sources.

Name	Pressure Type	Medium	Process	Features	Notes	Reference
DC glow discharge	low	vacuum	electron bombardment (plasma cleaning by etching)	the product is located on the anode	An additional power supply is required, discharge combustion stabilization system can lead to the creation of defects, an additional power supply is required, a system for stabilizing the burning of the discharge	[134]
			ion bombardment (surface sputtering)	the product is located on the cathode (E $\approx$ 10^2–10^3 eV)		
High frequency discharge	low	vacuum	plasma etching	the product is located on a grounded electrode	used for cleaning metals and dielectrics	[135]
			ion etching	the product is located on the power electrode	used for cleaning metals and dielectrics	[136]
			plasma etching (reduced ion energy)	the product is located on a separate electrode	"soft" mode of plasma etching	[137]
Glow discharges with a hollow cathode	low	vacuum	HF-discharge (plasma cleaning, etching)	high density of plasma	high energy efficiency, low temperature	[138,139]
			magnetron etching and sputtering	uses a strong magnetic field	application of magnetron type discharge	[140]
Super high frequency discharge	low	vacuum	microwave plasma with high radical density for cleaning and etching	high rate of chemical reactions	electronic cyclotron mode	[141]
Dielectric barrier discharge (DBD) systems	high	atmospheric	plasma filamentary discharge cleaning	frequency of 10–10^4 Hz, a continuous energy source must provide the required degree of ionization	use of high-velocity gas streams to remove cleaning products	[142]
Capillary barrier discharge	high	atmospheric	plasma cleaning by a filamentary discharge implemented by a device with a dielectric material with a number of small holes	the uniformity of the discharge depends on the location of the capillaries	provides a higher plasma density, the possibility of the appearance of unwanted substances on the substrate	[143]

Table 5. *Cont.*

Name	Pressure Type	Medium	Process	Features	Notes	Reference
Corona discharge	high	atmospheric	plasma cleaning and etching	use of a needle (wire) electrode use of a planar electrode	the addition of oxygen is required for the decomposition of organic contaminants, the possibility of surface oxidation, and the decomposition of cleaning products	[144]
High frequency and super high frequency discharges	high	atmospheric	HF-cleaning SHF-cleaning	the conditions are realized as in the case of a glow discharge; the plasma is generated in He, N_2, Helium with the addition of 1–3% gas (O_2, N_2, H_2, or CF_4). parameters within the arc and atmospheric pressure glow discharge	high energy efficiency, low temperature provides significant heating of the cleaned surface	[145]

High frequency (HF) discharges can carry out surface cleaning, microwave discharges, plasma flows, and dielectric barrier discharge (DBD) systems [146–150]. These sources are of low and high pressure operating in a vacuum and at atmospheric pressure.

Plasma processes at atmospheric pressure do not require the use of expensive vacuum equipment. In addition, time is not wasted during the implementation of the technological process to create a vacuum. However, the process of plasma cleaning at atmospheric pressure also has disadvantages [151,152]: cleaning with a plasma flow requires the use of shielding gases. The process is carried out in sealed volumes using high-velocity gas flows to prevent re-contamination of the surface with cleaning products.

Atmospheric pressure plasma sources used for surface cleaning include:

- Sources of dielectric barrier discharge (DBD);
- Sources realizing pulsed direct current discharges (corona discharge);
- HF and SHF frequency sources.

4.2. Sources of Dielectric Barrier Discharge

One of the promising sources of plasma at high pressures is an electric discharge controlled by dielectric barriers, the so-called dielectric barrier discharge [153].

The design of the Dielectric Barrier Discharge (DBD) source can be implemented according to the traditional scheme and using a dielectric electrode, which has a number of small-holed capillaries.

Operating principle. DBD sources generate nonequilibrium plasma with a relatively low gas temperature. AC power sources generate the discharge power with 10–104 Hz frequencies. The DBD source consists of two plane-parallel electrodes, one of which is covered with a dielectric material (Figure 4) [154,155]. There are design schemes when both electrodes are covered with a dielectric material. An alternating voltage of 50–20,000 Hz is used when generating a discharge. The amplitude significantly exceeds the breakdown voltage. Microdischarges that occur locally are evenly distributed over the interelectrode gap. The gap between the electrodes is more than 1.5 mm. The gap size varies and depends on the gas and its pressure. Glass, quartz, ceramics, and polymers are used as dielectric materials. Distinguish discharges by geometry, volumetric, and surface.

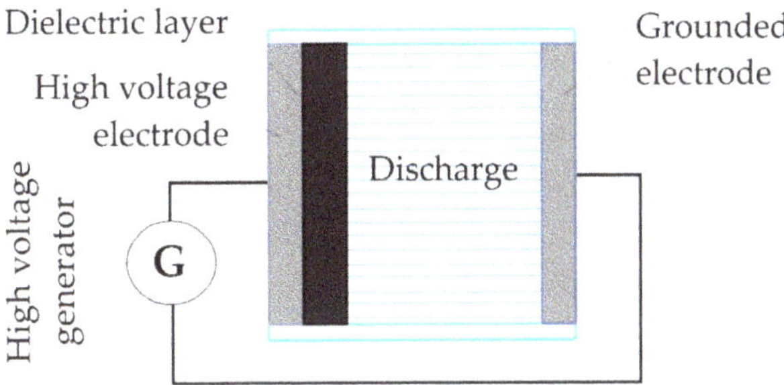

Figure 4. DBD device used for surface cleaning.

The charges are collected on the dielectric surface and discharged in microseconds during operation. The discharge exists at a certain degree of ionization. The gas filling the discharge gap causes the emission of a photon during the discharge, and the frequency and energy of the photon correspond to the type of gas [156].

Other devices for implementing the discharge have also been developed. For example, a device has been developed that implements the DBD method, characterized in that a high-resistance dielectric layer is used to cover one of the electrodes. Such a device generates a discharge called a resistive barrier, and a gallium arsenide (GaAs) semiconductor layer can be used as the dielectric layer.

The rate of purification by a device implementing DBD depends on the plasma gas composition. For example, adding a few percent of gaseous O_2 to the composition of the inert gas Ar makes it possible to intensify the process. However, a higher concentration of O_2 leads to slower removal of surface contamination or to the formation of an undesirable oxide layer. Nitrogen can also be additionally introduced into the composition of the mixture, which also contributes to a better purification process [157].

Source advantages are in the possibility of implementing DBD in technologically simple installations at atmospheric pressure, high electric field strength, soft impact on the treated surface. Disadvantages are limited length of the discharge gap, increased energy consumption.

In addition, in the practice of implementing cleaning with a dielectric barrier discharge, a device can be used in which the dielectric electrode has a number of small holes and capillaries. Such a device implements a capillary barrier discharge. Finally, a diagram of a device that implements a capillary barrier discharge is shown in Figure 5 [158].

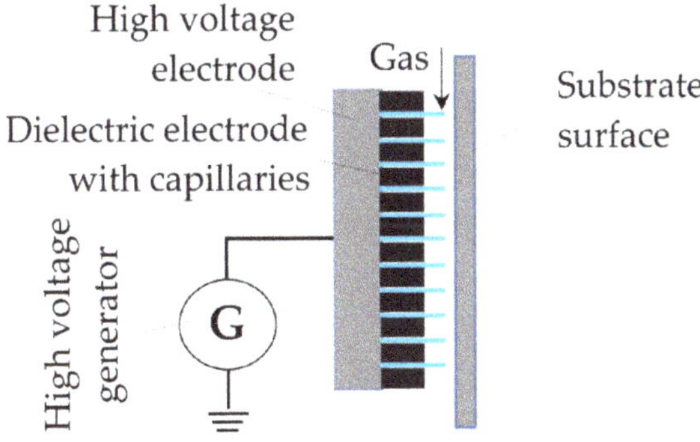

Figure 5. Capillary discharge device used for surface cleaning.

Operating principle. A discharge implemented by a DBD source ignited at atmospheric pressure in a nitrogen working gas environment or in air. As in the implementation of the above discharge, a dielectric barrier is installed between the electrodes, in which there are holes and capillaries. The temperature and density of the plasma of a capillary barrier discharge depend on the initial pressure of the gas, its chemical composition, the diameter of the hole (capillary), and the amplitude value of the current [159]. In addition, the high plasma density allows a higher cleaning power to be achieved.

In contrast to the discharge implemented by the DBD source, a capillary discharge produces stable plasma flows up to 4 cm long. The uniformity of the discharge, in turn, depends on the location of the holes in the dielectric electrode [151]. Source advantage is greater cleaning efficiency compared with traditional DBD. A flaw is in the possibility of surface damage and the deposition of undesirable substances on the substrate.

4.3. Sources of Corona Discharge

Plasma sources realizing a corona discharge operate in an electric field with high intensity, at high air pressure. Ionization processes during discharge generation occur only near the corona electrode. The diagram of the device is shown in Figure 4.

Operating principle. Using a corona discharge for cleaning the surface is possible using non-uniform fields that occur at electrodes with large surface curvature. Such an independent discharge is typical for electrodes in point, thin wires. Near such cathodes, a corona-shaped glow appears, and the field strength has higher values. The flow of a corona discharge is observed if an electron arises during the random ionization of a molecule (neutral). It accelerates in the electric field, acquiring energy enough to ionize another molecule. As a result of the process, a new negative electron and a positively charged ion are formed. Thus, a stream of charged particles is formed that then passes into an avalanche. The characteristics of a corona discharge depend on the type of corona (negative or positive) and the pulse applied to the corona electrode [160,161]. The scheme of the device that implements the corona discharge is shown in Figure 6.

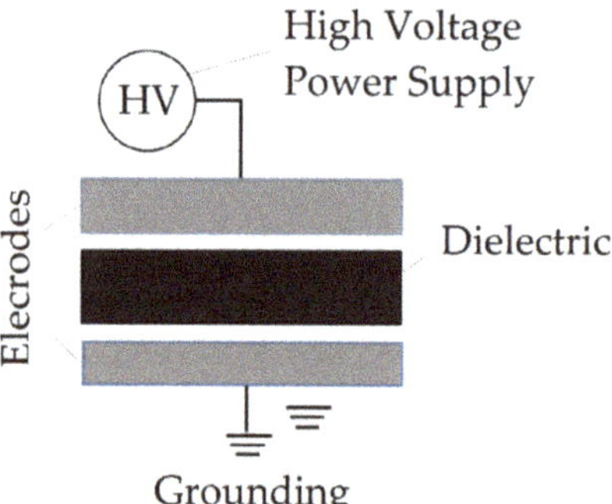

Figure 6. Corona devices used for surface cleaning.

Corona discharge device configurations use needle electrodes (thin wire) or planar electrodes. In this case, the energy is concentrated on the needle electrode, and the flat one is grounded. The addition of oxygen is necessary for the decomposition of organic pollutants. Voltage during processing is applied to the upper electrode. The dielectric in the system is located between two electrodes and air gaps. The bottom electrode is grounded. The gradually increasing voltage ionizes the air between the two electrodes and creates a corona discharge.

The main source advantage is in the ability to initiate a chemical reaction on the surface to be cleaned. Flaws are significant weakening of the plasma effect during processing, oxidation of the substrate surface due to ozone release, and the need for sophisticated ozone neutralization equipment.

4.4. High Frequency and Super High Frequency Discharges

For excitation in sources of High Frequency (HF) and Super High Frequency (SHF) (microwave) discharges, two electrodes must be in direct contact with the plasma. The direct current glow discharge is ignited and maintained, which is necessary to implement the cleaning process. However, such conditions can lead to disruption of the discharge due to the formation of a non-conductive film on the electrodes. The diagram of the device is shown in Figure 7 [162,163].

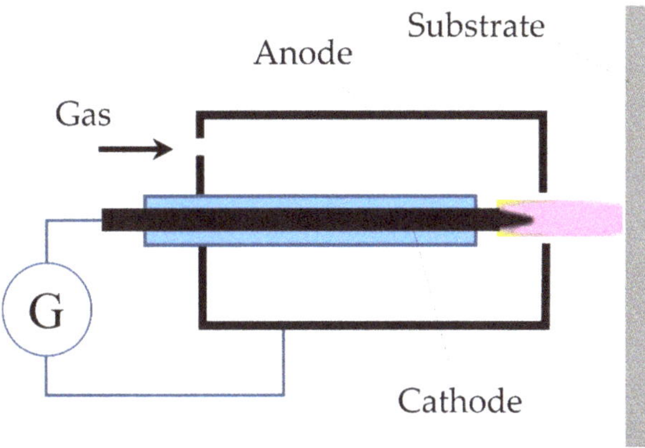

Figure 7. Plasma source for surface cleaning.

Atmospheric glow discharge is generated using a high-frequency discharge and microwave (microwave) at high pressure. Plasma is generated in He, N_2, air, and He with the addition of 1–3% molecular gas such as O_2, N_2, H_2, or CF_4, providing sufficient concentration of the plasma stream to clean the surface material effectively.

The atmospheric pressure plasma source is a device with an internal electrode and a grounded external electrode. The plasma gas flows between the electrodes at high speed and exits the nozzle. The plasma jet propagates from the anode (the anode is grounded). Plasma source operating modes: voltage of several hundred volts, power~1 kW. Surface treatment can be realized in HF and microwave modes. Plasma is generated in the resonant chamber and enters the processing zone through the opening-nozzle in the chamber [129,157]. The parameters of the plasma realizing the microwave discharge are close to the parameters of the arc and glow discharge of atmospheric pressure.

Device advantages are in the possibility of modifying the surface to be cleaned; the ability to control chemically active substances to improve the quality of cleaning.

Flaws are in limited areas of impact of HF and SHF discharge; there is significant heating of the substrate in the process of exposure to the source, and possible overheating of the surface. A plasma source based on atmospheric plasma based on a dielectric barrier discharge is most suitable for use as a source of plasma surface cleaning.

5. Conclusions

Today, the introduction of additive technologies, particularly LPBF in production, is hampered by the lack of scientifically based recommendations on the choice of process parameters for newly introduced metal-based materials and alloys. It is related to the lack of systematic studies of the LPBF, the influence of its factors on the quality of grown products, the formation of defects related to the presence of the oxide phase in the structure of the solids, and the lack of critical engineering solutions to minimize these types of structural defects and surface roughness. The analysis of oxide phase formation shows that the formation of structural and surface defects such as cracks, pores, oxide phases, etc., is determined by many process factors and related to the LPBF specifics depending on the chosen modes, material, and powder properties (reflectivity, sphericity, etc.).

Surface plasma cleaning methods demonstrate their ability to remove the oxide phase in metal products that can be used in laser powder bed fusion as a part of the unit. This provides conditions for reliable product growing that can be even more durable in extreme conditions by removing the oxide phase, smoothing the surface, and recrystallization by heat treatment (decrease in grain size). Systematization and comparative analysis of the existed plasma-based methods provided by different types of plasma sources for physical surface cleaning have shown that the plasma processing using additional ion bombardment by a plasma source based on an electric discharge with a dielectric barrier can be preferable to minimize the oxide phase and other structural and surface defects after LPBF.

Using a newly developed plasma source as a part of the LPBF unit can significantly reduce the labor intensity of production engineering products with the specific requirements to their physical and mechanical properties and surface quality and, in prospect, enlarge their exploitation ability and service life for working in cyclic loads.

Developing the LPBF production setup equipped with an atmospheric plasma source based on a dielectric barrier discharge could contribute to creating a new type of hybrid equipment for transition to the sixth technological paradigm associated with Kondratieff's waves.

6. Patents

Grigoriev Sergey Nikolaevich; Metel Alexander Sergeevich; Volosova Marina Alexandrovna; Melnik Yury Andreevich; Mustafaev Enver Serverovich. Device for processing dielectric products with fast atoms; #RU2752877C1, 2020-12-11.

Author Contributions: Conceptualization, S.N.G. and M.A.V.; methodology, S.V.F. and A.S.M.; software, S.V.F. and N.A.S.; validation, A.S.M. and S.R.S.; formal analysis, A.S.M. and A.A.O.; investigation, S.V.F.; resources, N.A.S.; data curation, S.R.S.; writing—original draft preparation, S.R.S. and N.A.S.; writing—review and editing, A.A.O. and M.A.V.; visualization, S.R.S. and N.A.S.; supervision, S.N.G.; project administration, S.N.G. and M.A.V.; funding acquisition, A.S.M. All authors have read and agreed to the published version of the manuscript.

Funding: This research was funded by the Russian Science Foundation, grant number No. 20-19-00620.

Institutional Review Board Statement: Not applicable.

Informed Consent Statement: Not applicable.

Data Availability Statement: Data are available in a publicly accessible repository.

Acknowledgments: The research was conducted at the Department of High-Efficiency Processing Technologies of MSTU Stankin.

Conflicts of Interest: The authors declare no conflict of interest.

References

1. Aversa, A.; Marchese, G.; Saboori, A.; Bassini, E.; Manfredi, D.; Biamino, S.; Ugues, D.; Fino, P.; Lombardi, M. New Aluminum Alloys Specifically Designed for Laser Powder Bed Fusion: A Review. *Materials* **2019**, *12*, 1007. [CrossRef] [PubMed]
2. Denti, L. Additive Manufactured A357.0 Samples Using the Laser Powder Bed Fusion Technique: Shear and Tensile Performance. *Metals* **2018**, *8*, 670. [CrossRef]
3. Sova, A.; Grigoriev, S.; Okunkova, A.; Smurov, I. Potential of cold gas dynamic spray as additive manufacturing technology. *Int. J. Adv. Manuf. Technol.* **2013**, *69*, 2269–2278. [CrossRef]
4. Sova, A.; Grigoriev, S.; Okunkova, A.; Bertrand, P.; Smurov, I. Cold spraying: From process fundamentals towards advanced applications. *Surf. Coat. Technol.* **2015**, *268*, 77–84.
5. Notenboom, G.; Maerten, O. Additive mass manufacturing of composite ceramic, metal and glass microparts and multilayers from nanosized particles, using inkjet and laser technology (Acerlink). In *5th European Conference on Advanced Materials and Processes and Applications (EUROMAT 97), Proceedings of the 5th European Conference on Advanced Materials and Processes and Applications: Materials, Functionality & Design, Maastricht, The Netherlands, 21–23 April 1997*; Sarton, L.A.J., Zeedijk, H.B., Eds.; Netherlands Soc Materials Science: Zwijndrecht, The Netherlands, 1997; Volume 3, p. 55.

6. Kumar, M.S. Rapid Prototyping-An overview of the present state-of-the-art. In Proceedings of the 13th National Convention of Mechanical Engineers, Modern Trends in Manufacturing Technology, Maastricht, The Netherlands, 6–8 November 1997; Chaturvedi, P., Tewari, N.K., Yadav, G.S., Rao, P.V., Eds.; Concept Publ Co.: New Delhi, India, 1998; pp. 123–130.

7. Aleshin, N.P.; Grigor'ev, M.V.; Shchipakov, N.A.; Prilutskii, M.A.; Murashov, V.V. Applying nondestructive testing to quality control of additive manufactured parts. *Russ. J. Nondestruct. Test.* **2016**, *52*, 600–609. [CrossRef]

8. Metel, A.S.; Grigoriev, S.N.; Tarasova, T.V.; Melnik, Y.A.; Volosova, M.A.; Okunkova, A.A.; Podrabinnik, P.A.; Mustafaev, E.S. Surface Quality of Metal Parts Produced by Laser Powder Bed Fusion: Ion Polishing in Gas-Discharge Plasma Proposal. *Technologies* **2021**, *9*, 27. [CrossRef]

9. Liu, D.; Lee, B.; Babkin, A.; Chang, Y. Research Progress of Arc Additive Manufacture Technology. *Materials* **2021**, *14*, 1415. [CrossRef]

10. Grigor'yants, A.G.; Shiganov, I.N. Development of Domestic Equipment for Laser Additive Technologies by Melting Metallic Powders. *Russ. Metall.* **2020**, *6*, 649–653. [CrossRef]

11. Ivanov, A.D.; Minaev, V.L.; Vishnyakov, G.N. A Shearograph for Nondestructive Testing of Products Obtained by Additive Technologies. *Instrum. Exp. Tech.* **2019**, *62*, 871–875. [CrossRef]

12. Kurbatov, A.S.; Orekhov, A.A.; Rabinskiy, L.N.; Tushavina, O.V.; Kuznetsova, E.L. Research of The Problem of Loss of Stability of Cylindrical Thin-Walled Structures Under Intense Local Temperature Exposure. *Period. Tche Quim.* **2020**, *17*, 884–891.

13. Khmyrov, R.S.; Grigoriev, S.N.; Okunkova, A.A.; Gusarov, A.V. On the possibility of selective laser melting of quartz glass. *Phys. Procedia* **2014**, *56*, 345–356. [CrossRef]

14. Khmyrov, R.S.; Protasov, C.E.; Grigoriev, S.N.; Gusarov, A.V. Crack-free selective laser melting of silica glass: Single beads and monolayers on the substrate of the same material. *Int. J. Adv. Manuf. Technol.* **2016**, *85*, 1461–1469. [CrossRef]

15. Grigoriev, S.; Peretyagin, P.; Smirnov, A.; Solis, W.; Diaz, L.A.; Fernandez, A.; Torrecillas, R. Effect of graphene addition on the mechanical and electrical properties of Al_2O_3-SiCw ceramics. *J. Eur. Ceram. Soc.* **2017**, *37*, 2473–2479. [CrossRef]

16. Sova, A.; Grigoriev, S.; Okunkova, A.; Smurov, I. Cold spray deposition of 316L stainless steel coatings on aluminium surface with following laser post-treatment. *Surf. Coat. Technol.* **2013**, *235*, 283–289. [CrossRef]

17. Klingaa, C.G.; Mohanty, S.; Hjermitslev, A.B.; Haahr-Lillevang, L.; Hattel, J.H. Towards a digital twin of laser powder bed fusion with a focus on gas flow variables. *J. Manuf. Process.* **2021**, *65*, 312–327. [CrossRef]

18. Kozior, T.; Bochnia, J. The Influence of Printing Orientation on Surface Texture Parameters in Powder Bed Fusion Technology with 316L Steel. *Micromachines* **2020**, *11*, 639. [CrossRef]

19. Asnafi, N. Application of Laser-Based Powder Bed Fusion for Direct Metal Tooling. *Metals* **2021**, *11*, 458. [CrossRef]

20. Grigor'ev, S.N.; Kozochkin, M.P.; Fedorov, S.V.; Porvatov, A.N.; Okun'kova, A.A. Study of Electroerosion Processing by Vibroacoustic Diagnostic Methods. *Meas. Tech.* **2015**, *58*, 878–884. [CrossRef]

21. Vila, C.; Siller, H.R.; Rodriguez, C.A.; Bruscas, G.M.; Serrano, J. Economical and technological study of surface grinding versus face milling in hardened AISI D3 steel machining operations. *Int. J. Prod. Econ.* **2012**, *138*, 273–283. [CrossRef]

22. Bekem, A.; Ozbay, B.; Bulduk, M.E. Effect of dendritic copper powder addition to polyamide 12 in selective laser sintering. *J. Fac. Eng. Archit. Gazi Univ.* **2021**, *36*, 421–431.

23. Qin, Y.L.; Sun, B.H.; Zhang, H.; Ni, D.R.; Xiao, B.L.; Ma, Z.Y. Development of Selective Laser Melted Aluminum Alloys and Aluminum Matrix Composites in Aerospace Field. *Chin. J. Lasers-Zhongguo Jiguang* **2021**, *48*, 1402002.

24. Wang, H.; Chen, J.Q.; Luo, H.I..; Wang, D.; Song, C.H.; Yao, X.Y.; Chen, P.; Yan, M. Bimetal printing of high entropy alloy/metallic glass by laser powder bed fusion additive manufacturing. *Intermetallics* **2022**, *141*, 107430. [CrossRef]

25. Sova, A.; Okunkova, A.; Grigoriev, S.; Smurov, I. Velocity of the Particles Accelerated by a Cold Spray Micronozzle: Experimental Measurements and Numerical Simulation. *J. Therm. Spray Technol.* **2012**, *22*, 75–80. [CrossRef]

26. Santecchia, E.; Spigarelli, S.; Cabibbo, M. Material Reuse in Laser Powder Bed Fusion: Side Effects of the Laser—Metal Powder Interaction. *Metals* **2020**, *10*, 341. [CrossRef]

27. Grigoriev, S.N.; Kozochkin, M.P.; Porvatov, A.N.; Volosova, M.A.; Okunkova, A.A. Electrical discharge machining of ceramic nanocomposites: Sublimation phenomena and adaptive control. *Heliyon* **2019**, *5*, e02629. [CrossRef] [PubMed]

28. Khorasani, M.; Ghasemi, A.; Rolfe, B.; Gibson, I. Additive manufacturing a powerful tool for the aerospace industry. *Rapid Prototyp. J.* **2022**, *28*, 87–100. [CrossRef]

29. Mindt, H.W.; Desmaison, O.; Megahed, M.; Peralta, A.; Neumann, J. Modeling of Powder Bed Manufacturing Defects. *J. Mater. Eng. Perform.* **2018**, *27*, 32–43. [CrossRef]

30. Metel, A.S.; Grigoriev, S.N.; Tarasova, T.V.; Filatova, A.A.; Sundukov, S.K.; Volosova, M.A.; Okunkova, A.A.; Melnik, Y.A.; Podrabinnik, P.A. Influence of Postprocessing on Wear Resistance of Aerospace Steel Parts Produced by Laser Powder Bed Fusion. *Technologies* **2020**, *8*, 73. [CrossRef]

31. Mahmood, M.A.; Ur Rehman, A.; Pitir, F.; Salamci, M.U.; Mihailescu, I.N. Laser Melting Deposition Additive Manufacturing of Ti6Al4V Biomedical Alloy: Mesoscopic In-Situ Flow Field Mapping via Computational Fluid Dynamics and Analytical Modelling with Empirical Testing. *Materials* **2021**, *14*, 7749. [CrossRef]

32. Dostovalov, A.V.; Korolkov, V.P.; Babin, S.A. Formation of thermochemical laser-induced periodic surface structures on Ti films by a femtosecond IR Gaussian beam: Regimes, limiting factors, and optical properties. *Appl. Phys. B-Lasers Opt.* **2017**, *123*, 30. [CrossRef]

33. Gusarov, A.V.; Grigoriev, S.N.; Volosova, M.A.; Melnik, Y.A.; Laskin, A.; Kotoban, D.V.; Okunkova, A.A. On productivity of laser additive manufacturing. *J. Mater. Process. Technol.* **2018**, *261*, 213–232. [CrossRef]
34. Yadroitsev, I.; Bertrand, P.H.; Antonenkova, G.; Grigoriev, S.; Smurov, I. Use of track/layer morphology to develop functional parts by selectivelaser melting. *J. Laser Appl.* **2013**, *25*, 052003. [CrossRef]
35. Kotoban, D.; Grigoriev, S.; Okunkova, A.; Sova, A. Influence of a shape of single track on deposition efficiency of 316L stainless steel powder in cold spray. *Surf. Coat. Technol.* **2017**, *309*, 951–958. [CrossRef]
36. Doubenskaia, M.; Pavlov, M.; Grigoriev, S.; Tikhonova, E.; Smurov, I. Comprehensive Optical Monitoring of Selective Laser Melting. *J. Laser Micro Nanoeng.* **2012**, *7*, 236–243. [CrossRef]
37. Glaziev, S.Y. The discovery of regularities of change of technological orders in the central economics and mathematics institute of the soviet academy of sciences. *Econ. Math. Methods* **2018**, *54*, 17–30. [CrossRef]
38. Korotayev, A.V.; Tsirel, S.V. A spectral analysis of world GDP dynamics: Kondratiev waves, Kuznets swings, Juglar and Kitchin cycles in global economic development, and the 2008–2009 economic crisis. *Struct. Dyn.* **2010**, *4*, 3–57. [CrossRef]
39. Perez, C. Technological revolutions and techno-economic paradigms. *Camb. J. Econ.* **2010**, *34*, 185–202. [CrossRef]
40. Wonglimpiyarat, J. Towards the sixth Kondratieff cycle of nano revolution. *Int. J. Nanotechnol. Mol. Comput.* **2011**, *3*, 87–100. [CrossRef]
41. Smurov, I.; Doubenskaia, M.; Grigoriev, S.; Nazarov, A. Optical Monitoring in Laser Cladding of Ti6Al4V. *J. Spray Technol.* **2012**, *21*, 1357–1362. [CrossRef]
42. Scotti, F.M.; Teixeira, F.R.; da Silva, L.J.; de Araujo, D.B.; Reis, R.P.; Scotti, A. Thermal management in WAAM through the CMT Advanced process and an active cooling technique. *J. Manuf. Process.* **2020**, *57*, 23–35. [CrossRef]
43. Duman, B.; Ozsoy, K. A deep learning-based approach for defect detection in powder bed fusion additive manufacturing using transfer learning. *J. Fac. Eng. Archit. Gazi Univ.* **2022**, *37*, 361–375.
44. Sing, S.L.; Yeong, W.Y.; Wiria, F.E.; Tay, B.Y.; Zhao, Z.Q.; Zhao, L.; Tian, Z.L.; Yang, S.F. Direct selective laser sintering and melting of ceramics: A review. *Rapid Prototyp. J.* **2017**, *23*, 611–623. [CrossRef]
45. Deirmina, F.; Davies, P.A.; Casati, R. Effects of Powder Atomization Route and Post-Processing Thermal Treatments on the Mechanical Properties and Fatigue Resistance of Additively Manufactured 18Ni300 Maraging Steel. *Adv. Eng. Mater.* **2021**, *24*, 2101011. [CrossRef]
46. Yan, J.; Zhou, Y.; Gu, R.; Zhang, X.; Quach, W.-M.; Yan, M. A Comprehensive Study of Steel Powders (316L, H13, P20 and 18Ni300) for Their Selective Laser Melting Additive Manufacturing. *Metals* **2019**, *9*, 86. [CrossRef]
47. Metel, A.; Tarasova, T.; Gutsaliuk, E.; Khmyrov, R.; Egorov, S.; Grigoriev, S. Possibilities of Additive Technologies for the Manufacturing of Tooling from Corrosion-Resistant Steels in Order to Protect Parts Surfaces from Thermochemical Treatment. *Metals* **2021**, *11*, 1551. [CrossRef]
48. Fomin, V.M.; Golyshev, A.A.; Malikov, A.G.; Orishich, A.M.; Filippov, A.A. Creation of A Functionally Gradient Material by the Selective Laser Melting Method. *J. Appl. Mech. Tech. Phys.* **2020**, *61*, 878–887. [CrossRef]
49. Akopyan, T.K.; Letyagin, N.V.; Avxentieva, N.N. High-tech alloys based on Al-Ca-La(-Mn) eutectic system for casting, metal forming and selective laser melting. *Non-Ferr. Met.* **2020**, *1*, 52–59. [CrossRef]
50. Mazeeva, A.K.; Staritsyn, M.V.; Bobyr, V.V.; Manninen, S.A.; Kuznetsov, P.A.; Klimov, V.N. Magnetic properties of Fe-Ni permalloy produced by selective laser melting. *J. Alloy. Compd.* **2020**, *814*, 152315. [CrossRef]
51. Samodurova, M.; Logachev, I.; Shaburova, N.; Samoilova, O.; Radionova, L.; Zakirov, R.; Pashkeev, K.; Myasoedov, V.; Trofimov, E. A Study of the Structural Characteristics of Titanium Alloy Products Manufactured Using Additive Technologies by Combining the Selective Laser Melting and Direct Metal Deposition Methods. *Materials* **2019**, *12*, 3269. [CrossRef]
52. Okunkova, A.; Peretyagin, P.; Vladimirov, Y.; Volosova, M.; Torrecillas, R.; Fedorov, S.V. Laser-beam modulation to improve efficiency of selecting laser melting for metal powders. In Proceedings of the Conference on Laser Sources and Applications II, Laser Sources and Applications II, Laser Sources and Applications II, Brussels, Belgium, 14–17 April 2014; Mackenzie, J.I., Jelinkova, H., Taira, T., Ahmed, M.A., Eds.; SPIE-Int Soc Optical Engineering: Bellingham, WA, USA, 2014; Volume 9135, p. 913524.
53. Gusarov, A.V.; Okun'kova, A.A.; Peretyagin, P.Y.; Zhirnov, I.V.; Podrabinnik, P.A. Means of Optical Diagnostics of Selective Laser Melting with Non-Gaussian Beams. *Meas. Tech.* **2015**, *58*, 872–877. [CrossRef]
54. Gu, R.N.; Wong, K.S.; Yan, M. Laser additive manufacturing of typical highly reflective materials-gold, silver and copper. *Sci. Sin. Phys. Mech. Astron.* **2020**, *50*, 034204.
55. Klotz, U.E.; Tiberto, D.; Held, F. Optimization of 18-karat yellow gold alloys for the additive manufacturing of jewelry and watch parts. *Gold Bull.* **2017**, *50*, 111–121. [CrossRef]
56. Chauveau, D. Review of NDT and process monitoring techniques usable to produce high-quality parts by welding or additive manufacturing. *Weld. World* **2018**, *62*, 1097–1118. [CrossRef]
57. Shi, Y.S.; Zhang, J.L.; Wen, S.F.; Song, B.; Yan, C.Z.; Wei, Q.S.; Wu, J.M.; Yin, Y.J.; Zhou, J.X.; Chen, R.; et al. Additive manufacturing and foundry innovation. *China Foundry* **2021**, *18*, 286–295. [CrossRef]
58. Yang, L.; Tang, S.Y.; Fan, Z.T.; Jiang, W.M.; Liu, X.W. Rapid casting technology based on selective laser sintering. *China Foundry* **2021**, *18*, 296–306. [CrossRef]
59. Abdul Rani, A.M.; Fua-Nizan, R.; Din, M.Y. Manufacturing methods for medical artificial prostheses—A review. *Malays. J. Fundam. Appl. Sci.* **2017**, *13*, 464–469.

60. Li, Y.; Li, D.C.; Lu, B.H.; Gao, D.J.; Zhou, J. Current status of additive manufacturing for tissue engineering scaffold. *Rapid Prototyp. J.* **2015**, *21*, 747–762. [CrossRef]

61. Ojo, O.O.; Taban, E. Defects and post-manufacturing processes of additively manufactured steels: A Review (Part 2). *Mater. Test.* **2020**, *62*, 835–848.

62. Filatova, A.; Tarasova, T.; Peretyagin, P. Developing processes for manufacturing metal aviation technology components using powder bed fusion methods. *MATEC Web Conf.* **2019**, *298*, 00116. [CrossRef]

63. Cortina, M.; Arrizubieta, J.I.; Calleja, A.; Ukar, E.; Alberdi, A. Case Study to Illustrate the Potential of Conformal Cooling Channels for Hot Stamping Dies Manufactured Using Hybrid Process of Laser Metal Deposition (LMD) and Milling. *Metals* **2018**, *8*, 102. [CrossRef]

64. Zhao, Y.; Li, F.; Chen, S.; Lu, Z.Y. Unit block-based process planning strategy of WAAM for complex shell-shaped component. *Int. J. Adv. Manuf. Technol.* **2019**, *104*, 3915–3927. [CrossRef]

65. Han, P. Additive Design and Manufacturing of Jet Engine Parts. *Engineering* **2017**, *3*, 648–652. [CrossRef]

66. Ghaffari, M.; Nemani, A.V.; Rafieazad, M.; Nasiri, A. Effect of Solidification Defects and HAZ Softening on the Anisotropic Mechanical Properties of a Wire Arc Additive-Manufactured Low-Carbon Low-Alloy Steel Part. *JOM* **2019**, *71*, 4215–4224. [CrossRef]

67. Sun, X.F.; Song, W.; Liang, J.J.; Li, J.G.; Zhou, Y.Z. Research and Development in Materials and Processes of Superalloy Fabricated by Laser Additive Manufacturing. *Acta Metall. Sin.* **2021**, *57*, 1471–1483.

68. Magerramova, L.; Isakov, V.; Shcherbinina, L.; Gukasyan, S.; Petrov, M.; Povalyukhin, D.; Volosevich, D.; Klimova-Korsmik, O. Design, Simulation and Optimization of an Additive Laser-Based Manufacturing Process for Gearbox Housing with Reduced Weight Made from AlSi10Mg Alloy. *Metals* **2022**, *12*, 67. [CrossRef]

69. Brailovski, V.; Kalinicheva, V.; Letenneur, M.; Lukashevich, K.; Sheremetyev, V.; Prokoshkin, S. Control of Density and Grain Structure of a Laser Powder Bed-Fused Superelastic Ti-18Zr-14Nb Alloy: Simulation-Driven Process Mapping. *Metals* **2020**, *10*, 1697. [CrossRef]

70. Ahsan, F.; Razmi, J.; Ladani, L. Process Parameter Optimization in Metal Laser-Based Powder Bed Fusion Using Image Processing and Statistical Analyses. *Metals* **2022**, *12*, 87. [CrossRef]

71. Gurin, V.D.; Kotoban, D.V.; Podrabinnik, P.A.; Zhirnov, I.V.; Peretyagin, P.Y.; Okunkova, A.A. Modeling of 3D technological fields and research of principal perspectives and limits in productivity improvement of selective laser melting. *Mech. Ind.* **2016**, *17*, 714. [CrossRef]

72. Zhirnov, I.V.; Podrabinnik, P.A.; Okunkova, A.A.; Gusarov, A.V. Laser beam profiling: Experimental study of its influence on single-track formation by selective laser melting. *Mech. Ind.* **2015**, *16*, 709. [CrossRef]

73. Yan, Q.; Song, B.; Shi, Y.S. Comparative study of performance comparison of AlSi10Mg alloy prepared by selective laser melting and casting. *J. Mater. Sci. Technol.* **2020**, *41*, 199–208. [CrossRef]

74. Albu, M.; Panzirsch, B.; Schröttner, H.; Mitsche, S.; Reichmann, K.; Poletti, M.C.; Kothleitner, G. High-Resolution Microstructure Characterization of Additively Manufactured X5CrNiCuNb17-4 Maraging Steel during Ex and In Situ Thermal Treatment. *Materials* **2021**, *14*, 7784. [CrossRef] [PubMed]

75. Tian, Z.; Zhang, C.; Wang, D.; Liu, W.; Fang, X.; Wellmann, D.; Zhao, Y.; Tian, Y. A Review on Laser Powder Bed Fusion of Inconel 625 Nickel-Based Alloy. *Appl. Sci.* **2020**, *10*, 81. [CrossRef]

76. Sotov, A.V.; Agapovichev, A.V.; Smelov, V.G.; Kokareva, V.V.; Zenina, M.V. Investigation of the Ni-Co-Cr alloy microstructure for the manufacturing of combustion chamber GTE by selective laser melting. *Int. J. Adv. Manuf. Technol.* **2019**, *101*, 3047–3053. [CrossRef]

77. Zegzulka, J.; Gelnar, D.; Jezerska, L. Characterization and fowability methods for metal powders. *Sci. Rep.* **2020**, *10*, 21004. [CrossRef]

78. Wang, L.; Li, E.L.; Shen, H.; Zou, R.P.; Yu, A.B.; Zhou, Z.Y. Adhesion effects on spreading of metal powders in selective laser melting. *Powder Technol.* **2020**, *363*, 602–610. [CrossRef]

79. Zakharov, O.V.; Brzhozovskii, B.M. Accuracy of centering during measurement by roundness gauges. *Meas. Tech.* **2006**, *49*, 1094–1097. [CrossRef]

80. Rezchikov, A.F.; Kochetkov, A.V.; Zakharov, O.V. Mathematical models for estimating the degree of influence of major factors on performance and accuracy of coordinate measuring machines. *MATEC Web Conf.* **2017**, *129*, 01054. [CrossRef]

81. Zakharov, O.V.; Balaev, A.F.; Kochetkov, A.V. Modeling Optimal Path of Touch Sensor of Coordinate Measuring Machine Based on Traveling Salesman Problem Solution. *Procedia Eng.* **2017**, *206*, 1458–1463. [CrossRef]

82. Grigoriev, S.N.; Metel, A.S.; Tarasova, T.V.; Filatova, A.A.; Sundukov, S.K.; Volosova, M.A.; Okunkova, A.A.; Melnik, Y.A.; Podrabinnik, P.A. Effect of Cavitation Erosion Wear, Vibration Tumbling, and Heat Treatment on Additively Manufactured Surface Quality and Properties. *Metals* **2020**, *10*, 1540. [CrossRef]

83. Lou, X.Y.; Andresen, P.L.; Rebak, R.B. Oxide inclusions in laser additive manufactured stainless steel and their effects on impact toughness and stress corrosion cracking behavior. *J. Nucl. Mater.* **2018**, *499*, 182–190. [CrossRef]

84. Zai, L.; Zhang, C.; Wang, Y.; Guo, W.; Wellmann, D.; Tong, X.; Tian, Y. Laser Powder Bed Fusion of Precipitation-Hardened Martensitic Stainless Steels: A Review. *Metals* **2020**, *10*, 255. [CrossRef]

85. Metel, A.S.; Stebulyanin, M.M.; Fedorov, S.V.; Okunkova, A.A. Power Density Distribution for Laser Additive Manufacturing (SLM): Potential, Fundamentals and Advanced Applications. *Technologies* **2019**, *7*, 5. [CrossRef]

86. Cheng, B.; Shrestha, S.; Chou, K. Stress and deformation evaluations of scanning strategy effect in selective laser melting. *Addit. Manuf.* **2016**, *12B*, 240–251.

87. Aota, L.S.; Bajaj, P.; Zilnyk, K.D.; Ponge, D.; Sandim, H.R.Z. The origin of abnormal grain growth upon thermomechanical processing of laser powder-bed fusion alloys. *Materialia* **2021**, *20*, 101243. [CrossRef]

88. Gong, H.; Nadimpalli, V.K.; Rafi, K.; Starr, T.; Stucker, B. Micro-CT Evaluation of Defects in Ti-6Al-4V Parts Fabricated by Metal Additive Manufacturing. *Technologies* **2019**, *7*, 44. [CrossRef]

89. Zhang, P.; He, A.N.; Liu, F.; Zhang, K.; Jiang, J.; Zhang, D.Z. Evaluation of Low Cycle Fatigue Performance of Selective Laser Melted Titanium Alloy Ti–6Al–4V. *Metals* **2019**, *9*, 1041. [CrossRef]

90. Loginova, I.S.; Bykovskiy, D.P.; Solonin, A.N.; Prosviryakov, A.S.; Cheverikin, V.V.; Pozdniakov, A.V.; Petrovskiy, V.N. Peculiarities of the Microstructure and Properties of Parts Produced by the Direct Laser Deposition of 316L Steel Powder. *Russ. J. Non-Ferr. Met.* **2019**, *60*, 87–94. [CrossRef]

91. Pelevin, I.A.; Ozherelkov, D.Y.; Chernyshikhin, S.V.; Nalivaiko, A.Y.; Gromov, A.A.; Chzhan, V.B.; Terekhin, E.A.; Tereshina, I.S. Selective laser melting of Nd-Fe-B: Single track study. *Mater. Lett.* **2022**, *315*, 131947. [CrossRef]

92. Tan, J.H.; Wong, W.L.E.; Dalgarno, K.W. An overview of powder granulometry on feedstock and part performance in the selective laser melting process. *Addit. Manuf.* **2017**, *18*, 228–255. [CrossRef]

93. Zhang, X.B.; Cheng, B.; Tuffile, C. Simulation study of the spatter removal process and optimization design of gas flow system in laser powder bed fusion. *Addit. Manuf.* **2020**, *32*, 101049. [CrossRef]

94. Vilanova, M.; Escribano-García, R.; Guraya, T.; San Sebastian, M. Optimizing Laser Powder Bed Fusion Parameters for IN-738LC by Response Surface Method. *Materials* **2020**, *13*, 4879. [CrossRef] [PubMed]

95. Ogawahara, M.; Sasaki, S. Effects of Defects and Inclusions on the Fatigue Properties of Inconel 718 Fabricated by Laser Powder Bed Fusion Followed by HIP. *Mater. Trans.* **2021**, *62*, 631–635. [CrossRef]

96. Aliprandi, P.; Giudice, F.; Guglielmino, E.; Sili, A. Tensile and Creep Properties Improvement of Ti-6Al-4V Alloy Specimens Produced by Electron Beam Powder Bed Fusion Additive Manufacturing. *Metals* **2019**, *9*, 1207. [CrossRef]

97. Fang, M.H.; Hu, F.G.; Han, Y.F.; Le, J.W.; Xi, J.J.; Song, J.W.; Ke, L.D.; Xiao, M.L.; Lu, W.J. Controllable mechanical anisotropy of selective laser melted Ti6Al4V: A new perspective into the effect of grain orientations and primary grain structure. *Mater. Sci. Eng. A-Struct. Mater. Prop. Microstruct. Process.* **2021**, *827*, 142031. [CrossRef]

98. Liu, J.; Liu, J.; Li, Y.; Zhang, R.; Zeng, Z.; Zhu, Y.; Zhang, K.; Huang, A. Effects of Post Heat Treatments on Microstructures and Mechanical Properties of Selective Laser Melted Ti6Al4V Alloy. *Metals* **2021**, *11*, 1593. [CrossRef]

99. Wang, C.; Yi, J.Z.; Qin, L.Y.; Wang, W.D.; Wang, X.M.; Yang, G. Effects of double annealing on microstructure and mechanical properties of laser melting deposition TA15 titanium alloy. *Mater. Res. Express* **2019**, *6*, 116526. [CrossRef]

100. Lin, D.Y.; Xu, L.Y.; Li, X.J.; Jing, H.Y.; Qin, G.; Pang, H.N.; Minami, F. A Si-containing FeCoCrNi high-entropy alloy with high strength and ductility synthesized in situ via selective laser melting. *Addit. Manuf.* **2020**, *35*, 101340. [CrossRef]

101. Miao, X.; Liu, X.; Lu, P.; Han, J.; Duan, W.; Wu, M. Influence of Scanning Strategy on the Performances of GO-Reinforced Ti6Al4V Nanocomposites Manufactured by SLM. *Metals* **2020**, *10*, 1379. [CrossRef]

102. Vrancken, B.; Ganeriwala, R.K.; Matthews, M.J. Analysis of laser-induced microcracking in tungsten under additive manufacturing conditions: Experiment and simulation. *Acta Mater.* **2020**, *194*, 464–472. [CrossRef]

103. Tian, Y.; Tomus, D.; Rometsch, P.; Wu, X. Influences of processing parameters on surface roughness of Hastelloy X produced by selective laser melting. *Addit. Manuf.* **2017**, *13*, 103–112. [CrossRef]

104. Tangsopha, W.; Thongsri, J.; Busayaporn, W. Simulation of ultrasonic cleaning and ways to improve the efficiency. In Proceedings of the 5th International Electrical Engineering Congress, Pattaya, Thailand, 8–10 March 2017; IEEE345: New York, NY, USA, 2017.

105. Fatyukhin, D.S.; Nigmetzyanov, R.I.; Prikhodko, V.M.; Sukhov, A.V.; Sundukov, S.K. A Comparison of the Effects of Ultrasonic Cavitation on the Surfaces of 45 and 40Kh Steels. *Metals* **2022**, *12*, 138. [CrossRef]

106. Gavrilov, N.V.; Mesyats, G.A.; Nikulin, S.P.; Radkovskii, G.V.; Elkind, A.; Perry, A.J.; Treglio, J.R. New broad beam gas ion source for industrial application. *J. Vac. Sci. Technol. A* **1996**, *14*, 1050–1055. [CrossRef]

107. Kolikov, V.; Bogomaz, A.; Budin, A. *Powerful Pulsed Plasma Generators: Research and Application*; Springer: Cham, Switzerland, 2018; Introduction Chapter; pp. 1–12.

108. Walton, S.G.; Cothran, C.D.; Amatucci, W.E. Electron Beam Propagation in Magnetic Fields. *IEEE Trans. Plasma Sci.* **2011**, *39*, 2574–2575. [CrossRef]

109. Yu, Y.C.; Bai, S.; Wang, S.T.; Hu, A.M. Ultra-Short Pulsed Laser Manufacturing and Surface Processing of Microdevices. *Engineering* **2018**, *4*, 779–786. [CrossRef]

110. Korkmaz, K.; Ribalko, A.V. Effect of pulse shape and energy on the surface roughness and mass transfer in the electrospark coating process. *Kov. Mater. Met. Mater.* **2011**, *49*, 265–270. [CrossRef]

111. Feng, S.; Xia, M.; Ge, C.C. Consecutive induction melting of nickel-based superalloy in electrode induction gas atomization. *Chin. Phys. Soc.* **2017**, *26*, 6. [CrossRef]

112. Antonov, D.; Strizhak, P. Explosive Disintegration of Two-Component Droplets in a Gas-Flow at Its Turbulization. *Therm. Sci.* **2019**, *23*, 2983–2993. [CrossRef]

113. Ivanov, Y.F.; Lopatin, I.V.; Petrikova, E.A.; Rygina, M.E.; Tolkachev, O.S.; Shimanskii, V.I. Structure and Properties of Silumin Surface after Vacuum Arc Plasma-Assisted Deposition of Coatings Irradiated by Low Energy High Current Pulsed Electron Beam. *Russ. Phys. J.* **2020**, *62*, 2106–2111. [CrossRef]

114. Bardos, L.; Barankova, H. Cold atmospheric plasma: Sources, processes, and applications. *Thin Solid Film.* **2010**, *518*, 6705–6713. [CrossRef]
115. Jiang, S.; Qiu, L.W.; Li, Z.; Zhang, L.; Rao, J.F. A New All-Solid-State Bipolar High-Voltage Multilevel Generator for Dielectric Barrier Discharge. *IEEE Trans. Plasma Sci.* **2020**, *48*, 1076–1081. [CrossRef]
116. Tatar, N.; Tuzlah, M.; Bahce, E. Investigation of the Lattice Production of Removable Dental Prostheses with CoCr Alloy Using Additive Manufacturing. *J. Mater. Eng. Perform.* **2021**, *30*, 6722–6731. [CrossRef]
117. Navinsek, B.; Panjan, P.; Milosev, I. PVD coatings as an environmentally clean alternative to electroplating and electroless processes. *Surf. Coat. Technol.* **1999**, *116*, 476–487. [CrossRef]
118. Kovalenko, V.A. Use of the vacuum arc discharge for the surface modification. *Fundam. Mech. Low-Energy-Beam-Modif. Surf. Growth Process.* **2000**, *585*, 129–134. [CrossRef]
119. Kikuchi, E.; Kikuchi, Y.; Hirao, M. Monitoring and Analysis of Solvent Emissions from Metal Cleaning Processes for Practical Process Improvement. *Ann. Occup. Hyg.* **2012**, *56*, 829–842. [PubMed]
120. Yoon, S.S.; Khang, D.Y. Facile and Clean Release of Vertical Si Nanowires by Wet Chemical Etching Based on Alkali Hydroxides. *Small* **2013**, *9*, 905–912. [CrossRef]
121. Razab, M.K.A.A.; Noor, A.M.; Jaafar, M.S.; Abdullah, N.H.; Suhaimi, F.M.; Mohamed, M.; Adam, N.; Yusuf, N.A.A.N. A review of incorporating Nd:YAG laser cleaning principal in automotive industry. *J. Radiat. Res. Appl. Sci.* **2018**, *11*, 393–402. [CrossRef]
122. Kuznetsov, A.P.; Buzinskij, O.I.; Gubsky, K.L.; Nikitina, E.A.; Savchenkov, A.V.; Tarasov, B.A.; Tugarinov, S.N. Development of a laser cleaning method for the first mirror surface of the charge exchange recombination spectroscopy diagnostics on ITER. *Phys. At. Nucl.* **2015**, *78*, 1677–1685. [CrossRef]
123. Metel, A.; Grigoriev, S.; Volosova, M.; Melnik, Y.; Mustafaev, E. Synthesis of aluminum nitride coatings assisted by fast argon atoms in a magnetron sputtering system with a separate input of argon and nitrogen. *Surf. Coat. Technol.* **2020**, *398*, 126078. [CrossRef]
124. Mizerov, A.M.; Timoshnev, S.N.; Nikitina, E.V.; Sobolev, M.S.; Shubin, K.Y.; Berezovskaia, T.N.; Mokhov, D.V.; Lundin, W.V.; Nikolaev, A.E.; Bouravleuv, A.D. On the Specific Features of the Plasma-Assisted MBE Synthesis of n^+-GaN Layers on GaN/c-Al$_2$O$_3$ Templates. *Semiconductors* **2019**, *53*, 1187–1191. [CrossRef]
125. Kamensky, E. Digital technologies in the Russians' everyday life: Analysis based on the opinion surveys. *Ann. XXI* **2020**, *186*, 134–142. [CrossRef]
126. Dai, Y.; Zhang, M.; Li, Q.; Wen, L.; Wang, H.; Chu, J. Separated Type Atmospheric Pressure Plasma Microjets Array for Maskless Microscale Etching. *Micromachines* **2017**, *8*, 173. [CrossRef]
127. Su, X.; Ji, C.; Xu, Y.; Li, D.; Walker, D.; Yu, G.; Li, H.; Wang, B. Surface Texture Evolution of Fused Silica in a Combined Process of Atmospheric Pressure Plasma Processing and Bonnet Polishing. *Coatings* **2019**, *9*, 676. [CrossRef]
128. Yen, T.; Chin, A.; Gritsenko, V. High Performance All Nonmetal SiNx Resistive Random Access Memory with Strong Process Dependence. *Sci. Rep.* **2020**, *10*, 2807. [CrossRef] [PubMed]
129. Solovyev, A.A.; Lebedynskiy, A.M.; Shipilova, A.V.; Ionov, I.V.; Smolyanskiy, E.A.; Lauk, A.L.; Remnev, G.E.; Maslov, A.S. Scale-up of Solid Oxide Fuel Cells with Magnetron Sputtered Electrolyte. *Fuel Cells* **2017**, *17*, 378–382. [CrossRef]
130. Zolotukhin, D.B.; Oks, E.M.; Tyunkov, A.V.; Yakovlev, E.V.; Yushkov, Y.G. Effect of a dielectric cavity on the ion etching of dielectrics by electron beam-produced plasma generated by a forevacuum plasma electron source. *Vacuum* **2021**, *192*, 110483. [CrossRef]
131. Marinov, D.; de Marneffe, J.F.; Smets, Q.; Arutchelvan, G.; Bal, K.M.; Voronina, E.; Rakhimova, T.; Mankelevich, Y.; El Kazzi, S.; Mehta, A.N.; et al. Reactive plasma cleaning and restoration of transition metal dichalcogenide monolayers. *NPJ 2D Mater. Appl.* **2021**, *5*, 17. [CrossRef]
132. Rakhimberlinova, Z.B.; Takibayeva, A.T.; Nazarova, O.G.; Iskakov, A.R.; Musina, G.N.; Kulakov, I.V. Activation Method of Cleaning Process Gas. *News Natl. Acad. Sci. Repub. Kazakhstan Ser. Chem. Technol.* **2020**, *3*, 73–79. [CrossRef]
133. Vizir, A.V.; Oks, E.M.; Yushkov, G.Y. Broad-beam high-current DC ion source based on a two-stage glow discharge plasma. *Rev. Sci. Instrum.* **2010**, *81*, 02B304. [CrossRef]
134. Gusev, V.V.; Dolgopolov, V.M.; Slovetskii, D.I.; Shelykhmanov, E.F. Spectral Features of a Hf Discharge Plasma in Etching Aluminum. *High Temp.* **1983**, *21*, 19–25.
135. Yang, S.Y.; Abelson, J.R. Amorphous-Silicon Alloys on C-Si-Influence of Substrate Cleaning and Ion-Bombardment on Film Adhesion and Microstructure. *J. Vac. Sci. Technol. A Vac. Surf. Film.* **1993**, *11*, 1327–1331. [CrossRef]
136. Ye, C.; Xu, Y.J.; Huang, X.J.; Ning, Z.Y. Effect of high-frequency on etching of SiCOH films in CHF3 dual-frequency capacitively coupled plasmas. *Thin Solid Film.* **2010**, *518*, 3223–3227. [CrossRef]
137. Denisov, V.V.; Akhmadeev, Y.H.; Koval, N.N.; Kovalsky, S.S.; Lopatin, I.V.; Ostroverkhov, E.V.; Pedin, N.N.; Yakovlev, V.V.; Schanin, P.M. The source of volume beam-plasma formations based on a high-current non-self-sustained glow discharge with a large hollow cathode. *Phys. Plasmas* **2019**, *26*, 123510. [CrossRef]
138. Lopatin, I.V.; Schanin, P.M.; Akhmadeev, Y.H.; Kovalsky, S.S.; Koval, N.N. Self-sustained low pressure glow discharge with a hollow cathode at currents of tens of amperes. *Plasma Phys. Rep.* **2012**, *38*, 585–589. [CrossRef]
139. Li, T.J.; Cui, S.H.; Liu, L.L.; Li, X.Y.; Wu, Z.C.; Ma, Z.Y.; Fu, R.K.Y.; Tian, X.B.; Chu, P.K.; Wu, Z.Z. Magnetic field optimization and high-power discharge characteristics of cylindrical sputtering cathode. *Acta Phys. Sin.* **2021**, *70*, 045202.

140. Raoux, S.; Tanaka, T.; Bhan, M.; Ponnekanti, H.; Seamons, M.; Deacon, T.; Xia, L.Q.; Pham, F.; Silvetti, D.; Cheung, D.; et al. Remote microwave plasma source for cleaning chemical vapor deposition chambers: Technology for reducing global warming gas emissions. *J. Vac. Sci. Technol. B* **1999**, *17*, 477–485. [CrossRef]
141. Watanabe, Y.; Elliott, S.; Firsov, A.; Houpt, A.; Leonov, S. Rapid control of force/momentum on a model ramp by quasi-DC plasma. *J. Phys. D Appl. Phys.* **2019**, *52*, 444003. [CrossRef]
142. Jiang, N.; Ji, A.L.; Cao, Z.X. Atmospheric pressure plasma jet: Effect of electrode configuration, discharge behavior, and its formation mechanism. *J. Appl. Phys.* **2009**, *106*, 013308. [CrossRef]
143. Sosnin, E.A.; Goltsova, P.A.; Panarin, V.A.; Skakun, V.S.; Tarasenko, V.F.; Didenko, M.V. Formation of Nitrogen Oxides in an Apokamp-Type Plasma Source. *Russ. Phys. J.* **2017**, *60*, 701–705. [CrossRef]
144. Es'kin, V.A.; Kudrin, A.V. Stationary Structure of a High-Frequency Discharge Maintained by a Distributed Electromagnetic Source in the Presence of an External Magnetic Field. *J. Exp. Theor. Phys.* **2010**, *110*, 701–709. [CrossRef]
145. Sun, D.L.; Hong, R.Y.; Wang, F.; Liu, J.Y.; Kumar, M.R. Synthesis and modification of carbon nanomaterials via AC arc and dielectric barrier discharge plasma. *Chem. Eng. J.* **2016**, *283*, 9–20. [CrossRef]
146. Kim, J.; Kim, S.-j.; Lee, Y.-N.; Kim, I.-T.; Cho, G. Discharge Characteristics and Plasma Erosion of Various Dielectric Materials in the Dielectric Barrier Discharges. *Appl. Sci.* **2018**, *8*, 1294. [CrossRef]
147. Li, Z.; Shi, Z.W.; Sun, Q.J.; Wei, C.Y.; Geng, X. Analysis of flow separation control using nanosecond-pulse discharge plasma actuators on a flying wing. *Plasma Sci. Technol.* **2018**, *20*, 115504. [CrossRef]
148. Nguyen, D.S.; Park, H.S.; Lee, C.M. Effect of cleaning gas stream on products in selective laser melting. *Mater. Manuf. Process.* **2019**, *34*, 455–461. [CrossRef]
149. Gabdrakhmanov, A.T.; Israphilov, I.H.; Shafigullin, L.N. The metal surface cleaning using a vapor-gas discharge. *IOP Conf. Ser. J. Phys. Conf. Ser.* **2018**, *1058*, 012009. [CrossRef]
150. Bolotov, M.G.; Prybytko, I.O. Application of Glow Discharge Plasma for Cleaning (Activation) and Modification of Metal Surfaces while Welding, Brazing, and Coating Deposition. *Prog. Phys. Met.* **2021**, *22*, 103–128. [CrossRef]
151. Sosnin, E.A.; Panarin, V.A.; Skakun, V.S.; Tarasenko, V.F.; Pechenitsin, D.S.; Kuznetsov, V.S. Source of an atmospheric-pressure plasma jet formed in air or nitrogen under barrier discharge excitation. *Tech. Phys.* **2016**, *61*, 789–792. [CrossRef]
152. Rao, J.F.; Liu, K.F.; Qiu, J. An Efficient All Solid-State Pulsed Generator for Pulsed Discharges. In Proceedings of the 2012 IEEE International Power Modulator and High Voltage Conference, San Diego, CA, USA, 3–7 June 2012; pp. 473–476.
153. Bocharnikov, V.M.; Volodin, V.V.; Golub, V.V.; Trifanov, I.V.; Serebryakov, D.G. Resonance Properties of a Low-Pressure Dielectric Barrier Discharge. *Tech. Phys.* **2020**, *65*, 1969–1974. [CrossRef]
154. Tao, X.P.; Lu, R.D.; Li, H. Electrical Characteristics of Dielectric-Barrier Discharges in Atmospheric Pressure Air Using a Power-Frequency Voltage Source. *Plasma Sci. Technol.* **2012**, *14*, 723–727. [CrossRef]
155. Erofeev, M.; Baksht, E.; Tarasenko, V. Miniaturized ultraviolet sources driven by dielectric barrier discharge and runaway electron preionized diffuse discharge. *Opt. Appl.* **2014**, *44*, 475–487.
156. Yang, L.Z.; Liu, Z.W.; Mao, Z.G.; Li, S.; Chen, Q. Formation and characteristics of patterns in atmospheric-pressure radio-frequency dielectric barrier discharge plasma. *Jpn. J. Appl. Phys.* **2017**, *56*, 01AC02. [CrossRef]
157. Brandenburg, R. Dielectric barrier discharges: Progress on plasma sources and on the understanding of regimes and single filaments. *Plasma Sources Sci. Technol.* **2018**, *27*, 079501. [CrossRef]
158. Timshina, M.; Eliseev, S.; Kalinin, N.; Letunovskaya, M.; Burtsev, V. Numerical investigation of dynamics and gas pressure effects in a nanosecond capillary sliding discharge. *J. Appl. Phys.* **2019**, *125*, 143302. [CrossRef]
159. Samaei, A.; Chaudhuri, S. Multiphysics modeling of metal surface cleaning using atmospheric pressure plasma. *J. Appl. Phys.* **2020**, *128*, 054903. [CrossRef]
160. Nasreen, A.; Shaker, K.; Nawab, Y. Effect of surface treatments on metal–composite adhesive bonding for high-performance structures: An overview. *Compos. Interfaces* **2021**, *28*, 1221–1256. [CrossRef]
161. Kuznetsov, V.S.; Sosnin, E.A.; Panarin, V.A.; Skakun, V.S.; Pechenitsyn, D.S.; Tarasenko, V.F. Observation of Streamer Coronas Preceding the Formation of an Apokampic Discharge. *Russ. Phys. J.* **2019**, *62*, 992–995. [CrossRef]
162. Emel'yanov, A.A.; Plotnikov, M.Y.; Rebrov, A.K.; Timoshenko, N.I.; Yudin, I.B. The Use of a Supersonic Jet of Gas Activated in a Microwave Discharge in Diamond Deposition. *Fluid Dyn.* **2021**, *56*, 106–115. [CrossRef]
163. Zorin, V.G.; Lazebnik, B.S.; Lunin, N.V.; Markov, G.A. Resonance Super-High Frequency Discharge in a Magnetic-Field. *Zhurnal Tekhnicheskoi Fiz.* **1975**, *45*, 1006–1009.

Review

Surface Laser Treatment of Cast Irons: A Review

Néstor Catalán [1,*], Esteban Ramos-Moore [2,3], Adrian Boccardo [4,5,6] and Diego Celentano [1,3,*]

[1] Departamento de Ingeniería Mecánica y Metalúrgica, Pontificia Universidad Católica de Chile, Av. Vicuña Mackenna 4860, Macul, Santiago 7820436, Chile

[2] Laboratorio de Películas Delgadas UC, Instituto de Física, Pontificia Universidad Católica de Chile, Casilla 306, Av. Vicuña Mackenna 4860, Macul, Santiago 7820436, Chile; evramos@uc.cl

[3] Centro de Investigación en Nanotecnología y Materiales Avanzados (CIEN-UC), Pontificia Universidad Católica de Chile, Av. Vicuña Mackenna 4860, Macul, Santiago 7820436, Chile

[4] Mechanical Engineering, School of Engineering, College of Science and Engineering, National University of Ireland (NUI), University Road, H91 HX31 Galway, Ireland; adrian.boccardo@nuigalway.ie

[5] I-Form Advanced Manufacturing Research Centre, NUI, University Road, H91 HX31 Galway, Ireland

[6] Instituto de Estudios Avanzados en Ingeniería y Tecnología (IDIT), CONICET-Universidad Nacional de Córdoba, Vélez Sarsfield 1611, Córdoba X5000, Argentina

* Correspondence: nacatalan@uc.cl (N.C.); dcelentano@ing.puc.cl (D.C.)

Abstract: Heat treatments are frequently used to modify the microstructure and mechanical properties of materials according to the requirements of their applications. Laser surface treatment (LST) has become a relevant technique due to the high control of the parameters and localization involved in surface modification. It allows for the rapid transformation of the microstructure near the surface, resulting in minimal distortion of the workpiece bulk. LST encompasses, in turn, laser surface melting and laser surface hardening techniques. Many of the works devoted to studying the effects of LST in cast iron are diverse and spread in several scientific communities. This work aims to review the main experimental aspects involved in the LST treatment of four cast-iron groups: gray (lamellar) cast iron, pearlitic ductile (nodular) iron, austempered ductile iron, and ferritic ductile iron. The effects of key experimental parameters, such as laser power, scanning velocity, and interaction time, on the microstructure, composition, hardness, and wear are presented, discussed, and overviewed. Finally, we highlight the main scientific and technological challenges regarding LST applied to cast irons.

Keywords: laser treatment; cast irons; microstructure; mechanical properties; wear

Citation: Catalán, N.; Ramos-Moore, E.; Boccardo, A.; Celentano, D. Surface Laser Treatment of Cast Irons: A Review. *Metals* **2022**, *12*, 562. https://doi.org/10.3390/met12040562

Academic Editor: Sergey N. Grigoriev

Received: 16 February 2022
Accepted: 24 March 2022
Published: 26 March 2022

1. Introduction

Over the last few decades, cast irons have been widely used in a variety of applications, such as for shafts, axles, engines, and gears in automobile parts or general industrial machinery [1–4], due to their high machinability, good mechanical properties, and low cost compared to other alloys.

Because of the increased production volume, an exhaustive characterization of the mechanical properties and fatigue behavior of cast irons has been extensively reported in the past literature. In addition, several tribological studies have been performed to address the wear mechanisms under diverse sliding conditions, as well as to establish a comparison of their wear rates against other industrial materials.

In general, intense loads in sliding tests reveal higher wear rates in cast irons than in steels. Thus, in extreme wear applications, e.g., mining, marine, aeronautic, or military industries, it is frequently required to improve their hardness and wear resistance in critical points of the material to meet requirements related to corrosion or fracture/fatigue failures. For this purpose, many conventional treatments have been developed and implemented to enhance the performance of cast irons in specific applications. Nonetheless, these techniques usually involve a full transformation of the workpiece, altering many of its beneficial bulk properties.

In this context, laser surface treatment (LST) has emerged as a novel opportunity to improve the useful life of cast irons. This technique is characterized by a precise, clean, and fast thermal process, where the heat input from a laser beam raises the temperature at the surface of the specimen, and then it is rapidly conducted into the rest of the material. The objective of LST is to achieve temperatures above the point of critical transformation (austenization or melting temperature), following a self-quenching process from the unaffected case bulk. Depending on the cooling and solidification rates to which they are exposed during the treatment, new phases can arise, directly influencing the new mechanical properties of the material.

Past articles and reviews have addressed, in general terms, the benefits and challenges of LST of a variety of cast irons. However, a thorough comparison and analysis of the importance of the initial microstructure and laser properties is yet to be performed. Therefore, this article presents a review of the literature according to LST of industrial iron castings, in order to understand the similarities and differences associated with these parameters and to serve as a guide to a reliable implementation and optimization of this technique.

First, a brief description of typical cast irons is provided, highlighting the different morphologies and conventional heat treatments that are commonly conducted in each one. Secondly, a more detailed theoretical basis of LST is presented, along with a summary of standard laser configurations used in metallic surface treatments, i.e., laser type, power, and scanning velocity. Moreover, the influence of other advanced parameters in the efficiency of LST is remarked. Afterward, an exploration of the results obtained in LST of gray cast irons (GIs) is offered to establish the influence of graphite morphology on the microstructural transformations that the treatment induces, as well as the resulting properties of the samples. Then the most relevant works for ductile irons (DIs) with predominantly pearlitic matrices are exposed, emphasizing the results as a function of laser configurations and characterization tests. Subsequently, investigations on laser-modified austempered ductile iron (ADI) castings are reviewed, mainly highlighting the differences in the scope of the treatment in terms of hardness, wear resistance, and depth of the affected area. Articles dealing with ferritic nodular cast iron treatments are then presented, addressing the main challenges of transforming the microstructure without melting the original matrix. To complete the review, a discussion of LST across all types of cast irons is given, highlighting the differences in achieved microstructure, hardness, and wear-related properties as a function of base material and laser operating parameters. At the same time, the main advantages and limitations of the treatments are presented, noting the improvement in overall performance compared to as-cast specimens. Finally, future technological, economic, and investigation areas related to the LST of cast irons are suggested.

2. Cast Irons

Cast irons comprise a large family of solid ferrous alloys. The main difference with steels is the higher carbon (C) and silicon (Si) content, with the richer carbon phase being critical in the microstructural transformations of the matrix. Additionally, significant amounts of manganese (Mn), phosphorus (P), and sulfur (S), as well as minor contents of molybdenum (Mo), chromium (Cr), and nickel (Ni), can be found in these castings. The most common classification for cast irons is based on microstructural features. The first feature is graphite morphology, where cast irons can be classified mainly as lamellar (gray) or nodular (spheroidal or ductile). On the other hand, the matrix phase can be used to identify cast irons as ferritic, pearlitic, martensitic, austenitic, or bainitic [5].

GIs are a broad class of casting alloys characterized by a lamellar (flake) graphite disposal, usually in a ferrous matrix. They are traditionally used in industrial applications, due to their good castability, flexible mechanical properties, and low cost compared to steel [6,7]. A variety of fatigue and crack growth models have been performed [8–10] to predict the fatigue life of GIs in standard applications, while other tribological studies [11–16] have identified the main wear mechanisms under dry and lubricated conditions, emphasizing the high variability in wear rates in different tribo-systems. Graphite plays a major lubricating role that helps reduce the wear coefficient. However, a cracking tendency is observed under high loads, and adhesion is the main wear mechanism in lubricated environments.

On the other hand, DI is a ferrous material containing graphite in a small and rounded shape, due to the nodulizing effect of magnesium, and it is present in higher concentrations. In comparison, DI exhibits higher strength and ductility, which is the main reason for its growth in industrial applications [5]. The effects of cooling rate and matrix type (i.e., pearlitic/ferritic) on the mechanical properties of DIs have been explored both experimentally and numerically [17–22], figuring that an increment of the ferritic or pearlitic phase can lead to very dissimilar properties. Thus, an appropriate balance of the alloying elements and the initial microstructure is needed to achieve the desired performance. In this regard, studies on crack growth have been conducted [23,24] that show a comparable behavior to that of GIs. However, under similar conditions, DI presents better fatigue resistance due to the higher stress intensity factor and the nodular shape of graphite that blunts the crack, while the lamellar shape of GI exhibits an anisotropic behavior that leads to stress concentrations at the interface between the matrix and the flake tips. Furthermore, wear analyses of DIs have been carried out to understand the underlying wear mechanisms and to compare the removal rates with those of industrial steels [4,25–28]. In mild wear environments, plowing is the predominant wear mechanism, and as the load applied increases, sliding wear becomes more severe, due to enhancement of the plastic deformation and initiation of cracks in the subsurface. Moreover, the influence of both bulk and matrix microhardness in the wear rates of ferrous alloys has been addressed, revealing that some initial cast irons microstructures could lead to tribological performances in the range of soft industrial steels, e.g., 52,100 and 1070.

Over the last decades, austempered ductile cast iron (ADI) has emerged as a novel option for achieving a good combination of mechanical properties. Because of its low cost and good machinability, ADIs have gained more interest in automotive and agricultural applications [2,3,29]. The characteristic ausferritic matrix is achieved by a pre-heating cycle, whose quality is primarily influenced by the processing window [30], i.e., the time interval between the ausferritic transformation and the carbide precipitation (Figure 1). Additional modifications of the austempering process have been explored to enhance the mechanical properties of ADIs [31,32].

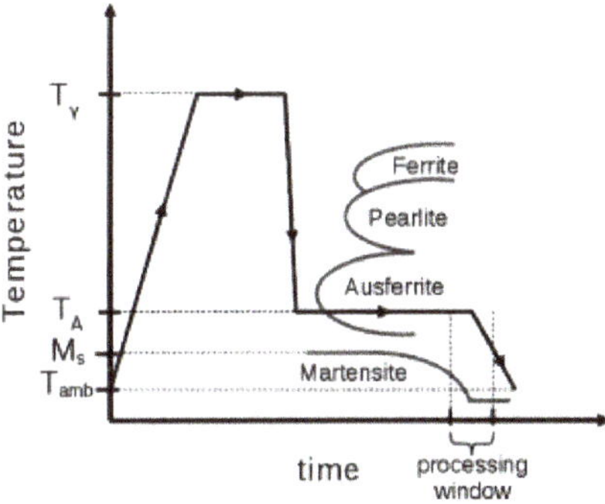

Figure 1. Phase transformations and variables of a conventional austempering heat treatment (Reprinted with permission from ref. [30]. 2015 Springer Nature).

The main limitations of ADI are its lower machinability and ductility when compared to as-cast ductile iron, due to higher austenitic content and the absence of pro-eutectoid ferrite, responsible for the work hardening of the material [29]. Nonetheless, tribological studies have demonstrated a superior ADI wear resistance in comparison to both GI and DI [33]. This difference is attributed to the higher toughness achieved by the bainitic microstructure of ADI, as it leads to severe plastic deformation in the direction of sliding with minimum material loss. Moreover, wear resistance can be further enhanced in ADIs based in the pre-austempered iron matrix, as graphite morphology plays a significant role in the control of crack propagation and thermal conductivity of the material and, therefore, of its thermal fatigue resistance.

Alternatively, the improvement of mechanical properties to fulfill industrial requirements has been achieved by applying conventional heat treatments on as-cast iron samples that can modify their entire primary microstructure. The most common treatments consist of normalizing, annealing, or quench-tempering, where furnaces are operated to provide uniform heating through the metallic piece until a critical temperature is reached, since, in the absence of chromium, iron carbides can be dissociated into austenite and graphite at annealing temperatures [34]. The microstructural transformations are also influenced by the cooling rate, which varies with the selected treatment.

The main results of normalizing are the enhancement of different properties, such as hardness, impact toughness, and tensile strength, and can be optimized with selected temperature, holding time, and cooling rate [35]. If the goal is to achieve high ductility and machinability at the expense of strength, then full annealing is conducted. The matrix transformations include the removal of cementite and decomposition of pearlite into graphite and ferrite [36]. Quench-tempering is used to mostly improve the wear resistance and strength of cast irons. The resulting microstructure of GIs consists of a retained austenite matrix with carbon-saturated martensite plates that becomes finer with higher alloying elements content [37], while a primary martensitic structure with small soft ferrite regions adjacent to the graphite nodules can be achieved in DIs [38].

3. Laser Processing of Cast Irons

As reviewed in the previous section, conventional heat treating of cast irons has been extensively used to enhance their performance to higher standards. However, it has been acknowledged that altering the properties of the entire bulk, e.g., to achieve a hardening effect, can be detrimental to the machinability advantages of these materials. Therefore, LST has appeared as a major opportunity to efficiently improve the overall performance of cast irons.

This technique is characterized by a highly localized, chemically clean, automatable, and fast thermal process, where only controlled regions of the material are hardened during the treatment. Steen and Mazumder [39] extensively detail the theoretical basis of LST, in which a laser beam is defocused or oscillated to irradiate a small area with high power density, and the relative motion between the beam and the sample, usually provided by a CNC table, allows users to precisely cover complex geometries (Figure 2). The laser heat supply raises the temperature at the surface of the specimen, which is transferred into the metallic body by thermal conduction.

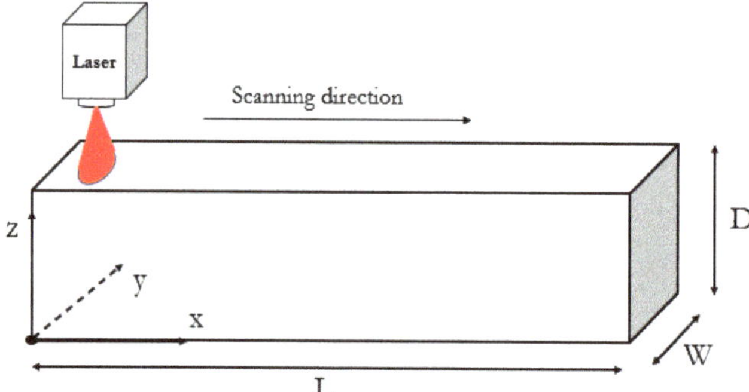

Figure 2. Schematic representation of LST in rectangular samples.

The objective of LST is to achieve temperatures above the point of critical transformation, which is the austenitizing temperature (solid-state transformation) in the case of laser surface hardening (LSH), or the melting temperature in the case of laser surface melting (LSM). Then a rapid self-quenching process takes place, because of the large volume of adjacent and unaffected material, resulting in a controlled low distortion of the workpiece. Since the heating rates are very high, the initial transformations take place under conditions far from equilibrium, while phase transformations on cooling can be addressed with the help of suitable continuous cooling transformation (CCT) diagrams. Depending on the cooling and solidification rates to which they are exposed during the treatment, the final microstructure directly influences the new mechanical properties of the material.

Laser surface modification of metallic samples, including cast irons, has been performed for over fifty years. As one of the first available tools, CO_2 lasers have been widely employed to achieve both LSH and LSM, usually aided by the application of a thin coating, in order to ensure higher absorption and thermal efficiency. However, modern breakthroughs have allowed the use of more efficient lasers for these purposes, such as Nd:YAG and diode lasers, where LST has been carried out with and without coatings with successful results. As shown in Table 1, other LST features include setting an appropriate scanning velocity, the laser power distribution, spot geometry and size, the number of scanned tracks (with or without overlapping), and the use of shielding gas to prevent oxidation. As some authors define, the linear energy density (or heat input) is a global parameter that includes the effect of both laser power and scanning speed by the following expression:

$$\overline{E} = \frac{P}{v},\tag{1}$$

where $\overline{E}$ is the heat input, P is the laser power, and v is the scanning velocity. In circular spots, this definition can be modified to consider the geometry of the irradiation zone and the energy distribution.

Table 1. Typical parameters and characterization tests used in LST of cast irons.

LST Parameter	Regular Value/Selection
Laser type	CO_2, Nd:YAG, fiber (diode)
Coating (optional)	Graphite, manganese/zinc phosphate
Shielding gas (optional)	Argon, nitrogen, helium
Laser power	Fixed, PID controller, linear ramp
Power distribution	Gaussian, uniform
Laser scanning velocity	Fixed, single, or multiple passes with overlapping
Spot geometry	Circular, elliptical, rectangular
Characterization tests	OM/SEM, macro/micro hardness, GDOES, XRD, wear/erosion resistance, residual stresses

Table 2 summarizes the main differences and challenges that LST poses for every reviewed as-cast iron. It is apparent that each material presents different limitations to the extent of LST; therefore, the laser parameters and conditions must be carefully selected to fulfill the technical requirements. In the following sections, a detailed review of the characterization and results of LST in cast irons is performed to delve into the differences and advantages of the technique as a function of the initial composition.

Table 2. Comparison of LST challenges for each iron casting.

Cast Iron	Main Challenges
GI	Graphite flakes act as stress raisers in sliding or rolling contact systems; Limited improvement in wear resistance.
ADI	Susceptible to tempering effect and crack propagation in LSM; Extended costs and processing time, due to pre-austempering stage.
Pearlitic DI	Matrix carbon enrichment during heat transfer lowers the melting point around graphite nodules and produces local melting; Reduced processing window for hardening without melting.
Ferritic DI	A lesser amount of pearlite further triggers local melting around graphite nodules; LSM cannot be avoided for significant hardened case depths (>250 μm); Wear resistance is commonly lower than pearlitic DIs under similar operating conditions.

4. Laser Surface Treatment of Gray Cast Irons

LST on GIs date back to the beginnings of this technology. CO_2 lasers are the most used for this purpose, but it is frequently necessary to apply thin coatings to increase absorptivity. The characterization of this technique is carried out mainly in terms of microstructure, hardness, and wear/erosion resistance, and the relevance of each work resides in the variation of some inputs of LST or the exploration of new approaches and tests for further understanding of the process.

Regarding the microstructural changes, when the laser parameters are sufficient to ensure reaching the solid-state transformation temperature without surpassing the melting temperature, it is reported that a martensitic structure is obtained from the fast cooling of the austenitic matrix. Hwang et al. [40] used a 5 kW CO_2 laser with fixed parameters, ranging from 1 to 4.5 kW and from 1 to 13 m/min, to modify the properties of as-cast GI samples with pearlite matrix and small amounts of cementite and steadite (iron phosphite, cementite, and ferrite). The purpose of their work was the improvement of the performance of GI piston rings used in marine diesel engines by LSH, and from the experimental results, the authors determined a proper heat input window from 30 to 45 J/mm to achieve the desired hardening effect.

This conclusion is supported by Liu and Previtali [41], as the authors agree that the feasibility window for hardening without melting (LSH) is very narrow since the melting point of cast irons is lower than in pure iron or steels. In this work, a Gaussian distribution diode laser with higher absorptivity was used, and two constant temperature levels were set on the surface of the samples (1000 and 1100 °C), varying the laser power using a proportional–derivative–integrative (PID) controller, to avoid the melting of the material during heating.

Furthermore, Wang et al. [42,43] compared the effects of LSH on five types of GIs: untreated (GI), quench-tempered (QTGI), austempered (AGI), quench-tempered and laser-hardened (LHQTGI), and laser-hardened austempered (LHAGI) samples. A continuous-wave Nd: YAG laser with Gaussian energy profile and 8 ms pulse duration was used, considering a 2 mm spacing between adjacent laser spots. Three similar zones were observed in the laser-treated cases: a central laser-hardened zone, where graphite flakes completely dissolve during heating and transform to a ledeburitic matrix; a heat-affected zone (HAZ) consisting of martensite, due to self-quenching, without changes in graphite morphology; and a typical substrate microstructure of tempered martensite below the HAZ. Figure 3 presents the microstructural transformations of LHQTGI as a function of the tempering temperature.

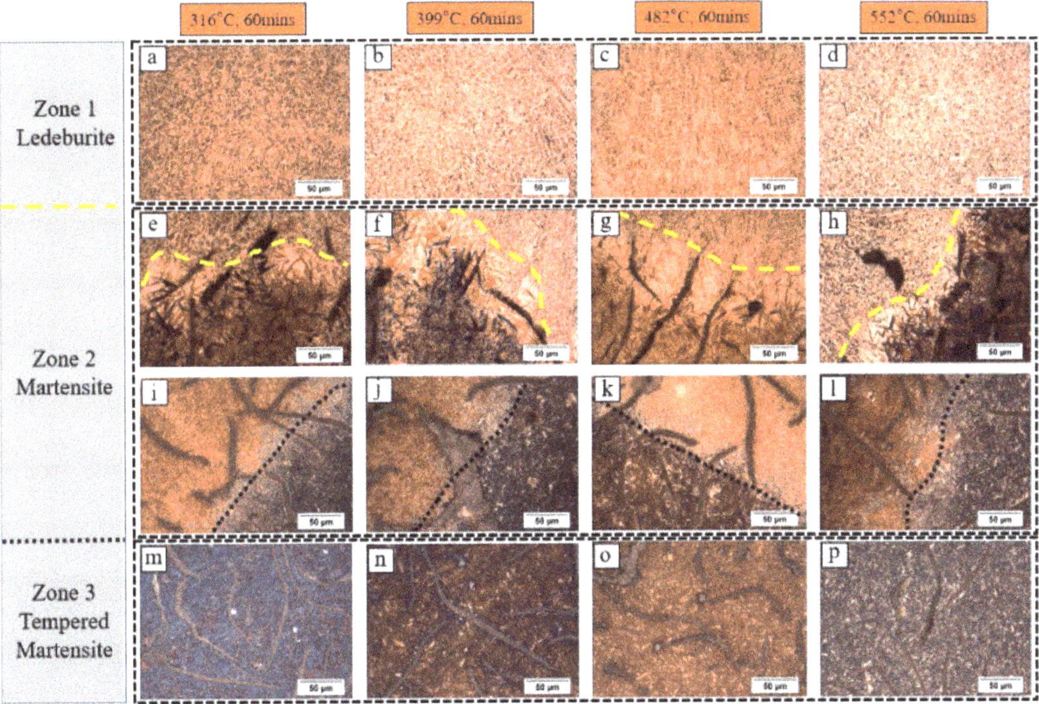

Figure 3. Microstructure of LHQTGI with different tempering temperatures: (**a**) laser hardened zone (Zone 1) at 316 °C, (**b**) Zone 1 at 399 °C, (**c**) Zone 1 at 482 °C, (**d**) Zone 1 at 552 °C, (**e**) interface between Zone 1 and Zone 2 (heat-affected zone) at 316 °C, (**f**) interface between Zone 1 and Zone 2 at 399 °C, (**g**) interface between Zone 1 and Zone 2 at 482 °C, (**h**) interface between Zone 1 and Zone 2 at 552 °C, (**i**) interface between Zone 2 and Zone 3 (substrate) at 316 °C, (**j**) interface between Zone 2 and Zone 3 at 399 °C, (**k**) interface between Zone 2 and Zone 3 at 482 °C, (**l**) interface between Zone 2 and Zone 3 at 552 °C, (**m**) Zone 3 at 316 °C, (**n**) Zone 3 at 399 °C, (**o**) Zone 3 at 482 °C, (**p**) Zone 3 at 552 °C (Reprinted with permission from ref. [43]. 2020 Elsevier).

On the other hand, Trafford et al. [44] made the first approach to LSM in GIs with different sets of fixed laser power and scanning velocity, established to ensure three different cases on the surface: hardening without fusion, with an incipient degree of fusion, and one with complete fusion on the laser path. In the case of LSM, three differentiated regions can be seen along the cross-section of the laser path: a fusion zone close to the surface, characterized by a fine ledeburite structure, which is caused by the rapid rate solidification of graphite sheets completely dissolved during heating; a deeper transition region, formed by a matrix composed of retained austenite and thick martensite plates, in which there is a partial degree of undissolved graphite sheets surrounded by a layer of ledeburite; and a base zone, where the action of the laser is not enough to produce the phase transformation. The authors also observed that cracks formed on the surface, with increased number and size according to the degree of fusion.

De Oliveira [45] also implemented a computer interface to control the temperature at the surface of GI samples, to ensure the transformation to austenite without melting, by linearly changing the heat input of a 2 kW Nd:YAG laser. However, this could not be fully achieved near the center of the laser spot, as three zones were identified by using microscopic observation: a melting zone next to the surface, with an arrange of needle-like martensite plates; a transformed region, in which graphite flakes remain unaltered and the pearlite matrix transforms to austenite during heating, and then to larger martensite plates after the cooling process; and the substrate material zone, where no transformation was observed. The author also highlighted the difference between the shape of martensite plates, due to the high thermal conductivity of the nearby graphite flakes and the amount of diffused carbon into austenite. Moreover, a continuous heating transformation (CHT) diagram was described to explain these transformations as a function of temperature and heating rates.

In terms of mechanical properties, the hardening effect is mostly measured with Vickers microhardness tests. Hwang et al. [40] reported an increment in the hardness from about 300 to 800–950 HV0.1 in the heat-affected zone. This satisfied the piston rings requirement of the minimum hardness of 450 HV0.1 in an effective depth of 300 μm. A tempering effect was also identified in overlapping zones, where hardness dropped to 470 HV0.1, still fulfilling the desired condition.

These results are in agreement with the work of Liu and Previtali [41]. A region of maximum hardness between 800 and 900 HV0.3 was generated in the non-overlapping area, while in the overlapping region, there was a tempering effect which reduced the hardness to 300–400 HV0.3 (Figure 4). Since overlapping is used to create a homogeneous modified surface in the case of a Gaussian laser, this unwanted effect is unavoidable. In this sense, the authors determined that an overlapping degree between 1 and 1.25 mm was acceptable to comply with industrial requirements, and this, based on their research, equaled a transformed zone with a minimum depth of 0.25 mm and hardness greater than 700 HV0.3. Wang et al. [42,43] also obtained similar hardness profiles with a Rockwell test (HRC), since the 2 mm spacing between adjacent laser tracks ensured avoiding the back-tempering effect.

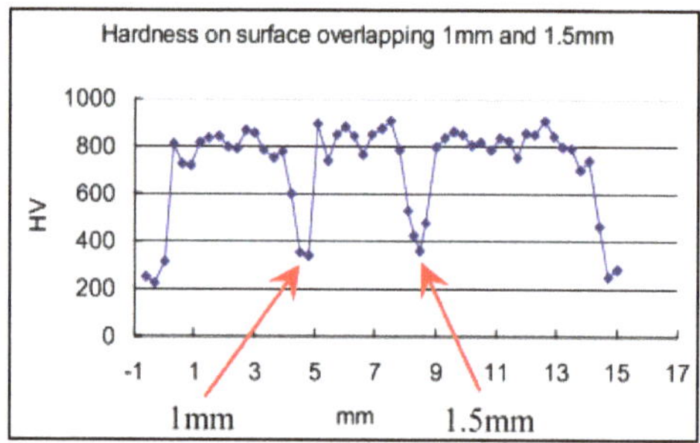

Figure 4. Microhardness profile as a function of overlapping size (Reprinted with permission from ref. [41]. 2010 Elsevier).

In the case of LSM, the improvement in hardness is comparable to LSH, but with a different magnitude. Trafford et al. [44] measured a maximum hardness of 850 HV0.3 for laser surface hardening without melting, and 950 HV0.3 for the LSM treatment. These values were reached in regions close to the surface, where the transformation is homogeneous, while hardness rapidly decays to the base value when the heat-affected zone (HAZ) is exceeded. De Oliveira [45] found a similar trend but with higher dispersion. The average hardness increased from 250 to about 600–900 HV, and the high variation was attributed to the randomness of the indentation location. Moreover, the higher hardness values were related to a higher content of martensite, as proved by X-ray diffraction (XRD) tests.

Concerning the improvement in wear resistance of GIs, Hwang et al. [40] first applied a pin-on-disc test under ASTM G99 conditions to evaluate the wear loss in both the pin (using LSH and untreated GI samples) and a conventional cast-iron disc. It was determined that LSH can double the wear life of the piston rings, which was supported with SEM observation to address the nature of the wear mechanisms. In untreated piston rings, adhesive wear features, such as plastic flow and tearing damage, were noted, while in laser-modified samples, only mild wear was observed around the remaining graphite flakes that acted as stress raisers.

Instead, Wang et al. [42,43] conducted ball-on-plate reciprocating sliding wear tests with a hard 4 mm–diameter alumina ball (45 HRC and surface roughness of 10 nm). Both laser-hardened specimens exhibited significantly lower mass loss, since they bared harder surfaces, with LHAGI displaying better performance (Figure 5). The worn surface of laser-treated cast irons revealed a polishing effect of the ceramic ball, while the severe damage in untreated specimens was related to the stress concentration around graphite flakes and the low fracture toughness of tempered martensite (zone 3), as in Hwang et al. [40]. The main wear mechanism was crack formation (Figure 6), which surged around the edge of graphite flakes on the surface or subsurface. Then, these cracks propagated along with the graphite flakes, conceiving either small-scale pits or large-scale spalls. From these observations, the authors suggested the possibility of using LHQTGI as a replacement for AGI, since its overall performance was enhanced, and the cost and energy consumption could be substantially reduced.

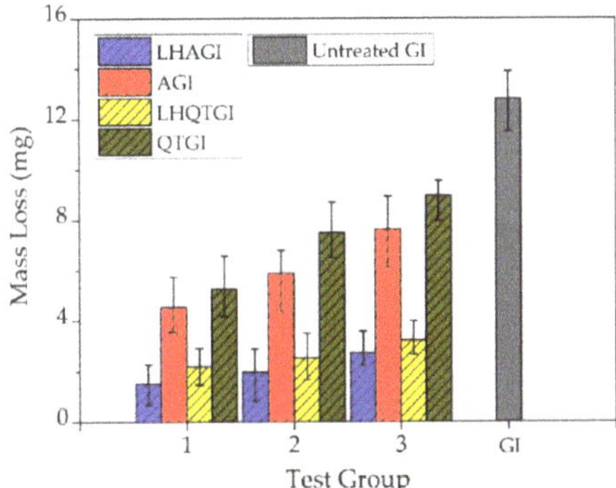

Figure 5. Mass loss of GI specimens after complete sliding wear test (Reprinted from ref. [42]).

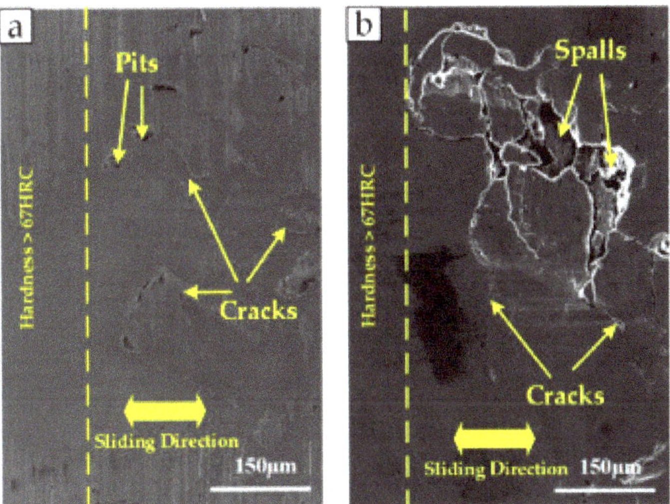

Figure 6. Worn surfaces of LHAGI and LHQTGI samples with different tempering temperatures: (**a**) 232 °C and (**b**) 316 °C (Reprinted from ref. [42]).

Trafford et al. [44] also compared the enhancement in wear resistance by LSH and LSM, using friction wear tests with a silicon carbide (SiC) platform, at a frequency of 2.5 Hz and a load of 5 kg, without lubrication. Wear rates, expressed as the amount of lost volume as a function of the total sliding distance and applied load, revealed that both LSH- and LSM-treated samples had higher resistance compared to the as-cast state, with less wear damage in the ledeburitic structure obtained by LSM.

As a summarizing work, Paczkowska [46] performed LST treatments by using various sets of laser powers and scanning velocities to determine the ranges of the surface energy density and interaction time that allowed the researcher to achieve different applications of the laser, such as tempering, hardening from the solid-state (LSH), fusion (LSM), or alloying. From the microscopic observation, as well as the measurements of temperature and hardness on the surface of the samples, the author defined the desired ranges based on a continuous wave CO_2 laser, as reported in Figure 7, highlighting how limited the processing windows are and the importance of finding the optimal laser parameters for each case.

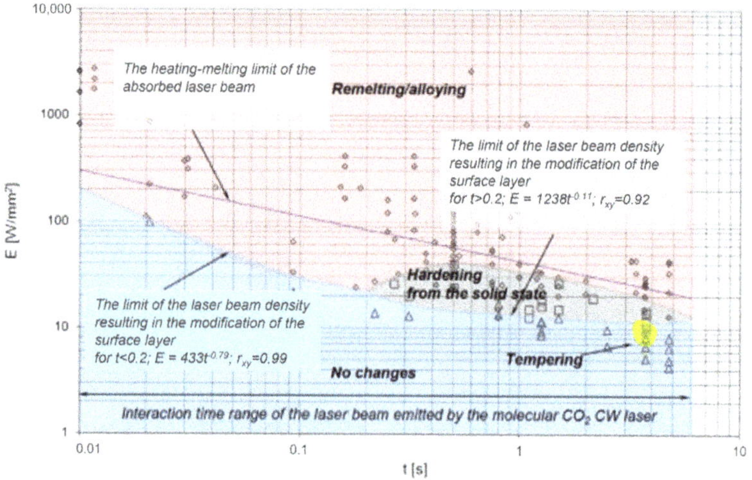

Figure 7. Energy density ranges for different heat treatments with CO_2 lasers on GI (Reprinted with permission from ref. [46]. 2016 Elsevier).

5. Laser Surface Treatment of Ductile Irons

LST on DIs has been also studied over the last decades to a similar extent, sharing the same transition between CO_2 to solid-state lasers, as well as most of the characterization tools with gray irons. However, the results indicate a strong dependence on the initial metallic matrix (e.g., pearlite or ferrite), which is responsible for the magnitude of diffusion times and cooling rates that play a major role in achieving LSH or LSM for a fixed laser input. Moreover, as Steen and Mazumder [39] proposed, many authors have verified the difficulties of achieving LSH on DIs due to the lowering of the melting point around the graphite nodules as carbon diffuses away from the graphite during the process.

5.1. Pearlitic Ductile Irons

One of the first works in the context of pearlitic DIs was carried out by Mathur and Molian [47]. The authors applied an LST treatment with a Gaussian CO_2 laser on both gray and ductile iron samples. The latter, named ASTM class 80-55-06, had a matrix with approximately 50% of pearlite. Three power levels (0.4, 0.8, and 1.2 kW) were defined, coupled with scanning velocities ranging from 4.2 to 169.3 mm/s. In addition, a manganese phosphate coating was used to increase the absorptivity of the material. The ductile iron changed its microstructure, initially pearlitic with graphite nodules surrounded by ferritic shells to one made up of two well-defined areas. The closest zone to the surface was identified as the fusion zone, where graphite nodules completely dissolved and solidified to give rise to a martensitic-like structure, while the HAZ was evidenced by graphite nodules surrounded by martensite rings. Furthermore, in the cross-sections to the laser passing direction, the shape of the transformed zone was parabolic, similar to the original laser distribution. In addition, the researchers qualitatively determined, from several attempts to fit the experimental data, the relationship between the variables of the experiment (laser power, diameter, and scanning velocity) and the size or depth of the transformed zone. It was concluded that the reached depth was directly proportional to the power of the laser and inversely proportional to the diameter of the laser and the scanning speed.

Moreover, Molian and Mathur [48] explored the differences of applying the same treatment but changing the laser circular shape to a square and elliptical type and making single or multiple passes through the working section. The authors were able to determine a linear correlation between the depth of the transformed area and the square-shaped laser parameters considered in the term $P/\sqrt{v}$. On the other hand, one-dimensional (1D) and three-dimensional (3D) heat-conduction models were used to simulate the temperature on the surface of the cast iron at constant power and different scanning velocities, with sufficiently high dispersions to conclude that the thermal response of the sample to the treatment could not be described by any simplified model.

The influence of the heat input and the solidification rates was further addressed by Chen et al. [49], where a 1 mm–diameter circular CO_2 laser was used at different powers and velocities to transform the structure of a pearlitic DI. For solidification rates greater than 5×10^4 K/s, the resulting microstructure consisted of primary austenite dendrites inserted in a continuous interdendritic cementite network, while martensitic microstructures with lamellar ferrite–cementite arrays were obtained for low rates.

These transformations are consistent with the work of Gadag et al. [50], where both a CO_2 laser at different levels of constant power (1 to 2.5 kW) and a 400 W Nd:YAG laser were used, coupled with different sets of scanning velocities. In this case, three zones were identified after LSM, aided by XRD patterns: a stationary region with a homogeneous microstructure consisting mainly of ledeburite, eutectic austenite, and cementite; a slowly decreasing section below the melting zone, with martensitic and fine pearlitic microstructure; and a rapid transition to the as-cast state. Moreover, a numerical simulation of the temperature during the treatment and the cooling rate at different locations was performed based on a 3D model of the heat equation that is written as follows:

$$\rho C_p \left(\frac{\partial T}{\partial t} \right) = \nabla (k \nabla T) - U \rho C_p \left(\frac{\partial T}{\partial x} \right), \tag{2}$$

where T represents the temperature, t is the time, ρ is the density of the material, Cp is the specific heat for ductile iron, k is its thermal conductivity, and U is the constant laser scanning velocity. The finite-difference solution for the temperature at different depths was determined as a function of time. With this dataset, a successful comparison with the experimental HAZ depth was achieved, both for its magnitude and parabolic shape, thus validating the considered model.

Fernández-Vicente et al. [51] focused on the formation of cracks during the application of LSH and LSM treatments on pearlitic and bainitic DI castings used in the production of hot industrial rolls. A surface energy density of 60 J/mm^2 was experimentally found to be the transition between LSH and LSM treatments, using scanning velocities from 2 to 14 mm/s and powers between 2.5 and 4.4 kW. Visible cracks were observed on the samples, as illustrated in Figure 8. Only for the lowest energy density LSH, the transformation occurred without the generation of microscopically observable cracks. The homogeneous structure of the fusion layer determined by Gadag et al. [50] was also present in this work, as evidenced by a matrix composed of primary martensite (M) dendrites, together with interdendritic cementite (C), residual blocks of retained austenite (A), plate-shaped carbides (D), and ledeburite (LD).

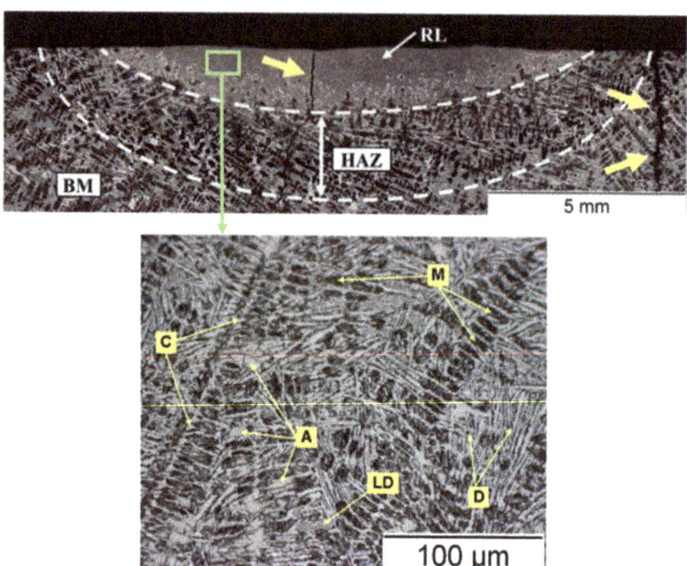

Figure 8. Microstructure of LSM ductile iron, formed by a parabolic remelted layer (RL), a heat-affected zone (HAZ), and the base material (BM), with visible cracks indicated by arrows (**top**); magnification of RL microstructure (**bottom**) (Reprinted with permission from ref. [51]. 2012 Elsevier).

A first approach on the hardening effect of LSH/LSM in pearlitic DIs was made in References [47,49], where Vickers tests were performed to measure an average hardness of 800–1000 HV in the fusion zone, compared to an original hardness between 220 and 290 HV, similar to the results found later in Molian and Baldwin [52]. However, the authors emphasized the difficulty of obtaining generalizable values in the HAZ, due to the influence of graphite nodules and their heterogeneous location in the microstructure. Moreover, in Reference [48], the effect of overlapping reported in References [40,41] for gray irons is verified for LSM.

A correlation between surface and microhardness as a function of LST parameters was also carried out by Gadag et al. [50], where it was evidenced that, the longer the interaction time (i.e., lower scanning velocities), the greater the HAZ depth. Moreover, the surface hardness increased with laser power (heat input). On the other hand, Nd:YAG produced greater surface hardness than CO_2 lasers under the same conditions, which was attributed to greater absorptivity and efficiency.

Recent investigations performed by Ghaini et al. [53] allowed us to develop a model that is able to address the absolute thermal efficiency of LSH on GGG-60 cast irons in terms of a hardening efficiency index (HEI) and a hardening ratio. A 600 W fiber laser with shielding argon gas was used at two different sets of power and four beam travel speeds to ensure a reasonable hardened area without melting. After microscopical observation and Vickers tests, the hardening efficiency index was defined as follows:

$$HEI = \frac{\Delta Hardness\ (average) \times Hardened\ area}{Heat\ Input}, \tag{3}$$

where the maximum achievable hardness was around 1020 HV0.3, the hardened area (or volume per unit length) was obtained from the microhardness profiles and the overall transformed region in cross-section micrographs, and the heat input was measured as the ratio between beam power and laser travel speed. A linear relationship between HEI and the term P^5/v was found, and it was compared to those of tool steels at similar operating parameters. Finally, the hardening ratio was established as follows:

$$\text{Hardening ratio} = \frac{A \times \rho \times (\int_{T_1}^{Ac3} C_p dT + \Delta H_a)}{P/v},$$ (4)

where A is the area of the hardened case in cross-section, ρ is the density, T_1 is the ambient temperature, $Ac3$ is the full austenite formation temperature, C_p is the specific heat, dT is the temperature differential, and ΔH_a is the austenization enthalpy. The results were correlated with the obtained values for HEIs, and it was determined that LST with 500 W laser power and 2 mm/s travel speed achieved the maximum thermal efficiency of 15.3%.

The wear mechanisms and resistance of laser-treated pearlitic DIs were initially evaluated by Chen et al. [54], based on their previous treatments [49]. From surface erosion, abrasion, and scratch tests, it was revealed that the rate of material loss from the laser-treated bodies was reduced by a factor of up to four times, depending on the surface hardness. Subsequently, transmission electron microscopy (TEM) and scanning electron microscopy (SEM) allowed them to conclude that, for continuous high applied loads, severe plastic deformation on the surface of the treated samples generated a state of internal stresses that was responsible for the growth of microcracks that propagated, favoring the wear rate of the material.

Gadag and Srinivasan [55] also provided evidence of the improvement in the behavior of DIs treated with LSM against cavitation erosion. The authors used three classes: a pearlitic type (600/3 class), a ferritic one (400/12 class), and a 600/3 DI treated with LSM. The researchers compared the corrosion resistance in different environments, and the rate of material loss was measured from observation by optical microscopy (OM) and SEM. For the erosion tests, the three samples were exposed to artificial seawater. A linear relationship between corrosion wear and exposure time was identified, while LSM reduced the erosion rate between 6 and 8 times compared to the as-cast specimens. This improvement was attributed to the transformation of the material into a finer grain structure (consisting of ledeburite in the melting region and martensite in the hardened area) with a uniform distribution of plastic deformations and a reduction in the free path for movement of dislocations.

Molian and Baldwin [52] suggested that the improvement in erosion resistance by LST could be attributed to the emergence of beneficial compressive residual stresses during the transformation to fine martensite and retained austenite microstructures. Fernández-Vicente et al. [51] further studied the nature and magnitude of both thermal and transformational stresses due to LSM. First, thermal stresses were estimated from the temperature variations during the cycle. In particular, the only contribution to the strain was the difference between the peak temperature of the heating cycle and the temperature after the irradiation ceased. These temperatures were also approximated by simplified solutions for the heat equation, and then the value of the thermal strain was measured as follows:

$$\varepsilon = (1 + v)\alpha \Delta T,$$ (5)

where ε is the thermal strain, v is the Poisson's ratio, α is the thermal expansion coefficient, and ΔT is the temperature variation. Then, following the linear elastic theory, the thermal stresses were estimated as follows:

$$\sigma = \frac{E}{2(1 + v)}\varepsilon,$$ (6)

where E is the elastic modulus, and σ is the thermal stress. On the other hand, the authors indicated that, for higher retained austenite contents (over 40%), the observed cracking in LSM treated samples was due to an increment in the tensile stresses from the lower specific volume of retained austenite compared to martensite or bainite (Figure 9). Finally, the retention of nucleated cracks in LSH specimens was associated with the relatively high fracture toughness of the Fe_3C carbides present in the microstructure, and this contributes to the absorption of the thermal and transformational stresses generated during the treatment.

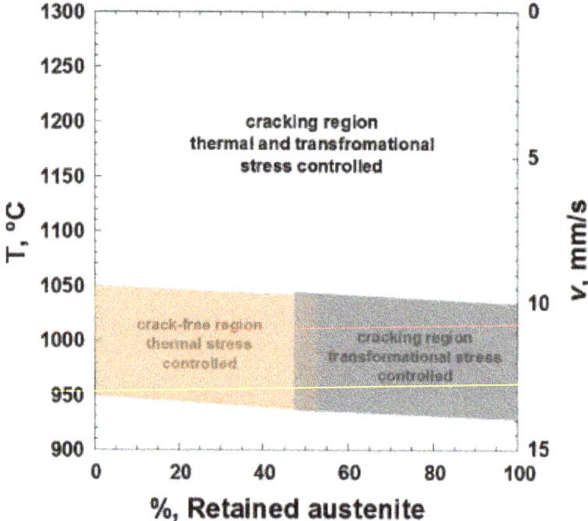

Figure 9. Laser processing window for LSH/LSM treatments as a function of retained austenite content, scanning velocity, and surface temperature (Reprinted with permission from ref. [51]. 2012 Elsevier).

5.2. Austempered Ductile Irons

Since austempering produces a unique initial matrix, LST can lead to different microstructural transformations and mechanical properties. Lu and Zhang [56] first carried out LSH with a CO_2 laser on bainitic ADI samples. The material followed a standard austempering process, with an austenitizing temperature of 890 °C, a cooling interval at 360 °C for 2 h, and a final quenching to ambient conditions. The initial austenitic bainite matrix also exhibited ferrite and austenite enriched in carbon, with a smaller portion of retained austenite and martensite. After laser hardening, the matrix changed to an acicular structure of martensite and retained austenite, with graphite spots close to the surface, as seen in other ductile iron samples.

An extensive numerical model of LST in ADIs was carried out by Roy and Manna [57], with the goal to predict the temperature in the vicinity of graphite nodules. The estimation is based on the heat equation for the case of a continuous wave CO_2 laser with Gaussian distribution profile. This equation is expressed as follows:

$$\nabla^2 T - \frac{1}{\alpha}\frac{\partial T}{\partial t} + \frac{q_r}{\lambda} = 0, \tag{7}$$

where T represents the temperature, α is the thermal diffusivity, q_r is the magnitude of the heat delivered by the laser per unit volume and time, λ is the thermal conductivity, and t is the time. Assuming that heat losses are negligible and that the thermal properties of the material do not depend on the temperature, the equation was solved analytically, and the temperature profile was used to determine the magnitude of carbon diffusion away from the graphite nodules, according to Fick's law, written as follows:

$$C(y, t) = \frac{C_f - C_e}{2}\left[1 - \mathrm{erf}\left(\frac{y}{2\sqrt{Dt}}\right)\right] + C_e, \tag{8}$$

which determines the carbon concentration toward the edges of the nodule. In this expression, C_f is defined as the carbon% in the austenitic phase at the matrix–nodule interface, C_e represents the carbon concentration in the matrix, y is the measured horizontal distance from the center of the nodule (assuming a perfect circular shape), D is the diffusion coefficient (which depends on temperature), and *t* is the time. This carbon concentration was then utilized to define, from the iron–carbon–silicon (Fe–C–Si) phase diagram, what the temperature is when fusion occurs. Then, by selecting the proper laser parameters, this model would allow us to achieve the material hardening without melting. The authors conclude that, for laser powers below 800 W and a fixed scanning velocity of 60 mm/s, the fusion width is negligible compared to the distribution of nodules in the matrix, and, therefore, the microstructure is predominantly martensitic.

Subsequently, Roy and Manna [58] compared the results of LSH and LSM on ADI samples, using sets of parameters to ensure melting and hardening without melting. A variable laser power between 0.8 and 1.5 kW and a speed between 20 and 1000 mm/s were selected for the case of LSM, and a power between 0.5 and 1 kW and a speed of 60 mm/s were chosen for LSH analysis. The depth of the transformed zone maintained a linear relationship with the power and an inverse correlation with the scanning velocity, regardless of the chosen parameters. Microscopic observation and X-ray diffraction analysis (XRD) made it possible to determine that LSM produced a higher proportion of retained austenite in the metallic matrix, while, in LSH, martensite was the predominant phase.

Zammit et al. [59] also compared the results of implementing LSH and LSM on ADI, considering a discrete spot laser in order to avoid the reported hardness reduction in overlapping zones. A 9 kW CO_2 laser with Gaussian energy deposition profile was used to generate stationary pulses, decreasing the total energy input and, thus, producing lower distortion and processing costs. First, it was determined that melting caused severe distortion on the surface, given by an increment in surface roughness from 0.15 to 1.34 μm (in arithmetic value). Moreover, the molten region was depleted of graphite nodules, forming soft phases, such as austenite and residual ferrite, along with low carbon martensite. On the other hand, LSH only increased the average surface roughness from 0.15 to 0.43 μm, while the microstructure was mainly composed of martensite with unaltered graphite nodules. Some nodules were surrounded by a bullseye ledeburitic structure, implying that the carbon diffusion lowered the melting point locally around them, as in Roy and Manna [57]. Soriano et al. [60] were then able to verify the microstructural features of LSH in ADI by using an Nd:YAG laser coupled with a PID controller to keep the temperature in the center of the laser constant at 975 °C, so that the melting point was not exceeded during the treatment, as in Liu and Previtali [41]. A hardened zone composed of coarse martensite in the near-surface, as well as a finer acicular structure below it, was distinguished, associated with the rapid solidification without melting of the material (Figure 10a). Moreover, as evidenced in Figure 10b, the mixture of retained austenite and upper and lower bainite structures below the coarse martensitic array verified the occurrence of a softer thermal cycle in this region.

Figure 10. Micrographs of (**a**) top and (**b**) medium zones of the laser hardened ADI sample (Reprinted with permission from ref. [60]. 2011 Elsevier).

Regarding the hardening effect of the LSH, the profile as a function of depth, measured by Lu and Zhang [56], with a Knoop test with 0.2 kg load, followed the trend of the results shown in Gadag et al. [50], with a homogeneous hardness of approximately 600 KHN in the transformed zone (~250 µm) and a subsequent rapid decay to the base value of the ADI specimen (300 KHN). Putatunda et al. [61] provided more evidence of the effects of LSH on the mechanical properties of ADIs. Hardness, yield stress, and ultimate tensile strength tests were conducted after initial austempering conditions similar to those of Lu and Zhang [56]. Changes in these mechanical properties were analyzed for four cases: (a) untreated ADI; samples treated with the same laser parameters, (b) initially without any type of coating; and then with two types of coating, namely (c) graphite and (d) manganese. LSH significantly increased most of the properties (except the ultimate tensile strength), obtaining the best performance under the use of coatings, since a greater energy absorption was responsible for a cooling rate that ensured the formation of a more resistant martensitic structure.

In Soriano et al. [60], the improvement of hardness was further backed up by the analysis of residual stresses (Figure 11) from XRD patterns, using the $\sin^2(\psi)$ method. The diffraction peak position of the (211) α-Fe phase was measured at nine different inclinations, ranging from $-45°$ to $45°$. The hardness of the heat-affected zone was quantified in the range of 700–800 HV, which decreased as the fraction of retained austenite increased in favor of the martensitic phase. Moreover, in the hardened region, compressive residual stresses were measured and explained by the volumetric increase associated with the transformation from austenite to harder martensite. As the depth increased, the degree of martensitic transformation decreased until the point where the matrix corresponds to the base material (0.9–1.5 mm), and, thus, the stresses become tensile. This result was also addressed by Zammit et al. [59] in their study, where it was found that compressive stresses take place near the surface, due to the 4% volumetric increase of austenite to martensite transformation. At approximately 160 µm, the stresses became tensile, due to the presence of retained austenite in the ausferritic bulk, therefore obtaining an estimation of the HAZ depth of the laser treatment.

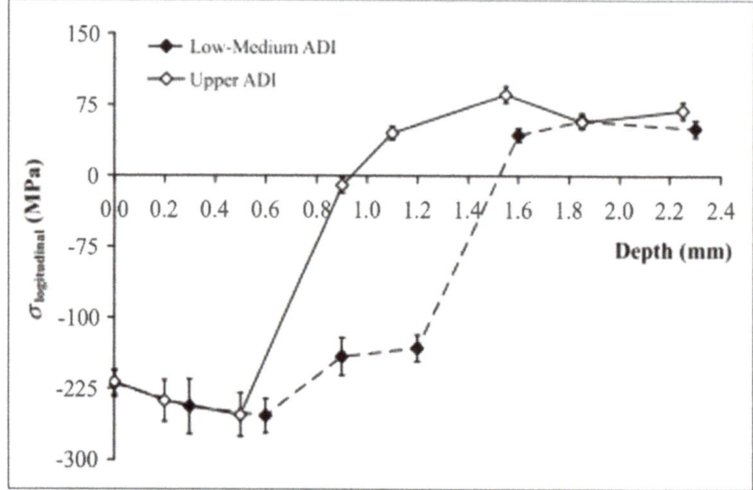

Figure 11. Evolution of residual stresses in the center of the laser track as a function of depth and ADI sample ((Reprinted with permission from ref. [60]. 2011 Elsevier).

The tribological performance of laser-modified ADIs was first assessed in Lu and Zhang [56], where the characteristics of sliding wear were analyzed from a 500-min test against a steel disc without lubrication, at a relative speed of 1.2 m/s and a variable load between 2 and 14 kgf. The results showed a linear relationship between the mass loss due to friction and the time of the test, as in Gadag and Srinivasan [55], but it was also determined that the wear rate, in both cases, consisted of two stages: a mild one (for loads less than 10 kgf) and a severe one (for loads greater than 10 kgf). These stages were related to microstructural changes during the wear test, where the austenite phase decreased as the applied load increased. The mild phase of wear is characterized by an oxidative mechanism near the surface, while the severe phase is dominated by lamination and delamination processes, the effect of which was observed by SEM in the formation of microcracks in the surface.

Roy and Manna [58] established a relationship between the hardness profiles and the wear resistance of ADIs. The hardness distribution in LSM was parabolic, while, for LSH, it was approximately constant within the transformed zone. Thus, the presence of a more homogeneous hardened zone in LSH provided the best tribological behavior, as observed in Figure 12. For a 5 kg load, three stages in the evolution of wear were identified. Initially, a high wear rate was associated with maximum contact area. Then a stationary stage was established, due to the adherence of surface asperities originated in the first phase, to finally give way to an accelerated wear stage due, to the formation of furrows and debris separation. The optimal performance obtained by LSH is supported by a lower extent of micro-fractures that is attributed to the lower probability of causing microcracks during friction contact. This result is in agreement with the deformation restrictions at the austenite–martensite interface that are caused by changes in the residual stresses during the transformation from austenite to martensite, as detailed in References [59,60].

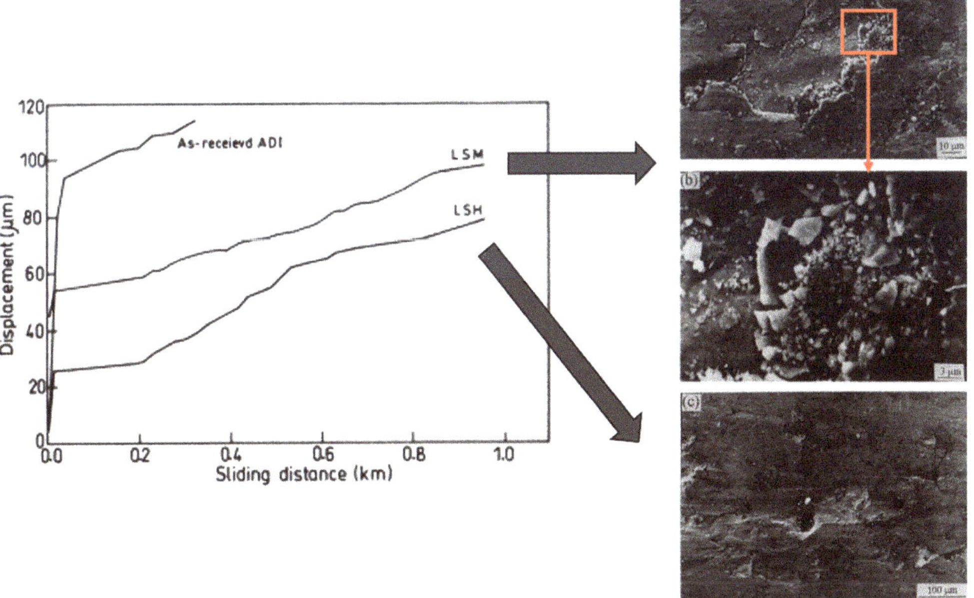

Figure 12. Evolution of displaced material as a function of sliding wear distance for as-received and laser-treated ADI samples (left); SEM micrographs of worn surfaces (right) for (**a**) LSM treated ADI, (**b**) magnified view of wear debris of (**a**), and (**c**) LSH-treated ADI (Reprinted with permission from ref. [58]. 2001 Elsevier).

The tribological characterization of LSH was completed in Zammit et al. [62]. Based on the results of Reference [59], the authors carried out discrete spot LSH treatments by using the same CO_2 laser at 600 W and 300 ms of pulse duration and considering three different arrays: laser spots separated by one spot diameter, adjacent spots, and 50% overlapped spots. Scuffing (pin-on-disc) and rolling contact fatigue (RCF) tests were performed to address the wear resistance and mechanisms. In the first test, as-austempered and laser-treated ADI pins were considered, with a hardened and oil-lubricated AISI D2 steel counterpart disc rotating at 1450 rpm under 10 MPa constant pressure. LSH treatment with adjacent spot tracks exhibited the highest sliding cycles to failure, followed by separated spots, showing that the back-tempering effect on hardness, due to overlapping, reduced the ADI wear resistance by an order of 10–100 times. Moreover, adjacent spot LSH treatment induced a higher martensite volume fraction, which is harder than the initial microstructure and is able to delay crack initiation and propagation. On the other hand, RCF tests were performed only on as-austempered and adjacent spot laser-treated ADIs. The LSH increased the number of cycles until fatigue failure over 10^6 times. This increment was also attributed to a harder microstructure and compressive residual stresses near the surface. It was concluded from SEM and pit observations that the main wear mechanism is governed by plastic deformation and propagation of cracks around graphite nodules, and the overall wear resistance is comparable to that of carburized steels. As a conclusion, the authors suggest that laser-treated ADIs could be suitable replacements for a variety of engineering components.

5.3. Ferritic Ductile Irons

Among all the ductile iron structures, LST in ferritic DIs has gained importance in the last decades because it lacks a significant pearlite amount (whose carbon content is higher due to the presence of cementite plates). This means that it is very difficult to find conventional ways of transforming the matrix to a martensitic type, and, on the other hand, it is difficult to achieve hardening without melting, since carbon diffusion from graphite nodules locally lowers the melting point, favoring the occurrence of melting near the surface [39].

The first modeling of the phase transformations during LST of ferritic DIs can be found in Grum and Sturm's study [63], where LSM was carried out with a low-power CO_2 laser. To ensure the fusion of the surfaces, an overlapping of 30% was considered. Three new regions arise after LSM: the melted zone, the hardened zone, and an intermediate transition zone. In the region where the liquidus temperature is exceeded, the predominantly ferritic matrix melts, and the graphite nodules diffuse toward the melted surface. This causes a strong dissolution of carbon in the liquid matrix, whose rapid solidification transforms it into austenite and ledeburite dendrites. The intermediate zone is characterized by a highly localized fusion process around the graphite nodules that, depending on the magnitude of the carbon diffusion toward their edges, can generate a ledeburite and/or martensite ring. The transformation scheme is shown in Figure 13. The differences between the transition zone and the hardened zone are that, although the matrix transforms into austenite during the heating cycle and is enriched in carbon due to diffusion away from the graphite nodules, this is not enough to locally reduce the melting point, and, therefore, the matrix becomes of martensitic type with residual austenite. In addition, due to the heterogeneous distribution of the nodules, there are areas with low carbon content where austenite reverts to ferrite during cooling.

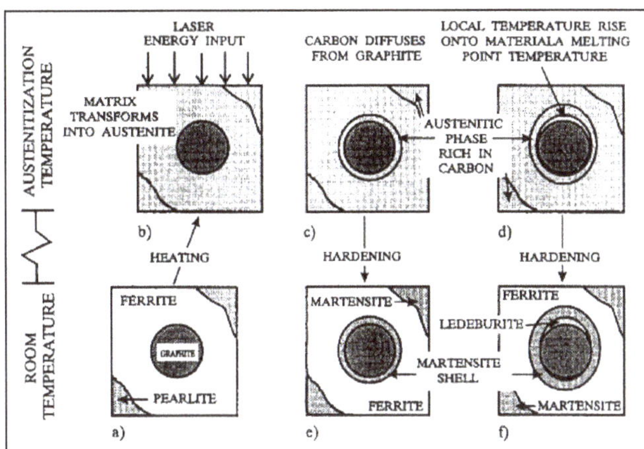

Figure 13. Schematic representation of phase changes around graphite nodules during LSM: (**a**) Initial DI microstructure, (**b**) transformation to nonhomogeneous austenite during heating, (**c**) carbon enrichment in austenitic matrix after diffusion, (**d**) local melting around graphite nodules, (**e**) formation of martensitic shells around graphite nodules after cooling, (**f**) formation of ledeburitic shells above graphite nodules that suffered local melting (Reprinted with permission from ref. [63]. 1996 Elsevier).

In Grum and Sturm [64], the analysis of LSM in ferritic DIs is completed, experimentally and numerically comparing the thickness of the martensite and ledeburite layers that surround the graphite nodules in transition and hardening zones. As described in Roy and Manna [57], from simplified models of temperature during the heating and cooling cycle, in conjunction with diffusion equations based on Fick's law, the authors determined, at different depths, if the temperature around the nodule is enough to produce fusion (and the formation of ledeburite when cooling), in addition to calculating the magnitude of the radial diffusion of the carbon atoms. As shown in Figure 14, the simulation precisely adjusted to what was experimentally determined, which validates the use of the aforementioned models. The average standard deviation obtained is less than 10% and can be explained by model simplifications, mainly in terms of diffusion, as average values that do not vary with temperature were used.

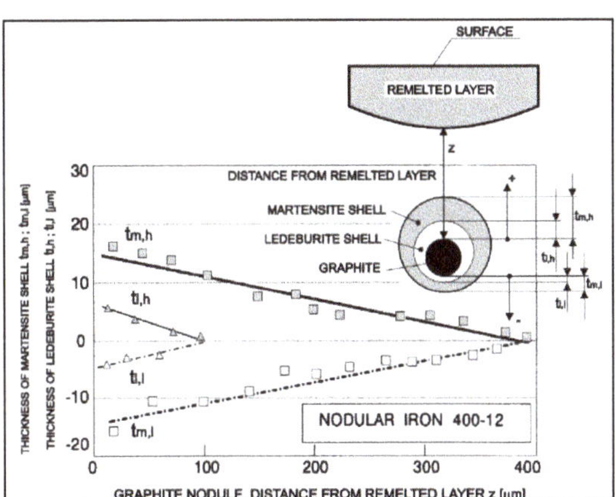

Figure 14. Comparison of the experimental and predicted thickness of martensite and ledeburite shells around graphite nodules (Reprinted with permission from ref. [64]. 2002 Elsevier).

This microstructural analysis was completed in Grum and Sturm [65], where the authors applied LSM to a similar DI sample to verify their previous models, as well as XRD and residual stress tests. The metallic surface was characterized by a high content of martensite (41%) and a portion of residual austenite (23%), and at higher depths, a reduction of the martensitic phase was observed in favor of the austenitic phase (24% and 42%, respectively, at 240 μm from the surface). From these results and the residual stresses measured at different points on the surface, it was determined that the residual austenite and martensite contents were the most influential on the nature of the internal stresses (Figure 15). The stresses on the surface always have a tensile character, since the transformations to austenitic phases involve a volume reduction of such magnitude that they cancel the effect of the transformation to martensite, thus implying a volumetric increase and, consequently, stresses of a compressive type.

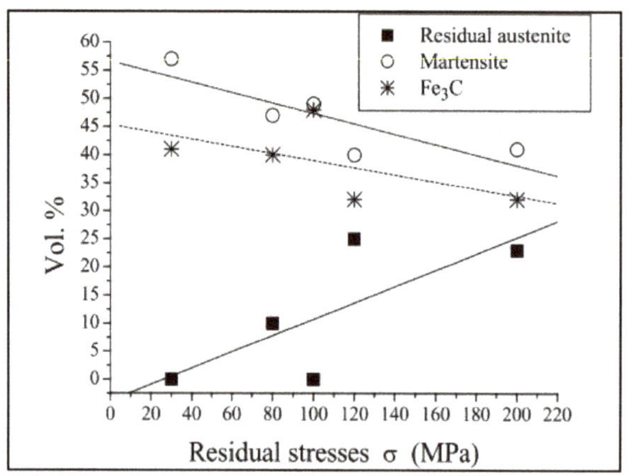

Figure 15. Evolution of residual stresses at the ductile iron surface as a function of volumetric % of phase constituents (reprinted from Reference (Reprinted with permission from ref. [65]. 2005 Inderscience Enterprises Ltd.).

Benyounis et al. [66] presented a new approach on the surface melting of DIs by conducting a comparison between the microstructural changes caused by two technologies: Nd:YAG laser and TIG. In the case of LSM, the laser was used at low power (0.1 kW) and speed of 1 mm/s with 50% overlapping, while in the case of TIG, the voltage was maintained at 50 V and the current was varied between 80 to 120 A to ensure fusion. The microstructure obtained by LSM consists of a fine dendritic structure of retained austenite, with residual percentages of martensite and cementite (Fe_3C). Although in TIG melting a dendritic structure was also observed in an austenite and cementite matrix, the main difference was the higher amount of retained austenite in LSM, which was attributed to a superior cooling rate.

The dendritic microstructure is also present in Alabeedi et al. [67], where LSM was carried out on an 83% ferrite, 11% pearlite, and 6% graphite ductile iron. A Gaussian distribution CO_2 laser was used at 3 kW power, with a scanning velocity of 10 mm/s. To achieve melting, a 50% overlapping was considered, and argon shielding gas was applied to also prevent oxidation and contamination of the specimen. The obtained microstructure was identified by SEM as austenite dendrites immersed in a cementite network, attributed to the high cooling rate during solidification. An additional transition region was observed below the melted zone; it was formed by a mixture of dendrites and thick martensite plates and added to some undissolved nodules surrounded by ledeburite layers.

Recent studies have further supported these results; for instance, Pagano et al. [68] carried out a complete analysis of the effects of the LSM technique on a ferritic DI, using an

Nd:YAG laser operating at 1 kW of power, with circular spot and Gaussian distribution. A parabolic transformed cross-section was obtained, coinciding with the laser distribution. In the area where fusion occurred, the microstructure became a network of austenite and ledeburite dendrites, with a small portion of martensite. Then a transition region with localized fusion was observed around the graphite nodules that did not melt. Finally, a hardened area was appreciated, where the graphite spheres are surrounded by a layer of martensite and immersed in a matrix of ferrite and residual austenite. These results corroborate the work of Grum and Sturm [63,64].

Another approach for LSH/LSM was conducted by Catalán et al. [69], where a fiber delivery diode laser with uniform energy distribution and rectangular shape was used. Two linear power ramps, along with two scanning velocities, were defined to achieve four different sets of linear energy densities in order to establish a relationship between the experimental results and the joint effect of all laser parameters, as well as to determine the transition between hardening from solid-state and melting. It was concluded that, for high-velocity settings (1000 mm/min), microstructural changes were not significant, whereas, for low-velocity cases (570 mm/min), severe changes could be observed. The melted zone depth increased directly proportional with the average linear energy (Figure 16), correlating with changes in the linear energy absorbed by the material and the thermal diffusion properties of the surface-modified samples. The chemical composition, as determined by GDOES, was strongly affected by the laser, with a significant increase for C and decrease for Si compared to the as-cast sample, due to the transformation of graphite nodules to carbides during the heating process. Regarding input parameters, at low energies, a Fe_3C phase was detected by XRD, whereas γ-Fe_2O_3 was raised at high energy densities, suggesting that surface oxidation occurred during the heating cycle.

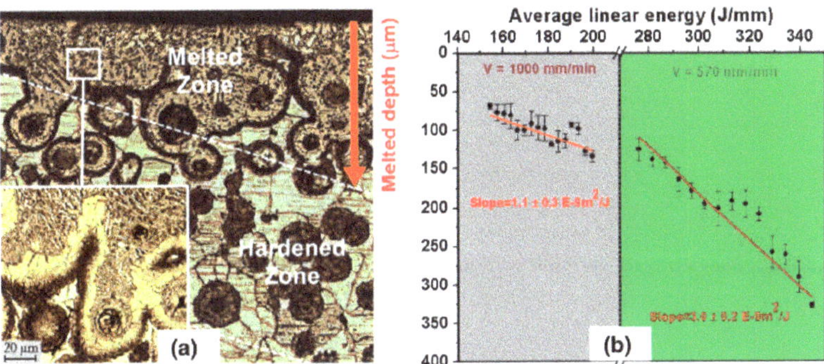

Figure 16. (**a**) Melted zone (MZ) and hardened zone (HZ) of laser surface-modified microstructure of DI; (**b**) melted depth as a function of the average linear energy of the laser (Reprinted with permission from ref. [69]. 2021 Springer Nature BV).

On the subject of hardness improvement, many of the reviewed works have quantitatively addressed the efficiency of LSM in DIs. For instance, Grum and Sturm [63] observed a uniform hardness distribution throughout the transformed layer (800–900 HV0.1), and when the limit of the hardened zone ($\approx$550 µm) was exceeded, it fell rapidly to the base value (around 250 HV0.1), and this is in agreement with the posterior measurements of Pagano et al. [68]. Benyounis et al. obtained a harder microstructure when using TIG melting over LSM (750 and 500–600 HV, respectively), since the supplied energy and the interaction time ensured less retained austenite content, implying that the tempering effect of melting was reduced.

In the melted zone, Catalán et al. [69] found a stable microhardness of 1000–1100 HV0.3, which was five times higher than the nominal value. This was attributed to the carbide/oxide formation in the transformed matrix, as observed in Figure 17, since the mentioned XRD analysis supported the existence of these harder phases. The main difference with other works is the uniformity in both hardness measurements and microstructure for the melted and hardened zones, since the laser energy distribution was uniform and rectangular shaped. Thus, overlapping of laser tracks was not needed and the tempering effect was avoided.

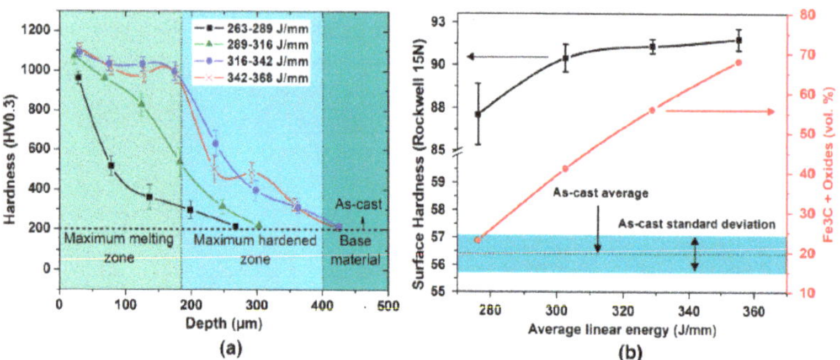

Figure 17. (**a**) Evolution of microhardness as a function of depth and linear energy. (**b**) Surface hardness and composition of as-cast and laser modified DIs (Reprinted with permission from ref. [69]. 2021 Springer Nature B.V.).

The wear response of laser melted DIs is broadly detailed in Pagano et al.'s [68] study, where a rotating steel cylinder for sliding wear tests was chosen to roll against the samples before and after being treated with LSM. The evolution of the coefficient of friction and material loss was recorded as a function of the total sliding distance. In this test, two stages were observed: initially, the coefficient of friction (COF) increases linearly with the sliding distance (run-in stage), and then it decreases to a stationary value (steady-state), due to the lubricating effect of graphite release. In this phase, since there is a partial dissolution of graphite in the hardened area, an increase in the adhesive component of the COF is generated by reducing the lubricating effect. Furthermore, the presence of hard phases, such as cementite and martensite, causes an increase in the abrasive component of the COF, which accounts for doubling the friction coefficient of the untreated DI. From Archard's equation for sliding wear, we obtain the following:

$$V = K\left(\frac{F_N \cdot S}{H}\right),$$ (9)

the authors verified that, for the same sliding distance (S), applied load (F_N), and wear factor (K), the greater hardness (H) of the LSM sample generated a lower volumetric loss V. This was supported by the fact that the original samples showed rapid plastic deformation during the test, while the presence of oxides on the surface of the specimens modified by LSM gave evidence of a moderate triboxidative wear regime.

Subsequently, Ceschini et al. [70] compared the tribological experiments carried out by Pagano et al. [61] for four different specimens: untreated GI (GJL300), as-cast DI (GJS400), low-energy melted DI (GJS400-LHV at 500 W), and DI treated with LSM at high energy (GJS400-HHV at 1000 W), using the same equipment and tests as the work in comparison. Due to different energy densities, the two LSM-treated DIs exhibited different surface hardness, with the GJS400-HHV being the hardest (1100 HV1, compared to 850 HV1 for the GJS400-LHV). The sample with the best performance was the GJS400-LHV, followed by the GJS400-HHV, as observed in Figure 18. In both cases, a slight triboxidative wear regime was observed, such that the differences reside in the presence of microcracks in the HHV cast iron, associated with its greater hardness and lower toughness. Furthermore, the results showed that graphite morphology in cast irons is a major factor in wear resistance, with the lamellar graphite shape (GJL300) being the one that acts as the greatest stress concentrator, favoring nucleation and crack growth during the sliding test.

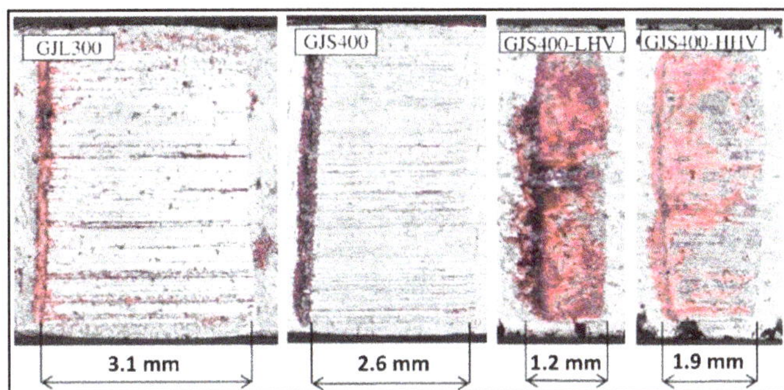

Figure 18. Worn surfaces of different cast irons for a 20 N applied load (Reprinted with permission from ref. [70]. 2016 Elsevier).

The erosion resistance enhancement due to LSM has been also assessed by Alabeedi et al. [67] in their study, where silica particles were bombarded at an average speed of 50 m/s at different incident angles. The amount of mass loss as a function of exposure time increased linearly, as in Gadag and Srinivasan [55], with a cumulative loss at the end of each trial (45 min) between 35 and 100 times less in the case of the treated samples. Furthermore, SEM examination evidenced that the wear mechanism in the untreated specimens begins with the deformation of the softer graphite nodules, which then favor the nucleation of cracks by acting as stress concentrators. Thus, the cracks form larger craters as they spread, so that the surface fractures and becomes dimpled. In contrast, wear in laser-modified samples is attributed to the initiation of fatigue cracks, which develop only into small craters. However, the hard structure formed in the process does not allow these cracks to propagate through the matrix and cause more damage, mainly due to the presence of retained austenite.

Finally, Boccardo et al. [71] developed a unidirectional coupled thermo-metallurgical model in order to predict the phase transformations of laser-treated DIs. In particular, the thermal model computed the temperature evolution during the process, while the metallurgical model predicted the phase evolution as a function of the temperature. The possible transformations during heating include the reverse eutectoid transformation (RET), homogenization of the austenite carbon content (HA), and melting transformation (MT), whereas the cooling process can comprise the eutectoid transformation (ET), solidification (ST), and martensitic transformations (MDT and MWT). To test the performance of this model, the authors considered the same DI as in Catalán et al. [69], with the four sets of linear power ramps and scanning velocities detailed in their work. Figure 19 shows the relationship between experimental and simulation results for 180 J/mm energy density. According to the presented metallurgical model, the observed thickness of the melted zone (Figure 19a) agrees with the computed temperature (Figure 19c), because it is greater than the temperature at which melting transformation ends (T_{Le}). For a depth of 100 µm, the material partially melted, because the temperature remains between the limits where melting transformation starts (T_{Ls}) and ends. Finally, for depths bigger than 300 µm, there is no phase transformation, since the temperature does not reach the point where RET starts. On the other hand, Figure 19b illustrates the evolution of the volumetric fraction of ledeburite, one of the main constituents of ductile iron during these transformations. The comparison between the computed ledeburite fraction and the thickness of melted, transition (hardened), and base material zones (Figure 19a) shows that the thermo-metallurgical model is able to identify these regions as a function of the ledeburite fraction in the transformed matrix. A linear correlation between measured and simulated martensite/ledeburite thickness layers was obtained, thus verifying that, for low thicknesses, the thermo-metallurgical model reasonably captures the trend of experimental results, and it slightly overestimates the thickness at higher magnitudes. Moreover, the thickness of the ledeburite–martensite layer depends on the laser parameters, and it is increased with the increment of laser power and the decrement of scanning velocity.

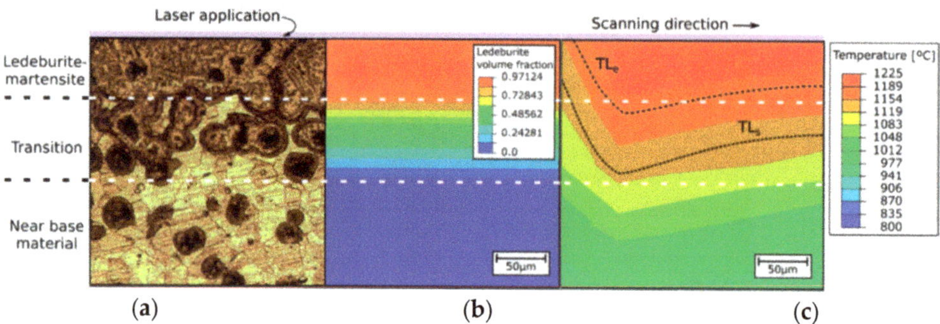

Figure 19. (**a**) Optical microscopy image of resulting microstructure of modified DI. (**b**) Computed ledeburite volume fraction at fixed laser parameters. (**c**) Computed temperature distribution at a fixed instant and parameters as a function of depth (Reprinted with permission from ref. [71]. 2021 Springer Nature B.V.).

5.4. Overall Facts of Surface Laser Treatment of Cast Irons

Considering the literature review, some similarities can be found in LST of cast irons, regardless of their initial microstructure. As presented in Figure 20, laser modification of iron castings generally leads to a martensitic or ledeburitic microstructure, depending on the amount of supplied energy and the subsequent phase transformations.

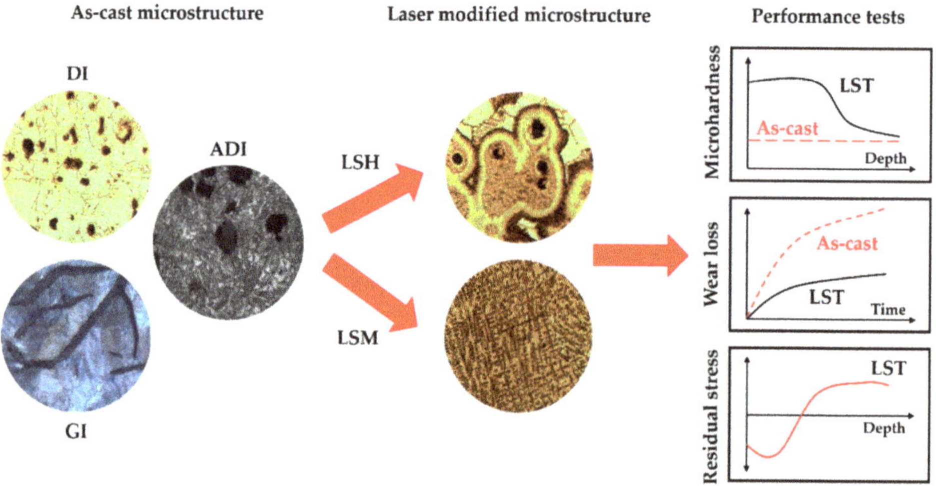

Figure 20. Summarizing diagram of microstructural transformations and overall response of laser-treated cast irons.

Moreover, mechanical/tribological characterization shows an improvement on hardness and wear resistance after LST, as explained by the formation of harder phases during rapid solidification. Microhardness curves tend to exhibit a stable, harder region, with a quasi-exponential decay to the as-cast value in zones with low degree transformation. In the case of tribological tests, mass loss is significantly reduced by both LSM and LSH; however, the wear rate is highly variable, and it depends on the conditions of the friction system. Finally, since LST encompasses volumetric transformations, the analysis of residual stresses is important, and it usually reveals a change from compressive stresses near the surface to tensile stresses further into the as-cast region.

6. Concluding Remarks

This review highlights the evolution of the characterization of the surface thermal treatment by laser in different iron castings, divided by the main constituents and structure of their matrix. Based on what is presented in this article, LST holds numerous advantages over conventional heat treatments, as it is a precise and highly localized process for which a harder and more wear-resistant microstructure is obtained without generating a significant distortion of the workpiece. The following main conclusions can be stated:

- The use of diode and Nd:YAG lasers without coatings allows for higher absorption coefficients of the radiated energy than when using conventional CO_2 lasers.
- Lasers with circular geometry and Gaussian energy distribution generate a parabolic heat-affected zone in the transverse section, where properties are not uniform within a constant layer depth.
- Single-pass LST with rectangular-shaped and uniform energy distribution lasers, as well as adjacent discrete laser spots, can be used to avoid a tempering effect of overlapping.
- Following laser modification, DIs and ADIs present lower wear damage under comparable conditions than GIs, since graphite flakes act as stress raisers, favoring crack nucleation and growth during friction, as in dry sliding tests.
- Regardless of the initial microstructure of the cast iron, the linear energy is the key parameter, since it considers the joint effect of experimental parameters, such as laser power, absorption layer thickness, and scanning velocity. It is suggested to apply surface hardening without melting on DIs and ADIs to achieve higher wear resistance, because of the nature of the residual stresses created during phase transformations (compressive in LSH and tensile in LSM).
- Further research challenges include the analysis of the effect of alloying elements, such as Mo, Cr, or Ni, on the thermal process and mechanical properties induced by LST. Moreover, extra features, such as grain growth and surface roughness, must be added into LST simulations in order to ensure a reliable validation and determination of the scope and precision of the model.
- Future technological challenges involve the analysis of costs and implementation of LST at the industrial level, especially in high-technology and -impact environments, to evaluate the performance of this treatment in real conditions. In this regard, the information presented in this review article is a guide to encourage future investigation prospects in cast irons and ferrous alloys in general.

Author Contributions: Conceptualization, E.R.-M., A.B. and D.C.; methodology, N.C., E.R.-M. and D.C.; investigation, N.C.; writing–original draft preparation, N.C.; writing–review and editing, E.R.-M., A.B. and D.C.; visualization, E.R.-M.; supervision, E.R.-M. and D.C.; project administration, E.R.-M. and D.C.; funding acquisition, E.R.-M., A.B. and D.C. All authors have read and agreed to the published version of the manuscript.

Funding: A. Boccardo thanks the financial support received by a research grant from Science Foundation Ireland (SFI) under grant number 16/RC/3872. E. Ramos-Moore thanks FONDECYT-ANID project 1180564, and D. Celentano thanks FONDECYT-ANID project 1180591.

Institutional Review Board Statement: Not applicable.

Informed Consent Statement: Not applicable.

Data Availability Statement: Any requirement about the data of the reviewed articles must be consulted directly to the corresponding authors.

Conflicts of Interest: The authors declare no conflict of interest.

References

1. Aliakbari, K.; Masoudi Nejad, R.; Akbarpour Mamaghani, T.; Pouryamout, P.; Rahimi Asiabaraki, H. Failure analysis of ductile iron crankshaft in compact pickup truck diesel engine. *Structures* **2022**, *36*, 482–492. [CrossRef]
2. Concli, F. Austempered Ductile Iron (ADI) for gears: Contact and bending fatigue behavior. *Procedia Struct. Integr.* **2018**, *8*, 14–23. [CrossRef]
3. Artola, G.; Gallastegi, I.; Izaga, J.; Barreña, M.; Rimmer, A. Austempered Ductile Iron (ADI) Alternative Material for High-Performance Applications. *Int. J. Met.* **2016**, *11*, 131–135. [CrossRef]
4. Chakrabarty, I. *Reference Module in Materials Science and Materials Engineering*; Elsevier: Amsterdam, The Netherlands, 2018; Alloy Cast Irons and Their Engineering Applications; pp. 1–25. ISBN 978-0-12-803581-8.
5. ASM International. Casting. In *ASM Handbook*, 9th ed.; ASM International: Materials Park, OH, USA, 1998; Volume 15, pp. 1–937.
6. Collini, L.; Nicoletto, G.; Konecná, R. Microstructure and mechanical properties of pearlitic gray cast iron. *Mater. Sci. Eng. A* **2008**, *488*, 529–539. [CrossRef]
7. Jabbari Benham, M.M.; Davami, P.; Varahram, N. Effect of cooling rate on microstructure and mechanical properties of gray cast iron. *Mater. Sci. Eng. A* **2010**, *528*, 583–588. [CrossRef]
8. Willidal, T.; Bauer, W.; Schumacher, P. Stress/strain behavior and fatigue limit of grey cast iron. *Mater. Sci. Eng. A* **2005**, *413–414*, 578–582. [CrossRef]
9. Baicchi, P.; Collini, L.; Riva, E. A methodology for the fatigue design of notched castings in gray cast iron. *Eng. Fract. Mech.* **2007**, *74*, 539–548. [CrossRef]
10. Damir, A.N.; Elkhatib, A.; Nassef, G. Prediction of fatigue life using modal analysis for grey and ductile cast iron. *Int. J. Fatigue* **2007**, *29*, 499–507. [CrossRef]
11. Cueva, G.; Sinatora, A.; Guesser, W.L.; Tschiptschin, A.P. Wear resistance of cast irons used in brake disc rotors. *Wear* **2003**, *255*, 1256–1260. [CrossRef]
12. Ferrer, C.; Pascual, M.; Busquets, D.; Rayón, E. Tribological study of Fe-Cu-Cr-graphite alloy and cast iron railway brake shoes by pin-on-disc technique. *Wear* **2010**, *268*, 784–789. [CrossRef]
13. Hirasata, K.; Hayashi, K.; Inamoyo, Y. Friction and wear of several kinds of cast irons under severe sliding conditions. *Wear* **2007**, *263*, 790–800. [CrossRef]
14. Prasad, B.K. Sliding wear response of cast iron as influenced by microstructural features and test condition. *Mater. Sci. Eng. A* **2007**, *456*, 373–385. [CrossRef]
15. Prasad, B.K. Sliding wear behaviour of a cast iron as affected by test environment and applied load. *Ind. Lubr. Tribol.* **2009**, *61*, 161–172. [CrossRef]
16. Ghasemi, R.; Elmquist, L. A study on graphite extrusion phenomenon under the sliding wear response of cast iron using microindentation and microscratch techniques. *Wear* **2014**, *320*, 120–126. [CrossRef]
17. Carazo, F.D.; Giusti, S.M.; Boccardo, A.D.; Godoy, L.A. Effective properties of nodular cast iron: A multi-scale computational approach. *Comput. Mater. Sci.* **2014**, *82*, 378–390. [CrossRef]
18. Gonzaga, R.A. Influence of ferrite and pearlite content on mechanical properties of ductile cast irons. *Mater. Sci. Eng. A* **2013**, *567*, 1–8. [CrossRef]
19. Li, N.; Xing, S.; Bao, P. Microstructure and Mechanical Properties of Nodular Cast Iron Produced by Melted Metal Die Forging Process. *J. Iron Steel Res. Int.* **2013**, *20*, 58–62. [CrossRef]
20. Ceschini, L.; Morri, A.; Morri, A.; Salsi, E.; Squatrito, R.; Todaro, I.; Tomesani, L. Microstructure and mechanical properties of heavy section ductile iron castings: Experimental and numerical evaluation of effect of cooling rates. *Int. J. Cast Met. Res.* **2015**, *28*, 365–374. [CrossRef]
21. González-Martínez, R.; de la Torre, U.; Ebel, A.; Lacaze, J.; Sertucha, J. Effects of high silicon contents on graphite morphology and room temperature mechanical properties of as-cast ferritic ductile cast irons. Part II—Mechanical properties. *Mater. Sci. Eng. A* **2018**, *712*, 803–811. [CrossRef]
22. Mourujärvi, A.; Widell, K.; Saukkonen, T.; Hänninen, H. Influence of chunky graphite on mechanical and fatigue properties of heavy-section cast iron. *Fatigue Fract. Eng. Mater. Struct.* **2009**, *32*, 379–390. [CrossRef]
23. Hosdez, J.; Limodin, N.; Najjar, D.; Witz, J.-F.; Charkaluk, E.; Osmond, P.; Forré, A.; Szmytka, F. Fatigue crack growth in compacted and spheroidal graphite cast irons. *Int. J. Fatigue* **2020**, *131*, 105319. [CrossRef]
24. Iacoviello, F.; Di Cocco, V.; Bellini, C. Fatigue crack propagation and damaging micromechanisms in Ductile Cast Irons. *Int. J. Fatigue* **2019**, *124*, 48–54. [CrossRef]
25. Mohamed, I.A.; Ibraheem, A.A.; Khashaba, M.I.; Ali, W.Y. Influence of Heat Treatment on Friction and Wear of Ductile Iron: II. Role of Chromium and Nickel. *Int. J. Control Autom. Syst.* **2014**, *3*, 10–16.
26. Ben Tkaya, M.; Mezlini, S.; El Mansori, M.; Zahouani, H. On some tribological effects of graphite nodules in wear mechanism of SG cast iron: Finite element and experimental analysis. *Wear* **2009**, *267*, 535–539. [CrossRef]
27. Pintaude, G.; Bernardes, F.G.; Santos, M.M.; Sinatora, A.; Albertin, E. Mild and severe wear of steels and cast irons in sliding abrasion. *Wear* **2009**, *267*, 19–25. [CrossRef]
28. Wei, M.X.; Wang, S.Q.; Cui, X.H. Comparative research on wear characteristics of spheroidal graphite cast iron and carbon steel. *Wear* **2012**, *274–275*, 84–93. [CrossRef]

29. Panneerselvam, S.; Putatunda, S.K.; Gundlach, R.; Boileau, J. Influence of intercritical austempering on the microstructure and mechanical properties of austempered ductile iron (ADI). *Mater. Sci. Eng. A* **2017**, *694*, 72–80. [CrossRef]

30. Boccardo, A.D.; Dardati, P.M.; Celentano, D.J.; Godoy, L.A.; Górny, M.; Tyraia, E. Numerical Simulation of Austempering Heat Treatment of a Ductile Cast Iron. *Metall. Mater. Trans. B* **2016**, *47*, 566–575. [CrossRef]

31. Putatunda, S.K. Development of austempered ductile cast iron (ADI) with simultaneous high yield strength and fracture toughness by a novel two-step austempering process. *Mater. Sci. Eng. A* **2001**, *315*, 70–80. [CrossRef]

32. Putatunda, S.K.; Gadicherla, P.K. Effect of Austempering Time on Mechanical Properties of a Low Manganese Austempered Ductile Iron. *J. Mater. Eng. Perform.* **2000**, *9*, 193–203. [CrossRef]

33. Ghaderi, A.R.; Nili Ahmadabadi, M.; Ghasemi, H.M. Effect of graphite morphologies on the tribological behavior of austempered cast iron. *Wear* **2003**, *255*, 410–416. [CrossRef]

34. ASM International. Heat Treating. In *ASM Handbook*, 9th ed.; ASM International: Materials Park, OH, USA, 1998; Volume 4, pp. 1–926.

35. Li, Y.; Song, R.; Zhao, Z.; Pei, Y. Effect of three-staged normalizing on the impact wear resistance of 3.23 mass% Cr-Mn-Cu-Si cast iron. *Wear* **2019**, *426–427*, 59–67. [CrossRef]

36. Wassilkowska, A.; Dabrowski, W. Silicon as a component of ferric oxide scale covering ductile iron pipes after annealing. *Eng. Fail. Anal.* **2021**, *125*, 105381. [CrossRef]

37. Vadiraj, A.; Balachandran, G.; Kamaraj, M.; Kazuya, E. Mechanical and wear behavior of quenched and tempered alloyed hypereutectic gray cast iron. *Mater. Des.* **2011**, *32*, 2438–2443. [CrossRef]

38. Wang, B.L.; Morris, D.S.; Farshid, S.; Lortz, S.; Ma, Q.; Wang, C.; Chen, Y.; Sanders, P. Rolling contact fatigue study of chilled and quenched/tempered ductile iron compared with AISI 1080 steel. *Wear* **2021**, *478–479*, 203890. [CrossRef]

39. Steen, W.M.; Mazumder, J. *Laser Material Processing*, 4th ed.; Springer: London, UK, 2010; pp. 295–317.

40. Hwang, J.H.; Lee, Y.S.; Kim, D.Y.; Youn, J.G. Laser Surface Hardening of Gray Cast Iron Used for Piston Ring. *J. Mater. Eng. Perform.* **2002**, *11*, 294–300. [CrossRef]

41. Liu, A.; Previtali, B. Laser surface treatment of grey cast iron by high power diode laser. *Phys. Procedia* **2010**, *5*, 439–448. [CrossRef]

42. Wang, B.; Barber, G.C.; Wang, R.; Pan, Y. Comparison of Wear Performance of Austempered and Quench-Tempered Gray Cast Irons Enhanced by Laser Hardening Treatment. *Appl. Sci.* **2020**, *10*, 3049. [CrossRef]

43. Wang, B.; Pan, Y.; Liu, Y.; Lyu, N.; Barber, G.C.; Wang, R.; Cui, W.; Qiu, F.; Hu, M. Effects of quench-tempering and laser hardening treatment on wear resistance of gray cast iron. *J. Mater. Res. Technol.* **2020**, *9*, 8163–8171. [CrossRef]

44. Trafford, D.N.H.; Bell, T.; Megaw, J.H.P.C.; Brandsen, A.S. Laser treatment of grey iron. *Met. Technol.* **1983**, *10*, 69–77. [CrossRef]

45. De Oliveira, U.O.B. Laser Treatment of Alloys: Processing, Microstructure and Structural Properties. Ph.D. Thesis, University of Groningen, Groningen, The Netherlands, 2007.

46. Paczkowska, M. The evaluation of the influence of laser treatment parameters on the type of thermal effects in the surface layer microstructure of gray irons. *Opt. Laser Technol.* **2016**, *76*, 143–148. [CrossRef]

47. Mathur, A.K.; Molian, P.A. Laser Heat Treatment of Cast Irons—Optimization of Process Variables: Part I. *J. Eng. Mater. Technol.* **1985**, *107*, 200–207. [CrossRef]

48. Molian, P.A.; Mathur, A.K. Laser Heat Treatment of Cast Irons—Optimization of Process Variables: Part II. *J. Eng. Mater. Technol.* **1986**, *108*, 233–239. [CrossRef]

49. Chen, C.H.; Ju, C.P.; Rigsbee, J.M. Laser surface modification of ductile iron: Part 1 Microstructure. *Mater. Sci. Technol.* **1988**, *4*, 161–166. [CrossRef]

50. Gadag, S.P.; Srinivasan, M.N.; Mordike, B.L. Effect of laser processing parameters on the structure of ductile iron. *Mater. Sci. Eng. A* **1995**, *196*, 145–154. [CrossRef]

51. Fernández-Vicente, A.; Pellizzari, M.; Arias, J.L. Feasibility of laser surface treatment of pearlitic and bainitic ductile irons for hot rolls. *J. Mater. Processing Technol.* **2012**, *212*, 989–1002. [CrossRef]

52. Molian, P.A.; Baldwin, M. Effects of single-pass laser heat treatment on erosion behavior of cast iron. *Wear* **1987**, *118*, 319–327. [CrossRef]

53. Ghaini, F.M.; Ameri, M.H.; Torkamany, M.J. Surface transformation hardening of ductile cast iron by a 600 w fiber laser. *Optik* **2020**, *203*, 163758. [CrossRef]

54. Chen, C.H.; Ju, C.P.; Rigsbee, J.M. Laser surface modification of ductile iron: Part 2 Wear mechanism. *Mater. Sci. Technol.* **1988**, *4*, 167–172. [CrossRef]

55. Gadag, S.P.; Srinivasan, M.N. Cavitation erosion of laser-melted ductile iron. *J. Mater. Process. Technol.* **1995**, *51*, 150–163. [CrossRef]

56. Lu, G.; Zhang, H. Sliding wear characteristics of austempered ductile iron with and without laser hardening. *Wear* **1990**, *138*, 1–12.

57. Roy, A.; Manna, I. Mathematical modelling of localized melting around graphite nodules during laser surface hardening of austempered ductile iron. *Opt. Lasers Eng.* **2000**, *34*, 369–383. [CrossRef]

58. Roy, A.; Manna, I. Laser surface engineering to improve wear resistance of austempered ductile iron. *Mater. Sci. Eng. A* **2001**, *297*, 85–93. [CrossRef]

59. Zammit, A.; Abela, S.; Betts, J.C.; Grech, M. Discrete spot laser hardening of austempered ductile iron. *Surf. Coat. Technol.* **2017**, *331*, 143–152. [CrossRef]

60. Soriano, C.; Leunda, J.; Lambarri, J.; García Navas, V.; Sanz, C. Effect of laser surface hardening on the microstructure, hardness and residual stresses of austempered ductile iron grades. *Appl. Surf. Sci.* **2011**, *257*, 7101–7106. [CrossRef]

61. Putatunda, S.; Bartosiewicz, L.; Hull, R.J.; Lander, M. Laser Hardening of Austempered Ductile Iron (ADI). *Mater. Manuf. Process.* **1997**, *12*, 137–151. [CrossRef]

62. Zammit, A.; Abela, S.; Betts, J.C.; Michalczewski, R.; Kalbarczyk, M.; Grech, M. Scuffing and rolling contact fatigue resistance of discrete laser spot hardened austempered ductile iron. *Wear* **2019**, *422–423*, 100–107. [CrossRef]

63. Grum, J.; Sturm, R. Microstructure analysis of nodular iron 400-12 after laser surface melt hardening. *Mater. Charact.* **1996**, *37*, 81–88. [CrossRef]

64. Grum, J.; Sturm, R. Comparison of measured and calculated thickness of martensite and ledeburite shells around graphite nodules in the hardened layer of nodular iron after laser surface remelting. *Appl. Surf. Sci.* **2002**, *187*, 116–123. [CrossRef]

65. Grum, J.; Sturm, R. Microstructure variations in the laser surface remelted layer of nodular iron. *Int. J. Microstruct. Mater. Prop.* **2005**, *1*, 11–23. [CrossRef]

66. Benyounis, K.Y.; Fakron, O.M.A.; Abboud, J.H.; Olabi, A.G.; Hashmi, M.J.S. Surface melting of nodular cast iron by Nd-YAG laser and TIG. *J. Mater. Process. Technol.* **2005**, *170*, 127–132. [CrossRef]

67. Alabeedi, K.F.; Abboud, J.H.; Benyounis, K.Y. Microstructure and erosion resistance enhancement of nodular cast iron by laser melting. *Wear* **2009**, *266*, 925–933. [CrossRef]

68. Pagano, N.; Angelini, V.; Ceschini, L.; Campana, G. Laser remelting for enhancing tribological performances of a ductile iron. *Procedia CIRP* **2016**, *41*, 987–991. [CrossRef]

69. Catalán, N.; Ramos-Moore, E.; Boccardo, A.; Celentano, D.; Alam, N.; Walczak, M.; Gunasegaram, D. Surface Laser Treatment on Ferritic Ductile Iron: Effect of Linear Energy on Microstructure, Chemical Composition, and Hardness. *Metall. Mater. Trans. B* **2021**, *52*, 755–763. [CrossRef]

70. Ceschini, L.; Campana, G.; Pagano, N.; Angelini, V. Effect of laser surface treatment on the dry sliding behavior of the EN-GJS400-12 ductile cast iron. *Tribol. Int.* **2016**, *104*, 342–351. [CrossRef]

71. Boccardo, A.D.; Catalán, N.; Celentano, D.J.; Ramos-Moore, E. A Thermo-metallurgical Model for Laser Surface Engineering Treatment of Nodular Cast Iron. *Metall. Mater. Trans. B* **2021**, *52*, 854–870. [CrossRef]

Article

Beam Shaping in Laser Powder Bed Fusion: Péclet Number and Dynamic Simulation

Sergey N. Grigoriev, Andrey V. Gusarov, Alexander S. Metel, Tatiana V. Tarasova, Marina A. Volosova, Anna A. Okunkova * and Andrey S. Gusev

Department of High-Efficiency Processing Technologies, Moscow State University of Technology STANKIN, 127055 Moscow, Russia; s.grigoriev@stankin.ru (S.N.G.); a.gusarov@stankin.ru (A.V.G.); a.metel@stankin.ru (A.S.M.); tarasova952@mail.ru (T.V.T.); m.volosova@stankin.ru (M.A.V.); gusev.angrey@bk.ru (A.S.G.)
* Correspondence: a.okunkova@stankin.ru; Tel.: +7-909-913-12-07

Abstract: A uniform distribution of power density (energy flux) in a stationary laser beam leads to a decrease in the overheating of the material in the center of the laser beam spot during laser powder bed fusion and a decrease in material losses due to its thermal ablation and chemical decomposition. The profile of the uniform cylindrical (flat-top) distribution of the laser beam power density was compared to the classical Gaussian mode (TEM_{00}) and inverse Gaussian (donut) distribution (airy distribution of the first harmonic, $TEM_{01*} = TEM_{01} + TEM_{10}$). Calculation of the Péclet number, which is a similarity criterion characterizing the relationship between convective and molecular processes of heat transfer (convection to diffusion) in a material flow in the liquid phase, shows that the cylindrical (flat-top) distribution ($TEM_{01*} + TEM_{00}$ mode) is effective in a narrow temperature range. TEM_{00} shows the most effective result for a wide range of temperatures, and TEM_{01*} is an intermediate in which evaporation losses decrease by more than 2.5 times, and it increases the absolute laser bandwidth when the relative bandwidth decreases by 24%.

Keywords: energy excess; heat diffusion; laser beam mode; laser powder bed fusion; numerical simulation; profiling; power density distribution; thermal conductivity

Citation: Grigoriev, S.N.; Gusarov, A.V.; Metel, A.S.; Tarasova, T.V.; Volosova, M.A.; Okunkova, A.A.; Gusev, A.S. Beam Shaping in Laser Powder Bed Fusion: Péclet Number and Dynamic Simulation. *Metals* **2022**, *12*, 722. https://doi.org/10.3390/met12050722

Academic Editors: Antonio Riveiro and Matteo Benedetti

Received: 1 March 2022
Accepted: 18 April 2022
Published: 24 April 2022

Publisher's Note: MDPI stays neutral with regard to jurisdictional claims in published maps and institutional affiliations.

1. Introduction

The well-known drawback of some laser material-processing technologies is non-uniform thermal conditions in the spot. The material is overheated in the center of the laser spot when an excess of the energy leads to intensive material evaporations and chemical decompositions [1–4], which is not characteristic of other additive technologies using alternative sources of concentrated energy flow [5,6]. Inversely, the material does not attain the necessary processing temperature at the periphery of the spot, and the energy is essentially lost by heat diffusion in the treated body (the target) [7–9]. Modern optics proposes shaping a laser beam that provides alternative laser power density distributions of transverse electromagnetic (TEM) mode:

- Airy distribution of the first harmonic (donut) $TEM_{01*} = TEM_{01} + TEM_{10}$;
- Uniform cylindrical (flat-top) distribution $TEM_{FT} = TEM_{01*} + TEM_{00}$.

These technical solutions have multiple laser powder bed fusion attempts but have never been researched theoretically with correction to the beam motion [10–12].

The lack of a reliable solution in terms of heat redistribution leads to the following disadvantages affecting the quality of parts obtained by laser-additive manufacturing and processing productivity (Figure 1) [13–17]:

- Local overheating, capturing the underlying layers, creating additional stresses during metal solidification (partially solved by subsequent heat treatment and preliminary heating of the substrate) [18–20];

- Active evaporation of the material and its chemical interaction with the atmosphere of the chamber (reduced due to the use of more gentle processing modes, which dramatically affects productivity) [21–23];
- Ejecting material from the processing area (reduces the surface quality of the part itself, damages the optics, and is reduced by gentle modes and preheating of the platform) [24–26].

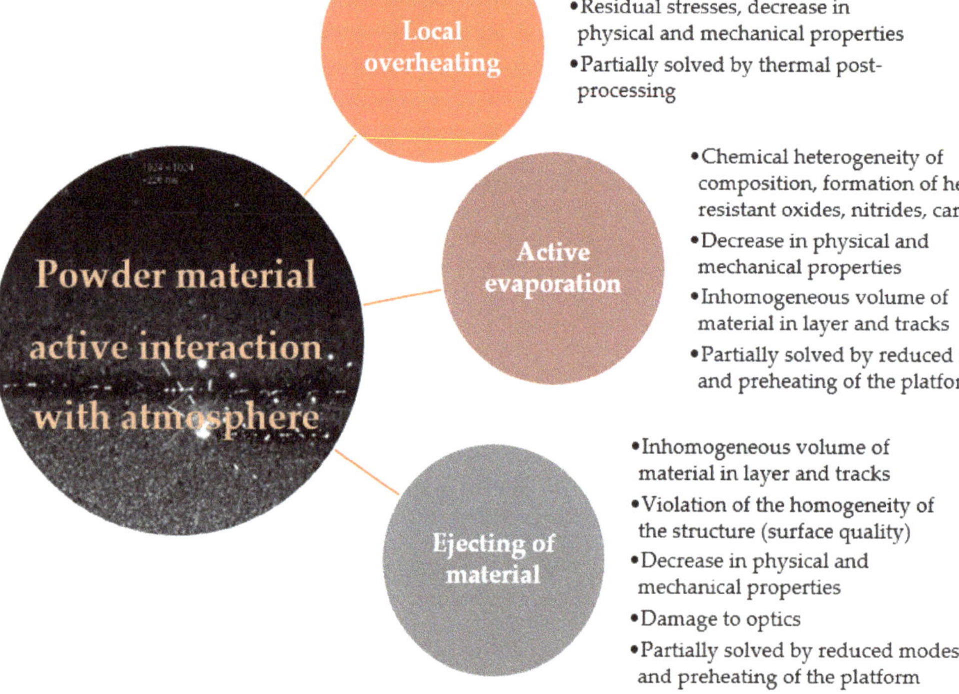

Figure 1. The main consequences of the active interaction of powder material with atmosphere and the existing ways of solving them.

An obvious disadvantage of using optical means for redistributing laser energy into the beam can be its expansion by 150–350%, which may not allow for obtaining more precision parts, but can become a significant advantage in the production of products with dimensions of more than 100 mm, for which the width of the heat-affected zone will be significantly reduced [27,28]. Figure 1 is based on the results of optical diagnostics and video monitoring described in detail in [27].

There are many factors that influence the final surface quality (roughness, uniformity, and dimensional accuracy) [29–33] such as:

- laser power, spot size, and laser power distribution among the laser system and optic parameters,
- scanning speed and strategy and hatch distance among strategy parameters,
- powder particle size, shape and morphology, and layer thickness among powder parameters,
- inertness of the atmosphere, impermeability of the chamber, dimensions of the part on the working platform (maximum angle of deviation of the beam from the vertical), and so on.

The conventional power (energy flux q, W/mm^2) density distribution in radius r of the laser focus is the classical bell-like one approximated by the normal Gauss distribution (Laguerre–Gaussian mode, circularly symmetric beam profile TEM$_{00}$) of the optical resonator as:

$$q = \frac{P}{\pi r_0^2} \exp\left(-\frac{r^2}{r_0^2}\right),$$

(1)

where P is the laser beam power, W and r_0 is the radius circle, mm.

In some laser-based technologies such as lithography (photo-activated processes) [34,35], laser scribing [36,37], and thin surface laser treatment (including medical purposes) [38–41], the optimal beam profile seems to be the flat-top (TEM$_{FT}$) one that provides the energy flux's uniformity (uniform laser power density distribution). The typical powder consolidation mechanisms in laser powder bed fusion are thermo-activated [42]. Then the objective is transferred from the uniform power density distribution (energy flux q, W/mm^2) to a radiation-induced uniform temperature field T ($^\circ$C).

Since the thermal energy is released on an adiabatic plane bounding a uniform conducting half-space inside a circle of radius r_0 (mm), with radial distribution [43]:

$$q = \frac{P}{2\pi r_0^2} \frac{1}{\sqrt{1 - r^2/r_0^2}},$$

(2)

the temperature rise over the circle:

$$T_0 = \frac{P}{4\lambda r_0},$$

(3)

where λ is the material thermal conductivity, W/mm·K. In this case, the laser radiation is absorbed by layered powder to heat a massive body with conduction as the principal heat transfer mechanism. Then profile (2) can be better for laser powder bed fusion and similar laser-based powder technologies. TEM$_{FT}$ profile (the cylindrical flat-top temperature distribution) is challenging to obtain because of a discontinuity at the beam boundary where $r = r_0$. Then the airy distribution of the first harmonic, (donut of the first overtone) TEM$_{01^*}$, seems to be a reasonable compromise [43]:

$$q = \frac{P}{\pi r_0^2} \frac{r^2}{r_0^2} \exp\left(-\frac{r^2}{r_0^2}\right),$$

(4)

In the thermo-activated processes, the laser beam scans the powder surface, resulting in a non-uniform temperature distribution over the laser spot for various laser beam profiles [44,45]. An inverse problem of heat diffusion for the scanning laser beam can be solved to find the ideal power density distribution. Still, the solution mainly depends on the scanning speed factor—its value and direction. The influence of direction on the absorbed energy flux shows that the laser beam profile would be asymmetric. Moreover, the laser beam scans quite fast (up to 400 mm/s) and changes direction rapidly. Therefore, it can be an even more complicated scientific and technical task never solved before, since most of the published work on beam profiling considers the symmetric beam for their calculations.

This work aims to compare three types of abovementioned laser beam profiles, research the influence of the scanning speed in a linear medium, and develop a non-linear model, including the material evaporation factor.

2. Numerical Simulations

2.1. Simulations and Influence of Scanning Speed

The powder layer on the target surface is considered thermally thin and is not taken into account. Laser radiation is supposed to be absorbed on the surface. In the case of partial reflection, the laser power in the equations mentioned above means the absorbed part of the laser beam radiation. In the coordinate system moving with the scanning speed, the steady-state heat diffusion equation is [43]:

$$\alpha \Delta T + u_s \frac{\partial T}{\partial x} = 0, \tag{5}$$

where u_s is scanning speed, m/s; α is the thermal diffusivity, m^2/s; and Δ is the Laplace operator. Equation (5) is solved by numerical or analytical methods where possible, with boundary condition:

$$T \to T_a \text{ at } x \to \pm\infty, \ y \to \pm\infty, \ z \to \infty, \tag{6}$$

where T_a is the ambient temperature. The target surface $z = 0$ is adiabatic, excluding the laser spot where

$$-\lambda \frac{\partial T}{\partial z} = q. \tag{7}$$

The temperature fields are presented in Figures 2 and 3.

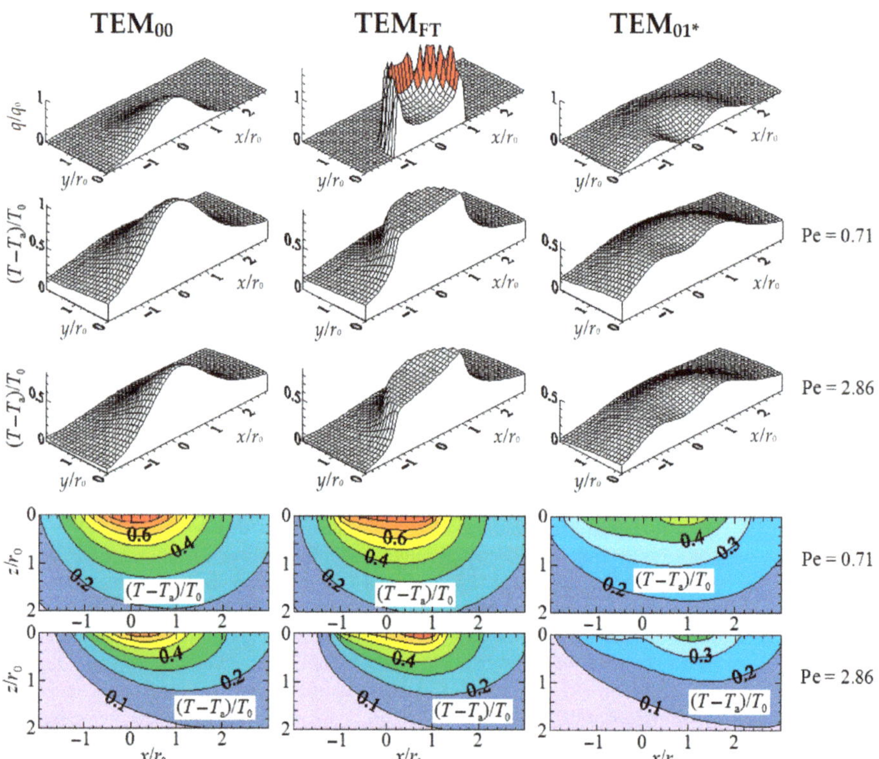

Figure 2. Normalized distributions: flux density of the absorbed laser energy q over the target surface $z = 0$ (top row); temperature T over the target surface (second and third rows); temperature T over the vertical plane of mirror symmetry $y = 0$ formed by the beam axis and the scanning line (two rows on the bottom). Red in the q/q_0 graph indicates the approach to the area of the discontinuity at the beam boundary where $r = r_0$.

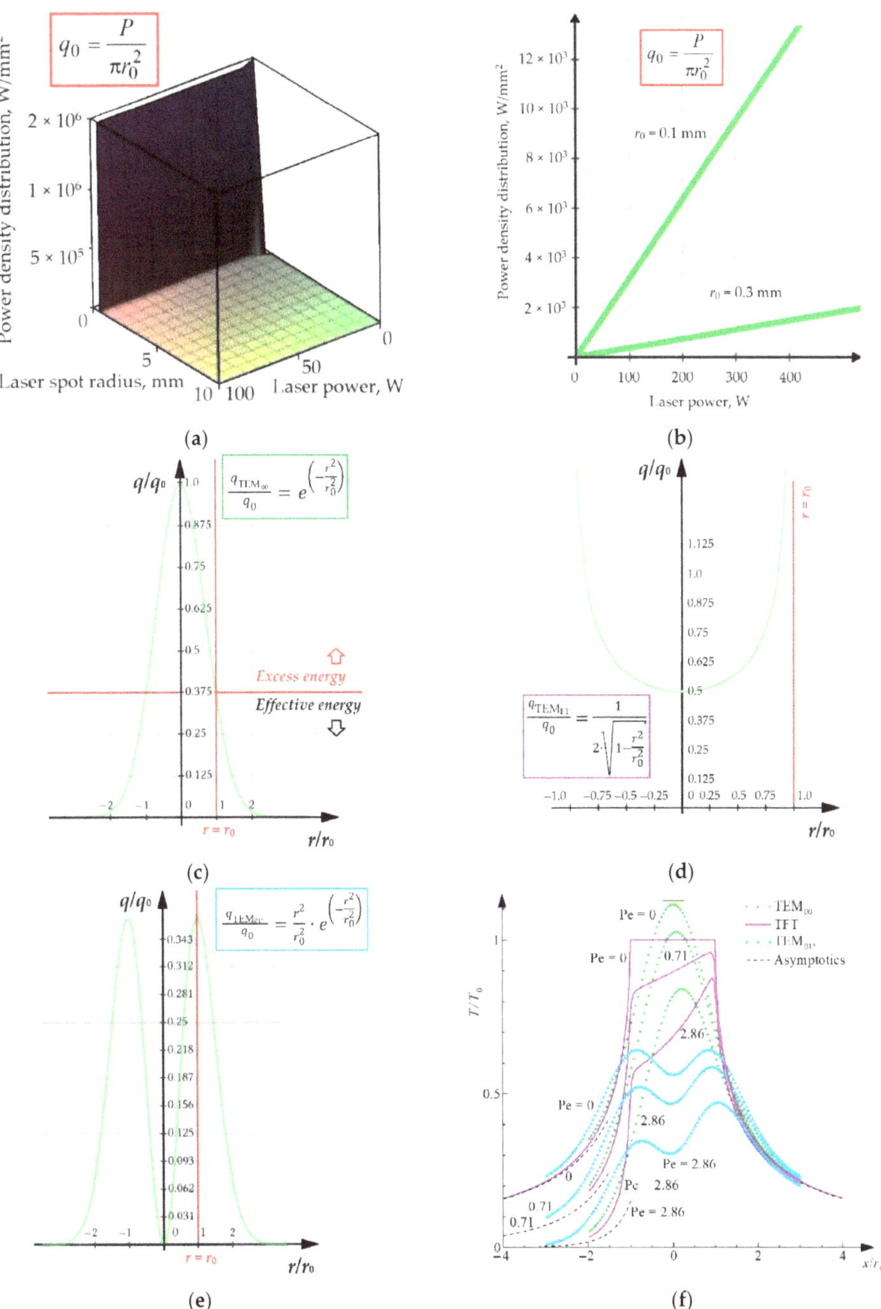

Figure 3. 3D plot of the implicit function q_0 (**a**); implicit function q_0 for various values of r_0 (**b**); normalized implicit function q/q_0 (TEM$_{00}$ profile) (**c**); normalized implicit function q/q_0 (TEM$_{FT}$ profile) (**d**); normalized implicit function q/q_0 (TEM$_{01*}$ profile) (**e**); and temperature distributions along the direction of the scanning speed on the surface, when $y = 0$, $z = 0$ (**f**). The beam boundary where $r = r_0$ is marked red in graphs (**c**–**e**).

The scanning speed is specified by the thermal Péclet number:

$$\text{Pe} = \frac{2r_0 u_s}{\alpha}. \tag{8}$$

The temperature rise relative to the ambience $(T - T_a)$ is normalized by T_0 specified by Equation (3). Normalizing coordinates by r_0 makes the obtained results universal for a linear conductive medium. The results significantly depend on the Péclet number. The top row in Figure 2 shows two-dimensional views of laser profiles (1), (2), and (4) normalized by [43] (Figure 3a,b):

$$q_0 = \frac{P}{\pi r_0^2}. \tag{9}$$

The normalized graphs of profiles are as follows (Figure 3c–e):

$$\frac{q_{\text{TEM}_{00}}}{q_0} = e^{\left(-\frac{r^2}{r_0^2}\right)}, \tag{10}$$

$$\frac{q_{\text{TEM}_{FT}}}{q_0} = \frac{1}{2 \cdot \sqrt{1 - \frac{r^2}{r_0^2}}}, \tag{11}$$

$$\frac{q_{\text{TEM}_{01*}}}{q_0} = \frac{r^2}{r_0^2} \cdot e^{\left(-\frac{r^2}{r_0^2}\right)}. \tag{12}$$

The other rows in Figure 2 are two-dimensional temperature distributions over two characteristic planes. The 3D plot of the implicit function is shown in Figure 3a.

Figure 3f shows all the obtained results as profiles of the surface temperature along line $y = 0$, $z = 0$. For all laser profiles, the temperature profiles decrease with the increase of Pe that corresponds to the increase of the scanning speed. The forward temperature front becomes sharper with the increase of Pe, and the backward temperature front is insensible to Pe, according to the well-known asymptotics:

$$\frac{T - T_a}{T_0} = \frac{2}{\pi} \frac{r_0}{R} \exp\left(\frac{\text{Pe}}{4} \frac{x - R}{r_0}\right), \tag{13}$$

with $R^2 = x^2 + y^2 + z^2$, shown by dashed lines in Figure 3f. In the case of mode TEM_{00}, all three numerically calculated temperature profiles are bell-like. At Pe = 0, the maximum is in the origin. The numerically obtained maximum value is about the analytical result $T_{\max}$,

$$\frac{T_{\max} - T_a}{T_0} = \frac{2}{\sqrt{\pi}}, \tag{14}$$

shown by a horizontal dash in Figure 3f. The increase of Pe slightly shifts the position of the temperature maximum in the direction opposite to that of the scanning speed vector that is explained by the thermal inertia of the target.

At Pe = 0, the flat-top laser beam profile forms steady-state temperature distribution

$$\frac{T - T_a}{T_0} = \frac{2}{\pi} \arcsin \frac{2r_0}{\sqrt{(r - r_0)^2 + z^2} + \sqrt{(r + r_0)^2 + z^2}}, \tag{15}$$

where $r^2 = x^2 + y^2$, with an exactly horizontal plate over the laser spot. When Pe increases, this plate inclines towards the scanning speed vector and slightly sags. In the case of donut mode, the surface temperature distribution inherits the ring-like ridge. The ridge becomes more asymmetric with the increase of Pe (Figure 3f).

2.2. Temperature and Energy Flux Profiles

Temperature distribution in a cross-section perpendicular to the scanning direction cannot objectively characterize the temperature conditions for laser powder bed fusion because retarding the maximum target temperature relative to the central cross-section $x = 0$. The retardation depends on the scanning speed value and the distance from the scanning axis (X). The most representative quantity is the maximum temperature along axis X for threshold-like and Arrhenius temperature dependencies of the process kinetics. Figure 4a shows the transverse profile of the quantity on the surface [43]:

$$\max_{x} T(x, y, 0), \tag{16}$$

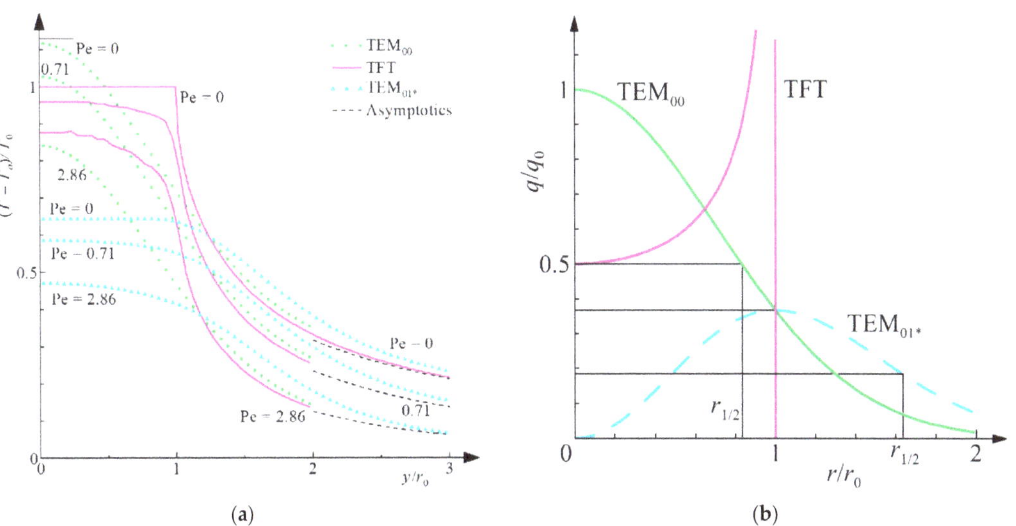

(a)

(b)

Figure 4. Maximum temperature T on the target surface $z = 0$ versus distance y from the scanning axis (**a**); the testing profiles (q/q_0) and estimation of their radii at half width $r_{1/2}$ (**b**).

The asymptotics at Pe = 0 are given by Equation (13) at $x = 0$. At Pe = 0.71 and Pe = 2.86, the asymptotics are obtained by numerical treatment of Equation (13) by Equation (16). The widths of the re-melted zone on the surface often estimate the contact's width between the consolidated powder, and the substrate can be deduced by this profile.

The transverse profiles of the surface temperature shown in Figure 4a present the thermal conditions for laser powder bed fusion. They cannot be compared with the tested laser beam profiles because all the obtained temperature profiles have different absolute maxima. The tentative laser-beam radius r_0 is not an objective measure of its width applicable to various beams' radial profiles. Thus, beam TEM$_{01*}$ in Figure 4b seems wider than beam TEM$_{00}$ at the same r_0. Let us estimate the width of a laser profile by its diameter at half-maximum $d_{\frac{1}{2}}$ that is conventional in laser technology applications. The scheme for estimating the corresponding radius at half-maximum $r_{\frac{1}{2}} = d_{\frac{1}{2}}/2$ is shown in Figure 4b and Table 1.

Table 1. Calculated absolute maximum of temperature $T_{\max}$ versus Péclet's number Pe.

Laser Beam Profile	Beam Radius at Half Maximum, $r_{1/2}/r_0$	$(T_{\max} - T_a)/T_0$		
		Pe = 0	Pe = 0.71	Pe = 2.86
TEM$_{00}$ (Gaussian)	$\sqrt{\ln 2} = 0.8326$	$2/\sqrt{\pi} = 1.128$	1.027	0.8417
TEM$_{01*}$ (donut)	1.6366	1.6453	0.5889	0.4735
TEM$_{FT}$ (flat-top)	1	1	0.9613	0.8819

It should be noted that temperatures above $T_{\max}$ are unallowable because of material evaporation or chemical decomposition. Temperatures below the minimum $T_{\min}$ are not sufficient to complete the specified physical or chemical processes. The boiling point is specified as $T_{\max}$, and the melting point is $T_{\min}$ for laser powder bed fusion of pure metals [46,47]. For alloys, $T_{\max}$ and $T_{\min}$ are determined by the component with the lowest boiling and melting points, correspondingly.

The temperature dependencies of the kinetic constants can be taken into account to define the laser powder bed fusion interval $(T_{\min}, T_{\max})$. The maximum temperature in the laser-processing zone and the width of the laser beam characterized by $d_{1/2}$ or $r_{1/2}$ can be effectively controlled by variation of the laser power or by laser beam expansion. The former quantity can be set at $T_{\max}$. The latter quantity can be set at the specified dimensional uncertainty.

Figure 5 shows the same temperature profiles as in Figure 4a to apply the chosen criterion for evaluating the laser beam profiles. However, these profiles are renormalized by their absolute maxima, height, laser beam radii at half maximum, and width. The normalizing constants for all the nine testing profiles are obtained from the data shown in Figure 4a,b and Table 1. A qualitative review of the temperature profiles shown in Figure 5 indicates that laser profile TEM$_{FT}$ results in the broadest top of the temperature profile, as expected. Laser profile TEM$_{00}$ results in the broadest base of the temperature profile. This means that evaluating the three tested laser profiles is not straightforward and depends on the acceptable temperature range of laser treatment $T_{\max} - T_{\min}$ relative to the maximum temperature increment $T_{\max} - T_a$. If the acceptable temperature range is narrow, the treated band of the surface is near the top level of the temperature profile. In this case, theoretically, the flat-top profile provides the widest laser-treated band, which means the most effective use of the laser energy. If the acceptable temperature range is wide, the most effective profile seems to be TEM$_{00}$.

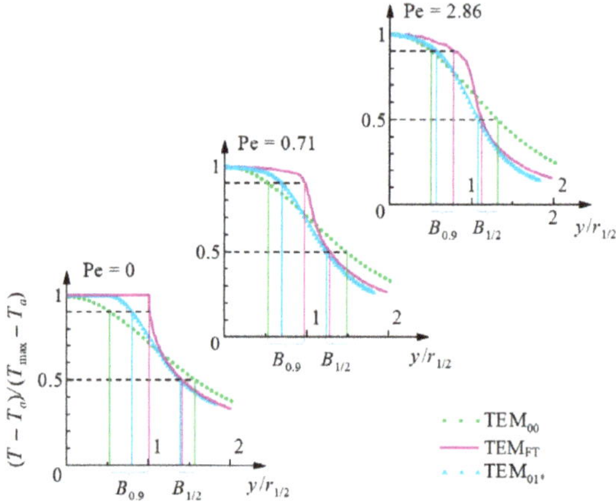

Figure 5. Normalized transverse profiles of the maximum surface temperature and the definition of the widths of laser-treated band ($B_{1/2}$ and $B_{0.9}$).

3. Model Evaluation

3.1. Quantitative Evaluation

Let us introduce the width of the laser-treated band B_n where non-dimensional parameter γ for quantitative evaluation of the laser beam profile characterizes the relative temperature range of the laser treatment [43]:

$$\gamma = \frac{T_{min} - T_a}{T_{max} - T_a}.$$ (17)

The definitions of $B_{1/2}$ and $B_{0.9}$ are shown in Figure 5. Band $B_{1/2}$ approximately corresponds to laser powder bed fusion of metals and alloys such as CoCr at the ambient temperature T_a with T_{max} equal to the boiling/decomposition point (~2800–3300 °C) and T_{min} equal to the melting point (~1250–1650 °C) [48]:

$$\gamma_{CoCr} = \frac{1458\ °C - 20\ °C}{3000\ °C - 20\ °C} \approx 0.4826.$$ (18)

The main properties of the cobalt-chromium alloy are shown in Table 2. The data presented in the table are taken from [49,50].

Table 2. Properties of the cobalt-chromium alloy (64–65% of Co, 29–30% of Cr).

Properties	Density, g/cm³	Melting Point, °C	Boiling Point, °C	Tensile Strength, kN/cm²	Yield Strength, kN/cm²	Young's Modulus, GPa	Coefficient of Thermal Expansion, $\times 10^{-6}$ °C^{-1}	Thermal Conductivity, W/(m·K)
CoCr alloy	8.0–8.4	1250–1650	2800–3000	$\geq$61.7–70	$\geq$50–64	210–250	11.2–14.2	13

Band $B_{0.9}$ corresponds to laser-additive manufacturing of oxide ceramics at the ambient temperature T_a with T_{max} equal to the temperature of chemical decomposition (~2900 °C) [51–55]. T_{min} should be chosen as high as possible because of the Arrhenius temperature dependence of the powder consolidation rate [56].

The calculated values of $B_{1/2}$ and $B_{0.9}$ versus Péclet's number for the laser beam profiles are shown in Table 3 and Figure 6. In the considered range of Péclet's numbers (Pe = 0–2.86), the conventional Gaussian profile of TEM$_{00}$ seems to be the most effective for the wide temperature range of laser treatment of $\frac{1}{2}$ (alloys, metals) when the flat-top profile can be significantly more advantageous for the narrow temperature range of 0.9 (mostly oxide ceramics). For $B_{1/2}$, profile TEM$_{01^*}$ seems to be the least effective one, and the flat-top is intermediate. For $B_{0.9}$, profile TEM$_{00}$ seems to be the least effective one, and TEM$_{01^*}$ is intermediate.

Table 3. Calculated widths of the laser-treated band $B_{1/2}$ and $B_{0.9}$ versus Péclet's number.

Laser Beam Profile	$B_{1/2}/d_{1/2}$			$B_{0.9}/d_{1/2}$		
	Pe = 0	Pe = 0.71	Pe = 2.86	Pe = 0	Pe = 0.71	Pe = 2.86
TEM$_{00}$ (Gaussian)	1.57	1.485	1.32	0.53	0.535	0.50
TEM$_{FT}$ (flat-top)	1.415	1.28	1.118	1.012	0.974	0.775
TEM$_{01^*}$ (donut)	1.39	1.24	1.07	0.80	0.70	0.565

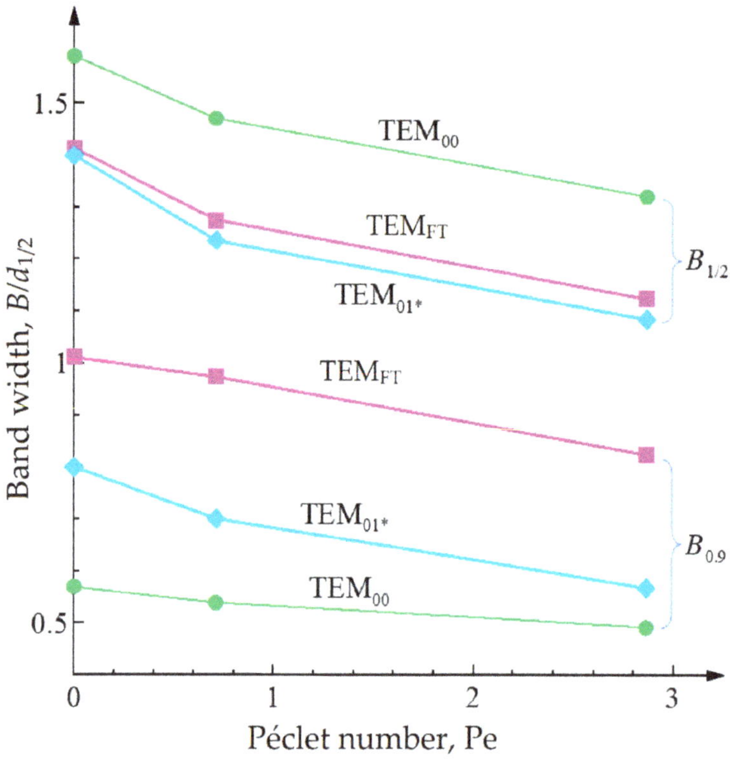

Figure 6. Widths of the laser treated band $B_{1/2}$ and $B_{0.9}$ versus Péclet's number.

3.2. Dynamic Evaluation

Let us calculate the steady temperature at the laser spot boundary for two laser modes and a laser power of 100 and 400 W. The experimental diameter of the laser spot will be approximately 100 μm (0.001 m) for the TEM$_{00}$ mode and 300 μm (0.003 m) for the TEM$_{01*}$ mode (Table 4) [57]. As can be seen, with an increase in the power of laser radiation to 400 W, due to excess heat, a multifaceted local overheating is predicted (the calculated temperature is 2.56 times higher than $T_{\max}$) at the boundary of the laser radiation of the Gaussian mode (as a result, active evaporation of metal from the processing zone). At the same time, when using the reverse Gaussian profile (donut), the temperature at the edge of the laser spot does not reach $T_{\min}$ (less than 2.34 times), which means that there is no sufficient heat to initiate the CoCr alloy granule fusion. The powder consolidation temperature can be closer to the melting temperature. Implicit graphs of the function of temperature on the radius for a cobalt-chromium alloy (λ = 13 W/(m·K)) depending on the power of laser radiation are shown in Figure 7 (Equation (3)). It should be noted that Figure 7a is an implicit graph of the temperature ($T_{\max} - T_a$) on the radius and laser power function for the material with the mentioned material thermal conductivity, where the solution area is marked red, since only values above zero can be taken into account for technological purposes, since other areas have no physical sense in the context of engineering.

Table 4. The steady temperature values at the laser spot boundary for two laser modes.

Laser Beam Profiles	Laser Spot Diameter, mm	Steady Temperature $(T_{max} - T_a)$, K	
		$P = 100$ W	$P = 400$ W
TEM$_{00}$ (Gaussian)	~0.1	1923	7692
TEM$_{01*}$ (donut)	~0.3	641.03	2564

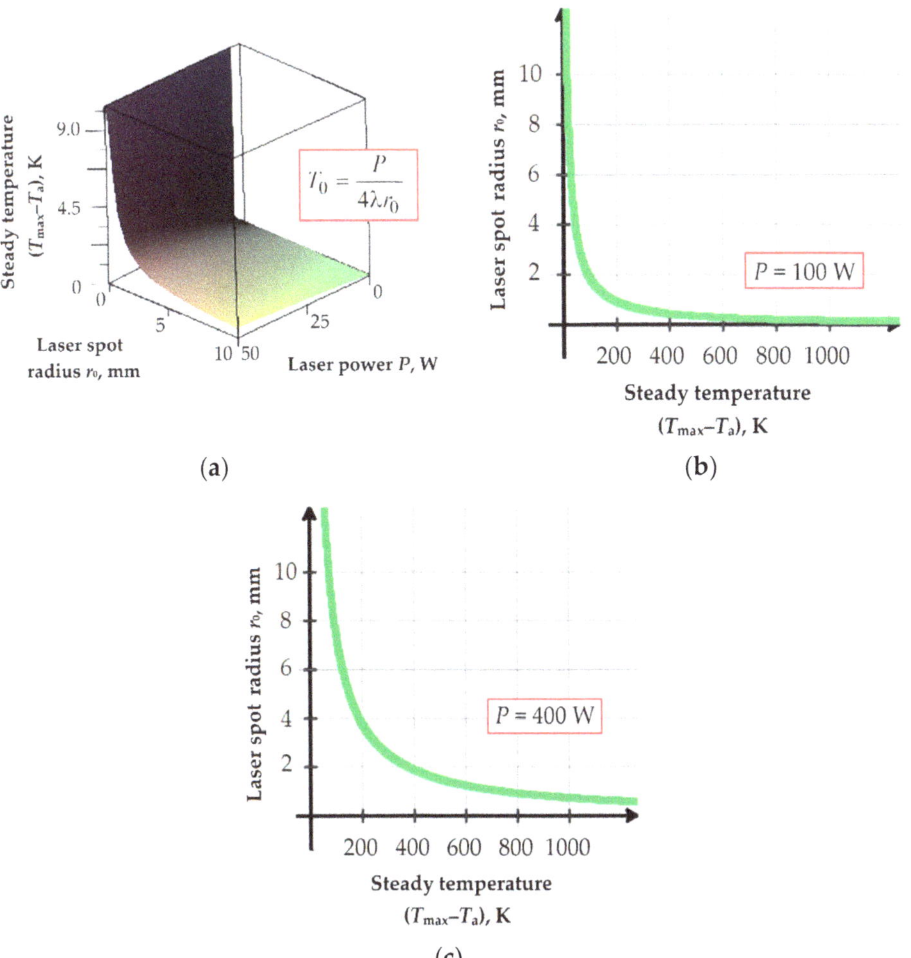

Figure 7. Implicit graphs of the function of temperature $(T_{max} - T_a)$ on the radius depending on the power of laser radiation for $\lambda = 13$ W/(m·K): (**a**) 3D-plot; (**b**) $P = 100$ W; (**c**) $P = 400$ W.

Table 5 presents two evaluated groups of laser beam parameters based on the experimental data obtained by optical achievements of the laser beam profiles using an expander and profiler installed in the LPBF setup and optical evaluation of the obtained profiles [28]. Specific energy contribution (J/m^2) was calculated by:

$$E = \frac{q_0}{u_s}. \tag{19}$$

Table 5. Parameters of laser powder bed fusion chosen for modeling.

Factor	Measuring Unit	Values	
Absorbed power of the beam, P	W	100	400
Laser beam radius, r_0	mm	~0.1/2	~0.3/2
Scanning velocity, u_s	m/s	0.0213	0.0286
Normalized power density distribution, q_0	W/m^2	0.320×10^8	0.142×10^8
Specific energy contribution, E	J/m^2	1.5×10^5	0.5×10^5
Péclet's number, Pe	-	0.71	2.86

Two numerical calculations for Gaussian (Equation (1)) and donut (Equation (4)) laser beam profiles are made for each group. Thermal diffusivity of CoCr alloy is presented in Table 6 [58,59]:

$$\alpha = \frac{\lambda}{\rho \cdot C_p},\tag{20}$$

where ρ is density, kg/m^3 and C_p is specific heat capacity, J/(kg·K). The dependence of the Péclet number on the laser spot radius and scanning speed for a cobalt-chromium alloy is shown in Figure 8.

Table 6. Thermal diffusivity α of CoCr alloy.

Thermal Diffusivity α, cm^2/s	
at 20 °C	at 500 °C
0.02–0.14	0.03–0.074

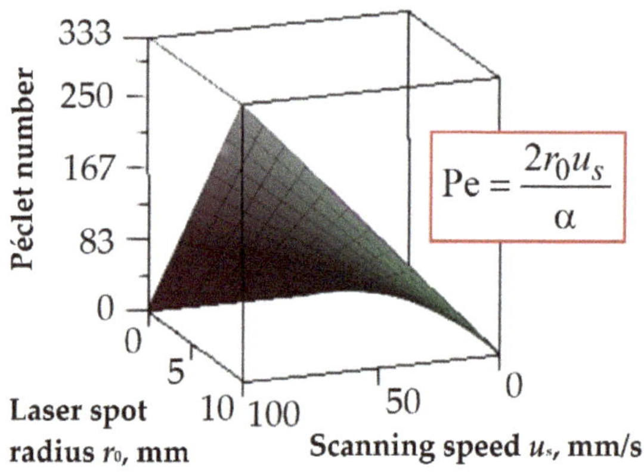

Figure 8. The implicit graph of the Péclet number on the laser spot radius and scanning speed for a cobalt-chromium alloy ($\alpha = 5.2 \times 10^{-6}$ m^2/s) (3D-plot).

Figure 9 shows the calculated temperature fields for two types of laser beam profiles: TEM$_{00}$ and TEM$_{01*}$ at laser powers of 100 and 400 W, correspondingly, when laser beam diameters are 0.109 and 0.310 mm, respectively. The difference from Figure 2 is that laser beam profiles are shown at the level of calculated steady temperatures (Table 4). Formation of the temperature plateau is explained by a small value of overheating sufficient for evaporation under the given conditions. In the case of mode TEM$_{01*}$, the characteristic temperature sink is still visible in the center. The energy losses for evaporation are listed in Table 7. The corresponding mass losses are proportional to the energy ones [43]. Comparison of values listed in Table 7 indicates that the change from mode TEM$_{00}$ to mode TEM$_{01*}$ decreases the evaporation loss for all four calculations made. Thus, the laser profile corresponding to mode TEM$_{01*}$ seems to provide more efficient laser power density distribution (Figure 10).

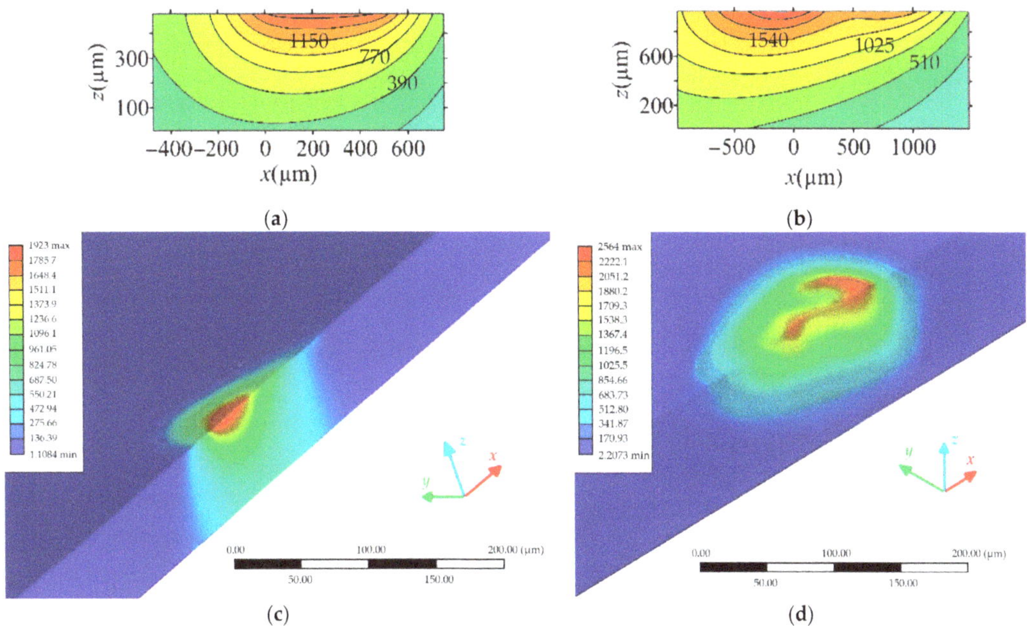

Figure 9. Calculated temperature distributions ($T_{\max} - T_{\mathrm{a}}$) in CoCr alloy: (**a**) over the vertical plane of mirror symmetry $y = 0$ formed by the beam axis and the scanning line for $P = 100$ W, Pe = 0.71; (**b**) over the vertical plane of mirror symmetry $y = 0$ formed by the beam axis and the scanning line for $P = 400$ W, Pe = 2.86; (**c**) results of temperature fields modeling for $P = 100$ W, Pe = 0.71 (cross-section); (**d**) results of temperature field modeling for $P = 400$ W, Pe = 2.86 (cross-section).

Table 7. Calculated values of power loss for evaporation P_v for CoCr alloy for the laser beam profiles.

Parameter	Evaporation Loss, P_v (W)	
	TEM$_{00}$, $P = 100$ W, Pe = 0.71	TEM$_{01^*}$, $P = 400$ W, Pe = 2.86
Max vapor velocity u_v, m/s	3.63	14.51
Max recoil pressure $p_{recoil} - p_0$, Pa	17.67	267.67
Mass loss rate L_{mass}, mg/s	144.30	520.22
Recoil force F_{recoil}, mN	0.55	7.57
Power loss for evaporation P_e, W	3.53	2.68

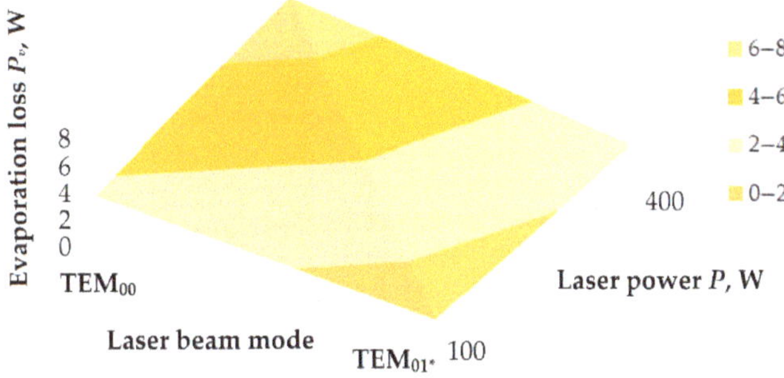

Figure 10. Calculated graphical presentation of power loss for evaporation P_v.

4. Discussion

It should be noted that the proposed dynamic model could not be used for precise data on the thermal history and simulation of the thermal stresses. The point was in researching an optimal laser power density distribution for the engineering tasks of LPBF. As known, the optimal melt pool configuration for the tasks of thick (more than 10 mm in thickness) material laser cutting or welding is torch-like (Figure 11) [25] and has a certain disadvantage when the laser power exceeds 100 W [26]. For laser scribing, surface treatment, and LPBF [40,41], the optimal one can be a more surface-like uniform distribution related to the following issues [60]:

a.   avoiding overheating in the centrum of the melt pool and consequences such as material loss on evaporation and ejecting granules from the melt pool of thermal heat with the laser power set at more than 100 W;

b.   avoiding secondary remelting and involvement of the previously solidified layers in the newly formed melt pool; and

c.   melt ejection under steam pressure.

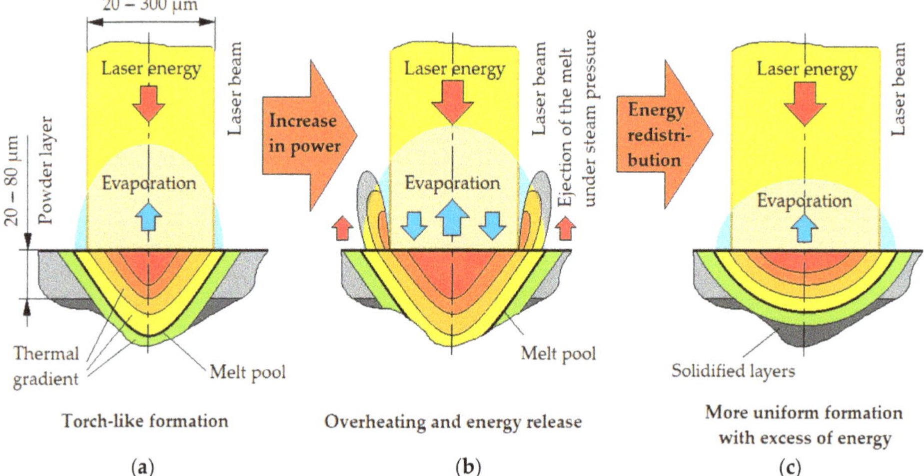

Figure 11. Melt pool formation: (**a**) torch-like; (**b**) torch-like with an increase of energy in the laser beam; (**c**) more uniform surface-like with an increase of redistributed energy in the melt pool.

The conducted research confirmed the effectiveness of the proposed approach not only for static modeling but for a dynamic one, as well. Achieved laser beam profiles are presented in Figure 12. As can be seen (Figure 12c), the flat-top profile is practically hard to be achieved close to the theoretical profile using the existed optical means [60]. The provided Figure 12d–f are reconstructed from the formed CoCr single tracks (Figure 12g–i) [61,62]. A detailed description of the developed LPBF setup equipped with an optical laser beam profiler and expander and optical diagnostics are presented in [26]. The experimental conditions are presented in [57]. Figure 13 presents the optical and modulation systems of LPBF setup.

The dynamic melt pool evaluation during experiments with metallic powders by optical diagnostic means [63,64] is expected for further research.

It should be noted that TEM_{FT} cannot be called a "desirable intensity distribution" since it was a theoretical proposal [50]. The idea was to achieve a more uniform energy density instead of peaks in the centrum of the laser beam spot. The picture of energy distribution in the laser beam spot and adsorbed energy by powder material is different. However, it can be even more varied, considering the dynamic factor (Pe number). Desirability can only be called a distribution that allows the achievement of uniform energy adsorption in the laser beam spot [57], taking into account the used material's thermal conductivity and dynamic factor. Definitely, it will be already varied for metallic [65,66] and ceramic [67–69] groups of materials. However, it can also vary depending on granulo-morphometric parameters of the powder, mainly shape and reflect ability [45,70,71], which was not considered in the article. The TEM_{00} + TEM_{01*} equation is the only way to achieve approximate TEM_{FT} by existing optical means [72,73].

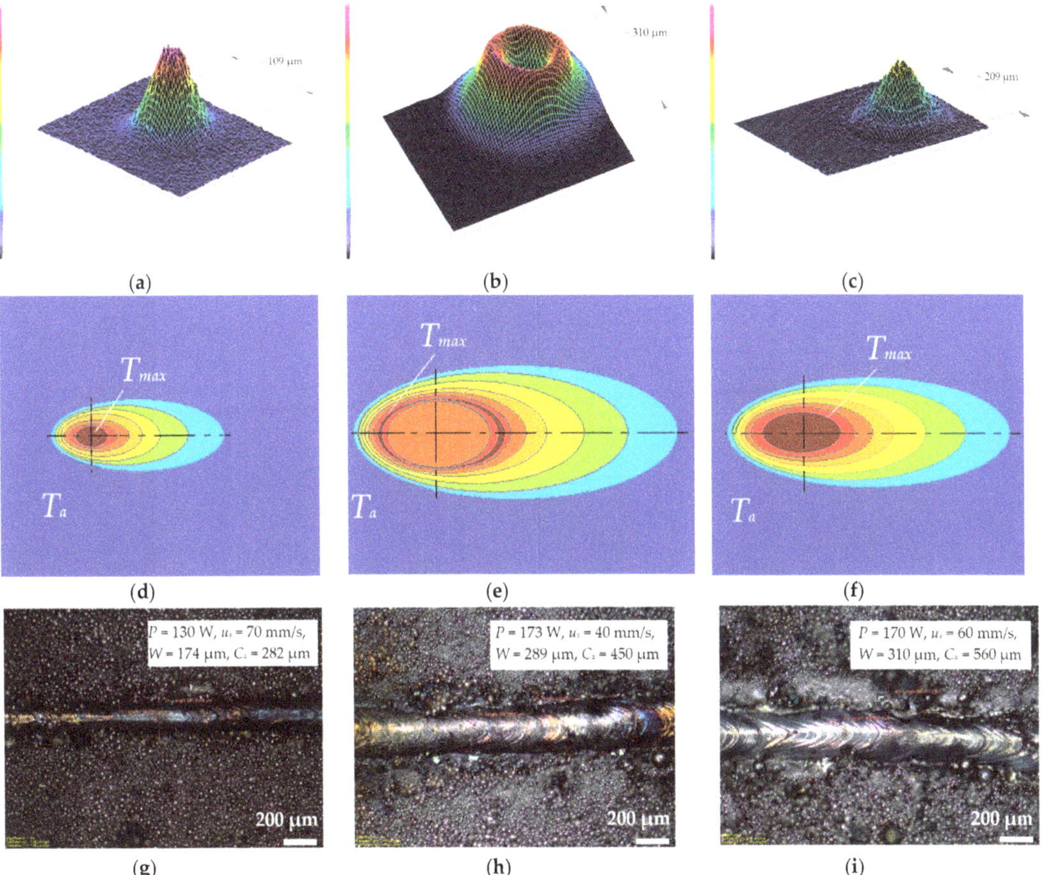

Figure 12. Laser beam profiles (objective control data achieved experimentally): (**a**) TEM_{00} (Gaussian); (**b**) TEM_{01*} (donut); (**c**) TEM_{FT} (flat-top); reconstruction of the temperature fields' features in the formed melt pools: (**d**) TEM_{00} (Gaussian); (**e**) TEM_{01*} (donut); (**f**) TEM_{FT} (flat-top); formed experimental tracks: (**g**) TEM_{00} (Gaussian); (**h**) TEM_{01*} (donut); (**i**) TEM_{FT} (flat-top), where W is a track's width, C_z is powder consolidation zone's width.

Comparing two radiation beams with different profiles is possible only with the different values for laser beam spot radii (Table 7). The same LPBF setup with a similar laser beam diameter provided technically and focused on a plane for all cases is practically used in the conditions of real production. Laser beam diameter corresponds to the main characteristics of the LPBF equipment (in our case, it is up to 100 μm) and cannot be changed quickly. The alternative laser beam profiles are experimentally achieved using a laser beam profiler and an expander and optically evaluated [60]. That was taken as a basis for theoretical evaluation of the dynamic factor to be closer to the common industrial conditions.

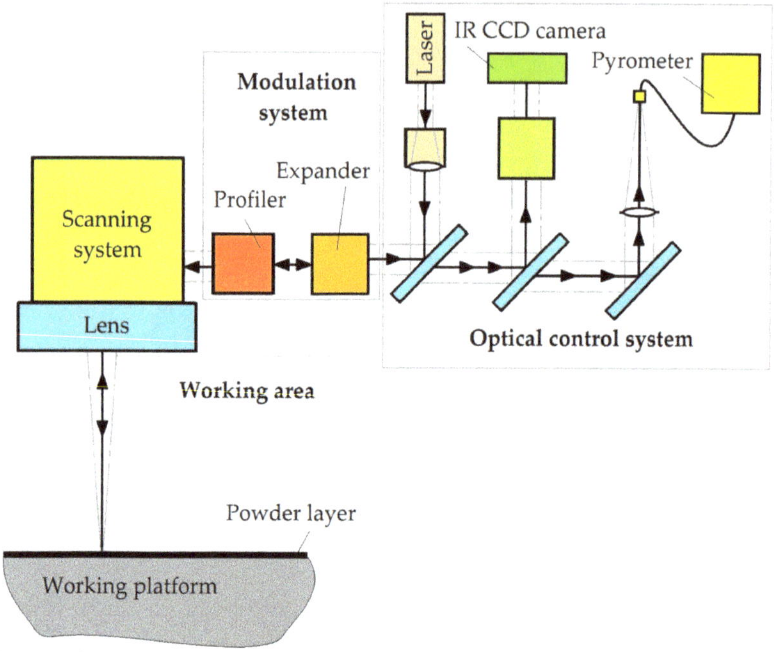

Figure 13. Modulation and optical control systems in-build into LPBF setup.

The average laser beam power distribution (E, J/m^2) will not be similar in these cases as it was previously evaluated and compared (Table 5). Still, the question is not in the energy density in the laser beam spot radii, but in the practically achievable profile that can be useful and implemented in standard or experimental LPBF equipment (Figure 13).

Practically, the achievable profile by mixing TEM$_{00}$ and TEM$_{01*}$ is far from the profile simulated based on Equation (2) due to the use available for market optical means. Moreover, as it was shown theoretically, the TEM$_{FT}$ profile is not the one that corresponds the most to the technological tasks of LPBF of metallic powder with the high material thermal conductivity (λ).

5. Conclusions

Three radial laser beam profiles of the power density distribution (energy flux) were compared for laser powder bed fusion. The uniform cylindrical (flat-top) distribution (TEM$_{01*}$ + TEM$_{00}$ mode) was compared with the standard Laguerre–Gaussian law distribution TEM$_{00}$ and the airy distribution of the first harmonic TEM$_{01*}$ (TEM$_{01}$ + TEM$_{10}$ mode).

The TEM$_{00}$ laser beam profile demonstrated the most effective result for a wide range of temperatures for thermos-activated processes such as laser powder bed fusion in the Péclet number range of 0–2.86, while the uniform cylindrical (flat-top) distribution is shown to be effective in a narrow temperature range. The inverse Gaussian (donut) laser beam distribution showed an interval result. With an increase in laser power, the transition from TEM$_{00}$ to TEM$_{01*}$ mode reduces the evaporation losses by more than 2.5 times, and it increases the absolute laser bandwidth when the relative bandwidth decreases by 24%.

The prospects of laser beam profiling for the purposes of increasing laser powder bed fusion productivity stay underestimated by the industry. However, they have a huge potential in the context of the switch to the sixth technological paradigm associated with Kondratieff's waves.

Author Contributions: Conceptualization, A.V.G. and S.N.G.; methodology, A.V.G.; software, A.S.G.; validation, A.S.M.; formal analysis, A.S.M. and T.V.T.; investigation, A.S.G. and T.V.T.; resources, T.V.T. and M.A.V.; data curation, A.A.O. and M.A.V.; writing—original draft preparation, A.A.O. and A.V.G.; writing—review and editing, A.V.G.; visualization, A.S.G., A.A.O. and A.V.G.; supervision, S.N.G.; project administration, S.N.G. and M.A.V.; funding acquisition, A.S.M. All authors have read and agreed to the published version of the manuscript.

Funding: This work was supported financially by the Ministry of Science and Higher Education of the Russian Federation (project No FSFS-2021-0003).

Institutional Review Board Statement: Not applicable.

Informed Consent Statement: Not applicable.

Data Availability Statement: Data are available in a publicly accessible repository.

Acknowledgments: The research was performed at the Department of High-Efficiency Processing Technologies of MSTU Stankin.

Conflicts of Interest: The authors declare no conflict of interest.

References

1. Lee, W.J.; Kim, E.A.; Woo, Y.J.; Park, I.; Yu, J.H.; Ha, T.; Choi, Y.S.; Lee, H.S. Effect of different WC particle shapes on laser-exposed microstructures during the directed energy deposition. *Powder Metall.* **2022**, *65*, 22–30. [CrossRef]
2. Ahsan, F.; Razmi, J.; Ladani, L. Process Parameter Optimization in Metal Laser-Based Powder Bed Fusion Using Image Processing and Statistical Analyses. *Metals* **2022**, *12*, 87. [CrossRef]
3. Gurin, V.D.; Kotoban, D.V.; Podrabinnik, P.A.; Zhirnov, I.V.; Peretyagin, P.Y.; Okunkova, A.A. Modeling of 3D technological fields and research of principal perspectives and limits in productivity improvement of selective laser melting. *Mech. Ind.* **2016**, *17*, 714. [CrossRef]
4. Khairallah, A.; Anderson, A.T.; Rubenchik, A.; King, W.E. Laser powder-bed fusion additive manufacturing: Physics of complex melt flow and formation mechanisms of pores, spatter, and denudation zones. *Acta Mater.* **2016**, *108*, 36–45. [CrossRef]
5. Sova, A.; Grigoriev, S.; Okunkova, A.; Smurov, I. Potential of cold gas dynamic spray as additive manufacturing technology. *Int. J. Adv. Manuf. Technol.* **2013**, *69*, 2269–2278. [CrossRef]
6. Sova, A.; Grigoriev, S.; Okunkova, A.; Sova, A.; Bertrand, P.; Smurov, I. Cold spraying: From process fundamentals towards advanced applications. *Surf. Coat. Technol.* **2015**, *268*, 77–84.
7. Monte, A.F.G.; Alves, G.A.; Marques, F.A.M. Thermal conductivity determination of erbium-doped crystals measured by spatially resolved confocal luminescence. *Appl. Opt.* **2018**, *57*, 7910–7914. [CrossRef]
8. Patel, S.; Reddy, P.; Kumar, A. A methodology to integrate melt pool convection with rapid solidification and undercooling kinetics in laser spot welding. *Int. J. Heat Mass Transf.* **2021**, *164*, 120575. [CrossRef]
9. Zhou, J.; Tsai, H.L. Modeling of transport phenomena in hybrid laser-MIG keyhole welding. *Int. J. Heat Mass Transf.* **2008**, *51*, 4353–4366. [CrossRef]
10. Doan, H.D.; Naoki, I.; Kazuyoshi, F. Laser processing by using fluidic laser beam shaper. *Int. J. Heat Mass Transf.* **2013**, *64*, 263–268. [CrossRef]
11. Litvin, I.A.; King, G.; Strauss, H. Beam shaping laser with controllable gain. *Appl. Phys. B-Lasers Opt.* **2017**, *123*, 1–5. [CrossRef]
12. Xia, X.P.; Cai, Z.B.; Yi, L. The splitted beam profile of laser beam in the interaction of intense lasers with overdense plasmas. *Laser Part. Beams* **2011**, *29*, 161–168. [CrossRef]
13. Gusarova, A.V.; Chumaevskii, A.V.; Osipovich, K.S.; Kalashnikov, K.N.; Kalashnikova, T.A. Regularities of Structural Changes after Friction Stir Processing in Materials Obtained by the Additive Method. *Nanosci. Technol.-Int. J.* **2020**, *11*, 195–205. [CrossRef]
14. Barro, Ó.; Arias-González, F.; Lusquiños, F.; Comesaña, R.; del Val, J.; Riveiro, A.; Badaoui, A.; Gómez-Baño, F.; Pou, J. Effect of Four Manufacturing Techniques (Casting, Laser Directed Energy Deposition, Milling and Selective Laser Melting) on Microstructural, Mechanical and Electrochemical Properties of Co-Cr Dental Alloys, Before and After PFM Firing Process. *Metals* **2020**, *10*, 1291. [CrossRef]
15. Sendino, S.; Gardon, M.; Lartategui, F.; Martinez, S.; Lamikiz, A. The Effect of the Laser Incidence Angle in the Surface of L-PBF Processed Parts. *Coatings* **2020**, *10*, 1024. [CrossRef]
16. Nalivaiko, A.Y.; Arnautov, A.N.; Zmanovsky, S.V.; Gromov, A.A. Al-Si-Cu and Al-Si-Cu-Ni alloys for additive manufacturing: Composition, morphology and physical characteristics of powders. *Mater. Res. Express* **2019**, *6*, 086536. [CrossRef]
17. Zavalov, Y.N.; Dubrov, A.V. Short Time Correlation Analysis of Melt Pool Behavior in Laser Metal Deposition Using Coaxial Optical Monitoring. *Sensors* **2021**, *21*, 8402. [CrossRef]
18. Chen, J.H.; Zhao, C.C.; Li, K.L.; Shen, Z.J.; Liu, W. Formability and Controlling of Cracks in Laser Powder Bed Fusion of Tungsten-5% Tantalum Carbide Alloys. *Chin. J. Lasers-Zhongguo Jiguang* **2021**, *48*, 1502006.
19. Alberti, E.A.; Bueno, B.M.P.; D'Oliveira, A.S.C.M. Additive manufacturing using plasma transferred arc. *Int. J. Adv. Manuf. Technol.* **2016**, *83*, 1861–1871. [CrossRef]

20. Kishore, V.; Ajinjeru, C.; Nycz, A.; Post, B.; Lindahl, J.; Kunc, V.; Duty, C. Infrared preheating to improve interlayer strength of big area additive manufacturing (BAAM) components. *Addit. Manuf.* **2017**, *14*, 7–12. [CrossRef]

21. Fateri, M.; Kaouk, A.; Cowley, A.; Siarov, S.; Palou, M.V.; Gonzalez, F.G.; Marchant, R.; Cristoforetti, S.; Sperl, M. Feasibility study on additive manufacturing of recyclable objects for space applications. *Addit. Manuf.* **2018**, *24*, 400–404. [CrossRef]

22. Sleiman, K.; Rettschlag, K.; Jaschke, P.; Capps, N.; Kinzel, E.C.; Overmeyer, L.; Kaierle, S. Material loss analysis in glass additive manufacturing by laser glass deposition. *J. Laser Appl.* **2021**, *33*, 042050. [CrossRef]

23. Żrodowski, Ł.; Wróblewski, R.; Choma, T.; Morończyk, B.; Ostrysz, M.; Leonowicz, M.; Łacisz, W.; Błyskun, P.; Wróbel, J.S.; Cieślak, G.; et al. Novel Cold Crucible Ultrasonic Atomization Powder Production Method for 3D Printing. *Materials* **2021**, *14*, 2541. [CrossRef] [PubMed]

24. Harkin, R.; Wu, H.; Nikam, S.; Quinn, J.; McFadden, S. Analysis of Spatter Removal by Sieving during a Powder-Bed Fusion Manufacturing Campaign in Grade 23 Titanium Alloy. *Metals* **2021**, *11*, 399. [CrossRef]

25. Zubko, M.; Loskot, J.; Świec, P.; Prusik, K.; Janikowski, Z. Analysis of Stainless Steel Waste Products Generated during Laser Cutting in Nitrogen Atmosphere. *Metals* **2020**, *10*, 1572. [CrossRef]

26. Metel, A.S.; Grigoriev, S.N.; Tarasova, T.V.; Melnik, Y.A.; Volosova, M.A.; Okunkova, A.A.; Podrabinnik, P.A.; Mustafaev, E.S. Surface Quality of Metal Parts Produced by Laser Powder Bed Fusion: Ion Polishing in Gas-Discharge Plasma Proposal. *Technologies* **2021**, *9*, 27. [CrossRef]

27. Okunkova, A.; Peretyagin, P.; Vladimirov, Y.; Volosova, M.; Torrecillas, R.; Fedorov, S.V. Laser-beam modulation to improve efficiency of selecting laser melting for metal powders. In Proceedings of the Conference on Laser Sources and Applications II, Laser Sources and Applications II, Brussels, Belgium, 14–17 April 2014; Mackenzie, J.I., Jelinkova, H., Taira, T., Ahmed, M.A., Eds.; SPIE-Int Soc Optical Engineering: Bellingham, WA, USA, 2014; Volume 9135, p. 913524.

28. Zhirnov, I.V.; Podrabinnik, P.A.; Okunkova, A.A.; Gusarov, A.V. Laser beam profiling: Experimental study of its influence on single-track formation by selective laser melting. *Mech. Ind.* **2015**, *16*, 709. [CrossRef]

29. Wang, S.D.; Zhang, X.; Zheng, Y.; Li, B.W.; Qin, H.T.; Li, Q. Similarity evaluation of 3D surface topography measurements. *Meas. Sci. Technol.* **2021**, *32*, 125003. [CrossRef]

30. Sagbas, B.; Gumus, B.E.; Kahraman, Y.; Dowling, D.P. Impact of print bed build location on the dimensional accuracy and surface quality of parts printed by multi jet fusion. *J. Manuf. Processes* **2021**, *70*, 290–299. [CrossRef]

31. Zakharov, O.V.; Brzhozovskii, B.M. Accuracy of centering during measurement by roundness gauges. *Meas. Tech.* **2006**, *49*, 1094–1097. [CrossRef]

32. Rezchikov, A.F.; Kochetkov, A.V.; Zakharov, O.V. Mathematical models for estimating the degree of influence of major factors on performance and accuracy of coordinate measuring machines. *MATEC Web Conf.* **2017**, *129*, 01054. [CrossRef]

33. Zakharov, O.V.; Balaev, A.F.; Kochetkov, A.V. Modeling Optimal Path of Touch Sensor of Coordinate Measuring Machine Based on Traveling Salesman Problem Solution. *Procedia Eng.* **2017**, *206*, 1458–1463. [CrossRef]

34. Wu, J.; Tao, K.; Miao, J.M. Production of centimeter-scale sub-wavelength nanopatterns by controlling the light path of adhesive photomasks. *J. Mater. Chem. C* **2015**, *3*, 6796–6808. [CrossRef]

35. Sima, C.; Gates, J.C.; Rogers, H.L.; Mennea, P.L.; Holmes, C.; Zervas, M.N.; Smith, P.G.R. Phase controlled integrated interferometric single-sideband filter based on planar Bragg gratings implementing photonic Hilbert transform. *Opt. Lett.* **2013**, *38*, 727–729. [CrossRef]

36. Raciukaitis, G.; Stankevicius, E.; Gecys, P.; Gedvilas, M.; Bischoff, C.; Jager, E.; Umhofer, U.; Volklein, F. Laser Processing by Using Diffractive Optical Laser Beam Shaping Technique. *J. Laser Micro Nanoeng.* **2011**, *6*, 37–43. [CrossRef]

37. Nammi, S.; Vasa, N.J.; Balaganesan, G.; Gupta, S.; Mathur, A.C. Influence of pulsed Nd3+: YAG laser beam profile and wavelength on microscribing of copper and aluminum thin films. *J. Micro-Nanolithography Mems Moems* **2015**, *14*, 044503. [CrossRef]

38. Zhu, R.; Zhang, Y.K.; Lin, C.H.; Chen, Y. Residual stress distribution and surface geometry of medical Ti13Nb13Zr alloy treated by laser shock peening with flat-top laser beam. *Surf. Topogr.-Metrol. Prop.* **2020**, *8*, 045026. [CrossRef]

39. Yen, W.C.; Huang, C.T.; Liu, H.P.; Lee, L.P. A Nd:YAG laser with a flat-top beam profile and constant divergence. *Opt. Laser Technol.* **1997**, *29*, 57–61. [CrossRef]

40. Wu, T.N.; Wu, Z.P.; He, Y.C.; Zhu, Z.; Wang, L.X.; Yin, K. Femtosecond laser textured porous nanowire structured glass for enhanced thermal imaging. *Chin. Opt. Lett.* **2022**, *20*, 033801. [CrossRef]

41. Yin, K.; Wu, Z.P.; Wu, J.R.; Zhu, Z.; Zhang, F.; Duan, J.A. Solar-driven thermal-wind synergistic effect on laser-textured superhydrophilic copper foam architectures for ultrahigh efficient vapor generation. *Appl. Phys. Lett.* **2021**, *118*, 211905. [CrossRef]

42. Aleksandrova, M.; Nikolov, N.; Pandiev, I. Thermo-Stimulation of Charges by Peltier Element for Trap Analysis in Polymer Layers. *Int. J. Polym. Anal. Charact.* **2011**, *16*, 221–227. [CrossRef]

43. Gusarov, A.V. Radiative transfer, absorption, and reflection by metal powder beds in laser powder-bed processing. *J. Quant. Spectrosc. Radiat. Transf.* **2020**, *257*, 107366. [CrossRef]

44. Yu, H.; Hayashi, S.; Kakehi, K.; Kuo, Y.-L. Study of Formed Oxides in IN718 Alloy during the Fabrication by Selective Laser Melting and Electron Beam Melting. *Metals* **2019**, *9*, 19. [CrossRef]

45. Cao, Y.; Bai, P.; Liu, F.; Hou, X. Investigation on the Precipitates of IN718 Alloy Fabricated by Selective Laser Melting. *Metals* **2019**, *9*, 1128. [CrossRef]
46. Qin, Y.; Liu, J.; Chen, Y.; Wen, P.; Zheng, Y.; Tian, Y.; Voshage, M.; Schleifenbaum, J.H. Influence of Laser Energy Input and Shielding Gas Flow on Evaporation Fume during Laser Powder Bed Fusion of Zn Metal. *Materials* **2021**, *14*, 2677. [CrossRef]
47. Gokcekaya, O.; Ishimoto, T.; Todo, T.; Suganuma, R.; Fukushima, R.; Narushima, T.; Nakano, T. Effect of Scan Length on Densification and Crystallographic Texture Formation of Pure Chromium Fabricated by Laser Powder Bed Fusion. *Crystals* **2021**, *11*, 9. [CrossRef]
48. Grigoriev, S.N.; Metel, A.S.; Tarasova, T.V.; Filatova, A.A.; Sundukov, S.K.; Volosova, M.A.; Okunkova, A.A.; Melnik, Y.A.; Podrabinnik, P.A. Effect of Cavitation Erosion Wear, Vibration Tumbling, and Heat Treatment on Additively Manufactured Surface Quality and Properties. *Metals* **2020**, *10*, 1540. [CrossRef]
49. Wang, J.-H.; Ren, J.; Liu, W.; Wu, X.-Y.; Gao, M.-X.; Bai, P.-K. Effect of Selective Laser Melting Process Parameters on Microstructure and Properties of Co-Cr Alloy. *Materials* **2018**, *11*, 1546. [CrossRef]
50. Metel, A.S.; Stebulyanin, M.M.; Fedorov, S.V.; Okunkova, A.A. Power Density Distribution for Laser Additive Manufacturing (SLM): Potential, Fundamentals and Advanced Applications. *Technologies* **2019**, *7*, 5. [CrossRef]
51. Kopec, M.; Jóźwiak, S.; Kowalewski, Z.L. A Novel Microstructural Evolution Model for Growth of Ultra-Fine Al_2O_3 Oxides from SiO_2 Silica Ceramic Decomposition during Self-Propagated High-Temperature Synthesis. *Materials* **2020**, *13*, 2821. [CrossRef]
52. Bulina, N.V.; Makarova, S.V.; Baev, S.G.; Matvienko, A.A.; Gerasimov, K.B.; Logutenko, O.A.; Bystrov, V.S. A Study of Thermal Stability of Hydroxyapatite. *Minerals* **2021**, *11*, 1310. [CrossRef]
53. Kuzin, V.V.; Grigor'ev, S.N.; Volosova, M.A. Effect of a TiC Coating on the Stress-Strain State of a Plate of a High-Density Nitride Ceramic under Nonsteady Thermoelastic Conditions. *Refract. Ind. Ceram.* **2014**, *54*, 376–380. [CrossRef]
54. Volosova, M.A.; Grigor'ev, S.N.; Kuzin, V.V. Effect of Titanium Nitride Coating on Stress Structural Inhomogeneity in Oxide-Carbide Ceramic. Part 4. Action of Heat Flow. *Refract. Ind. Ceram.* **2015**, *56*, 91–96. [CrossRef]
55. Grigoriev, S.N.; Kozochkin, M.P.; Porvatov, A.N.; Volosova, M.A.; Okunkova, A.A. Electrical discharge machining of ceramic nanocomposites: Sublimation phenomena and adaptive control. *Heliyon* **2019**, *5*, e02629. [CrossRef]
56. Kimura, H.; Myung, W.N.; Kobayashi, S.; Suzuki, M.; Toda, K.; Yuine, T. Consolidation of Mechanically Alloyed Amorphous Conbzr Powder by HIP. *J. Jpn. Inst. Met.* **1992**, *56*, 833–841. [CrossRef]
57. Gusarov, A.V.; Grigoriev, S.N.; Volosova, M.A.; Melnik, Y.A.; Laskin, A.; Kotoban, D.V.; Okunkova, A.A. On productivity of laser additive manufacturing. *J. Mater. Process. Technol.* **2018**, *261*, 213–232. [CrossRef]
58. Ponomarev, S.V.; Bulanov, E.V.; Bulanova, V.O.; Divin, A.G. Minimization of Measurement Errors of the Coefficients of Heat Conductivity and Thermal Diffusivity of Thermal Insulating Materials by the Plane Pulsed Heat Source Method. *Meas. Tech.* **2019**, *61*, 1203–1208. [CrossRef]
59. Sova, A.; Okunkova, A.; Grigoriev, S.; Smurov, I. Velocity of the Particles Accelerated by a Cold Spray Micronozzle: Experimental Measurements and Numerical Simulation. *J. Therm. Spray Technol.* **2012**, *22*, 75–80. [CrossRef]
60. Gusarov, A.V.; Okun'kova, A.A.; Peretyagin, P.Y.; Zhirnov, I.V.; Podrabinnik, P.A. Means of Optical Diagnostics of Selective Laser Melting with Non-Gaussian Beams. *Meas. Tech.* **2015**, *58*, 872–877. [CrossRef]
61. Yadroitsev, I.; Bertrand, P.; Antonenkova, G.; Grigoriev, S.; Smurov, I. Use of track/layer morphology to develop functional parts by selectivelaser melting. *J. Laser Appl.* **2013**, *25*, 052003. [CrossRef]
62. Kotoban, D.; Grigoriev, S.; Okunkova, A.; Sova, A. Influence of a shape of single track on deposition efficiency of 316L stainless steel powder in cold spray. *Surf. Coat. Technol.* **2017**, *309*, 951–958. [CrossRef]
63. Doubenskaia, M.; Pavlov, M.; Grigoriev, S.; Tikhonova, E.; Smurov, I. Comprehensive Optical Monitoring of Selective Laser Melting. *J. Laser Micro Nanoeng.* **2012**, *7*, 236–243. [CrossRef]
64. Smurov, I.; Doubenskaia, M.; Grigoriev, S.; Nazarov, A. Optical Monitoring in Laser Cladding of Ti6Al4V. *J. Spray Tech.* **2012**, *21*, 1357–1362. [CrossRef]
65. Sova, A.; Grigoriev, S.; Okunkova, A.; Smurov, I. Cold spray deposition of 316L stainless steel coatings on aluminium surface with following laser post-treatment. *Surf. Coat. Technol.* **2013**, *235*, 283–289. [CrossRef]
66. Metel, A.; Tarasova, T.; Gutsaliuk, E.; Khmyrov, R.; Egorov, S.; Grigoriev, S. Possibilities of Additive Technologies for the Manufacturing of Tooling from Corrosion-Resistant Steels in Order to Protect Parts Surfaces from Thermochemical Treatment. *Metals* **2021**, *11*, 1551. [CrossRef]
67. Khmyrov, R.S.; Grigoriev, S.N.; Okunkova, A.A.; Gusarov, A.V. On the possibility of selective laser melting of quartz glass. *Phys. Procedia* **2014**, *56*, 345–356. [CrossRef]
68. Khmyrov, R.S.; Protasov, C.E.; Grigoriev, S.N.; Gusarov, A.V. Crack-free selective laser melting of silica glass: Single beads and monolayers on the substrate of the same material. *Int. J. Adv. Manuf. Technol.* **2016**, *85*, 1461–1469. [CrossRef]
69. Grigoriev, S.; Peretyagin, P.; Smirnov, A.; Solis, W.; Diaz, L.A.; Fernandez, A.; Torrecillas, R. Effect of graphene addition on the mechanical and electrical properties of Al2O3 -SiCw ceramics. *J. Eur. Ceram. Soc.* **2017**, *37*, 2473–2479. [CrossRef]
70. Li, E.L.; Wang, L.; Yu, A.B.; Zhou, Z.Y. A three-phase model for simulation of heat transfer and melt pool behaviour in laser powder bed fusion process. *Powder Technol.* **2021**, *381*, 298–312. [CrossRef]

71. Pelevin, I.A.; Ozherelkov, D.Y.; Chernyshikhin, S.V.; Nalivaiko, A.Y.; Gromov, A.A.; Chzhan, V.B.; Terekhin, E.A.; Tereshina, I.S. Selective laser melting of Nd-Fe-B: Single track study. *Mater. Lett.* **2022**, *315*, 131947. [CrossRef]
72. Gorshkov, B.G.; Danileiko, Y.K.; Nikolaev, V.N.; Sidorin, A.V. An Effect of Multiple Exposure in Laser Damage of Optical-Materials. *Kvantovaya Elektron.* **1983**, *10*, 640–643.
73. Baburin, N.V.; Galagan, B.I.; Danileiko, Y.K.; Il'ichev, N.N.; Masalov, A.V.; Molchanov, V.Y.; Chikov, V.A. Two-frequency mode-locked lasing in a monoblock diode-pumped Nd3+: GGG laser. *Quantum Electron.* **2001**, *31*, 303–304. [CrossRef]

Article

Influence of Specific Energy on Microstructure and Properties of Laser Cladded FeCoCrNi High Entropy Alloy

Leilei Wang [1,*], Zhuanni Gao [1], Mengyao Wu [1], Fei Weng [2], Ting Liu [1] and Xiaohong Zhan [3,*]

1   College of Materials Science and Technology, Nanjing University of Aeronautics and Astronautics, Nanjing 211106, China; 18795996358@163.com (Z.G.); wumengyao0521@126.com (M.W.); liuting_nuaa123@126.com (T.L.)
2   Singapore Institute of Manufacturing Technology, Singapore 637662, Singapore; weng_fei@simtech.a-star.edu.sg
3   National Key Laboratory of Science and Technology on Helicopter Transmission, Nanjing University of Aeronautics and Astronautics, Nanjing 210016, China
*   Correspondence: meleileiwang@gmail.com (L.W.); xiaohongzhan_nuaa@126.com (X.Z.)

Received: 21 September 2020; Accepted: 29 October 2020; Published: 2 November 2020

Abstract: Specific energy is a key process parameter during laser cladding of high entropy alloy (HEA); however, the effect of specific energy on the microstructure, hardness, and wear resistance of HEA coating has not been completely understood in the literature. This paper aims at revealing the influence of specific energy on the microstructure and properties of laser cladded FeCoCrNi high entropy alloy on the Ti6Al4V substrate, and further obtains feasible process parameters for preparation of HEA coating. Results indicate that there are significant differences in the microstructure and properties of the coatings under different specific energy. The increase of specific energy plays a positive role in coarsening the microstructure, promoting the diffusion of Ti from the substrate to HEA coating, and subsequently affects the hardness of samples. The HEA coating is mainly composed of the face-centered cubic phase and body-centered cubic phase, precipitating a small amount of Fe-Cr phase and Laves phase. Metallurgical bonding is obtained between the base metal and the coatings of which the bonding region is mainly composed of columnar crystal and shrinkage cavities. The microhardness of the HEA coating reaches 1098 HV, which is about 200% higher than that of the TC4 substrate, and the wear resistance is significantly improved by the HEA coating.

Keywords: laser cladding; high entropy alloy; specific energy; phase transformation; wear resistance

---

1. Introduction

Titanium alloy has been widely used in the aerospace industry due to its superior advantages of low density, high strength, and anti-corrosion performance [1–4] However, the service life of the titanium alloy structural part is restricted by the insufficient wear resistance and hardness. Therefore, the surface strengthening of titanium alloy has attracted massive attention from researchers worldwide [5,6]. High entropy alloy (HEA), as a promising multi-component alloy, has drawn considerable attention to the repair of seriously worn aircraft flap slide due to its outstanding comprehensive properties, which is attributed to its effect of high entropy, lattice distortion, and slow diffusion [7–9]. HEA rapidly attracts massive attention in material science since it was first reported [10–12]. HEA has created an unexplored area of alloy compositions and exploited the potential to influence solid solution phase stability through the controlling of configurational entropy.

Preparation of HEA coating on titanium alloy is a feasible method to improve surface hardness and wear resistance of titanium alloy. At present, the main technologies used to prepare HEA coating include thermal spraying [13,14], laser/plasma cladding [15–18], physical vapor deposition [19], and powder metallurgy [20,21]. Among the above HEA coating methods, laser cladding has advantages such as high efficiency, reliable metallurgical bonding, high material utilization ratio, and excellent performance [22,23]. Therefore, laser cladding of HEA coating has been widely adopted for repairing and strengthening the titanium alloy structural parts.

The mechanism of surface strengthening by HEA was reported in the literature. Huang et al. [24,25] investigated the structure and properties of high entropy alloys; results indicated that the wear resistance of the cladding layer was significantly improved due to the second phase strengthening. The effect of the cladding process on the microstructure and properties of the HEA coatings was reported in the literature. Joseph et al. [26] conducted a comparative study between arc melting and laser melting; the microstructure and properties of the HEA coatings were found to be significantly different under different processes, but the compression experimental results were not much different. The effect of chemical compositions on the microstructure and properties of the HEA coatings was reported in the literature. Jiang et al. [27] conducted a comparative study on microstructure evolution and wear behavior of the laser cladding $CoFeNi_2V_{0.5}Nb_{0.75}$ and CoFeNi2V0.5Nb HEA coatings. Cai et al. [28] studied the alloying elements and dilution rates on the microstructure and properties of high entropy alloy cladding layers; results indicated that the migration of the Fe element from the matrix to the cladding layer could make the mixing entropy of CoCrNi coating close to the theoretical mixing entropy of FeCoCrNi high entropy alloy. The effect of laser power on the microstructure and properties of the HEA coatings was reported in the literature. Shu et al. [29] studied the effect of laser power on microstructure, mechanical and chemical properties of the CoCrBFeNiSi HEA coating; results indicated that the amorphous content in the coatings had a significant influence on microhardness, wear resistance, and corrosion resistance. Sui et al. [30] investigated the effect of specific energy on the microstructure and properties of laser cladded $TiN/Ti_3AlN-Ti_3Al$ composite coating; results revealed that the dilution rate of the coating increased with the increase of specific energy, the microhardness of the composite coating was approximately three times higher than that of the substrate, and the wear resistance was improved remarkably under optimum specific energy 58.3 J/mm^2. In summary, previous studies mainly focus on the thermal stability, oxidation resistance, microstructure evolution, mechanical properties, and wear behavior of the HEA coating, however, the effect of specific energy on microstructure, hardness, and wear resistance of HEA coating has not been completed understood in the literature.

Specific energy (E) is a key process parameter during laser cladding of HEA on the TC4 substrate. Specific energy is calculated by using the Equation: $E = P/(V \times D)$, where P (W) is the laser power, V (mm/s) is the scanning speed, and D (mm) is the laser spot diameter [30]. In this study, the influence of specific energy on the microstructure, phase transformation, hardness, and wear resistance of laser cladded FeCoCrNi HEA on the TC4 substrate will be systematically investigated, and the mechanism of surface strengthening lying in the laser cladding process will also be revealed, feasible process parameters for preparation of HEA coating will be obtained.

2. Materials and Methods

In this research, the experimental setup for the laser cladding process consists of a laser system (made in Shanghai, China), a 6-axis KUKA robot (made in MS, USA), a water-cooling machine to ensure the normal operation of the laser system, and a protection chamber to control the argon environment with less than 60 ppm oxygen content. Overall, the equipment of laser cladding is shown in Figure 1.

During the laser cladding process, the high entropy alloy powder and the surface substrate are subjected to the radiation of a high-energy laser beam, which quickly melts, diffuses, and solidifies to form a cladding layer that combines well with the substrate. The detailed schematic diagram of laser cladding is shown in Figure 2.

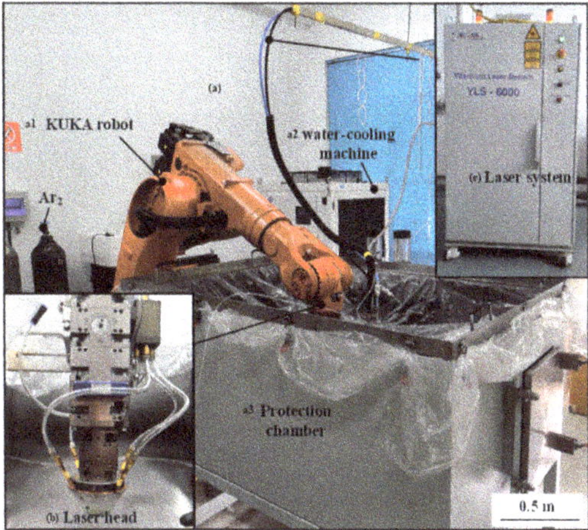

Figure 1. Equipment of laser cladding experiments: (**a**) laser cladding system with a KUKA robot, protection chamber, and water-cooling machine; (**b**) laser head; (**c**) laser system.

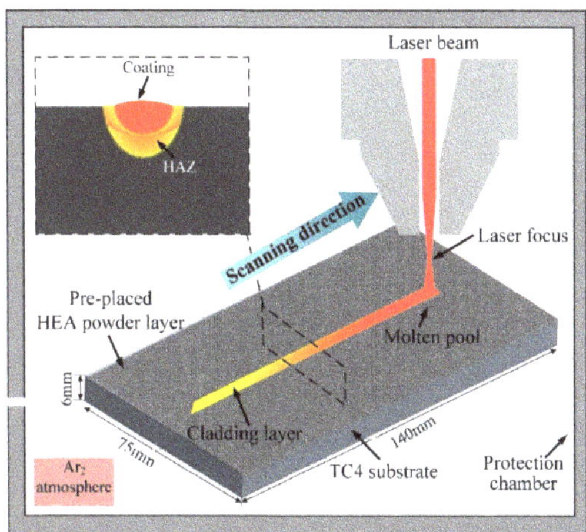

Figure 2. Schematic diagram of the experimental setup for the process.

In this paper, TC4 is used as the matrix material, and self-developed FeCoCrNi high entropy alloy is used as the powder material. The spherical pre-melted FeCoCrNi powders of 15~53 μm were produced by vacuum gas atomization under the Ar atmosphere. The element compositions of the substrate and HEA powder are shown in Table 1.

Table 1. Chemical compositions of the substrate and FeCoCrNi powder (wt%).

Materials	Ti	Fe	Co	Cr	Ni	Al	V
TC4 Substrate	Bal.	≤0.15	–	–	–	4.96	3.70
FeCoCrNi powder	–	24.44	26.18	22.43	Bal.	–	–

Meanwhile, reasonable control of the process parameters is essential for achieving the desired performance. However, the properties of laser cladded coating are affected by numerous factors, such as the laser scanning speed and laser power. Therefore, this paper introduces the concept of specific energy, which can reflect the changes in laser power and scanning speed at the same time. The selected laser cladding parameters are shown in Table 2.

Table 2. Laser cladding experiment parameters.

Case	Laser Power (W)	Scanning Speed (mm/s)	Beam Diameter (mm)	Specific Energy (J/mm^2)
1	900	11	2	40.9
2	1050	9	2	58.3
3	1250	7	2	89.3

After laser cladding, specimens were cut along the transverse sections, then they were ground and polished to 0.06 μm using colloidal silica, and subsequently they were electrolytically etched using the standard Kroll's reagent (HF:HNO$_3$:H$_2$O = 1:2:6) with 10 s. The crystal structure and lattice parameters of the laser cladding products and TC4 substrate were discussed from the corresponding XRD patterns. The XRD was tested on a Bruker D8 ADVANCE (made in Karlsruhe, Germany) Bragg-Brentano diffractometer with an X-ray wavelength of 1.54 Å. The microstructural research and chemical analysis of the laser cladded samples are performed using an FEI Inspect-F Scanning Electron Microscope (made in Hillsboro, OR, USA) equipped with an energy-dispersive spectroscopy (EDS) detector. The microhardness tests were performed using an HVS50 hardness tester, with 15 s load application time and under loads of 1000 g. Meanwhile, the wear experiments were carried out at room temperature using CFT-I type fretting friction and wear machine (made in Shanghai, China) with a normal load of 30 N, a motor speed of 300 t/m, and a wear time of 10 min. Besides, Si$_3$N$_4$ ceramic balls with a radius of 3 mm were used as the friction pair.

3. Results and Discussions

3.1. Macro Morphology

Based on the selected experimental parameters in Table 2, a high entropy alloy laser cladding experiment was carried out. The macro morphology of the cladding layer under different specific energies is shown in Figure 3. It can be seen from Figure 3 that as the specific energy increases, the surface quality of the cladding layer changes significantly. In order to investigate the influence of specific energy on the cladding layer, the cross-sections of each high entropy alloy laser cladding layers are shown in Figure 4.

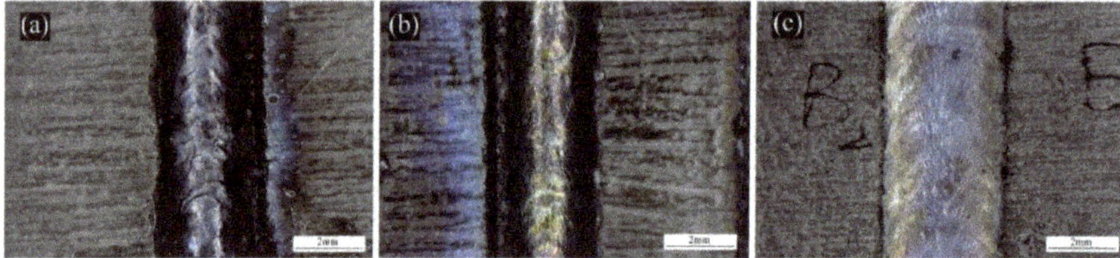

Figure 3. Surface morphology of high entropy alloy laser cladding layer under different specific energies. (**a**) 40.9 J/mm^2; (**b**) 58.3 J/mm^2; (**c**) 89.3 J/mm^2.

During the laser cladding process, affected by the influence of surface tension and wetting, a parabolic morphology was formed during the rapid solidification process. It can be seen from Figure 4 that the specific energy of laser cladding has a significant impact on the cladding layer morphology, specifically related to the cladding layer morphology size, that is, the cladding layer reinforcement (H), width (W), and melting depth (h). The test results are shown in Table 3. The results show that the specific energy has an important effect on the volume of the molten pool, which is specifically expressed in the reinforcement, width, and melting depth. With the increase of specific energy, the size of the molten pool will increase accordingly. The penetration and width of the molten pool increase with higher specific energy, while the reinforcement does not show any regularity with higher specific energy.

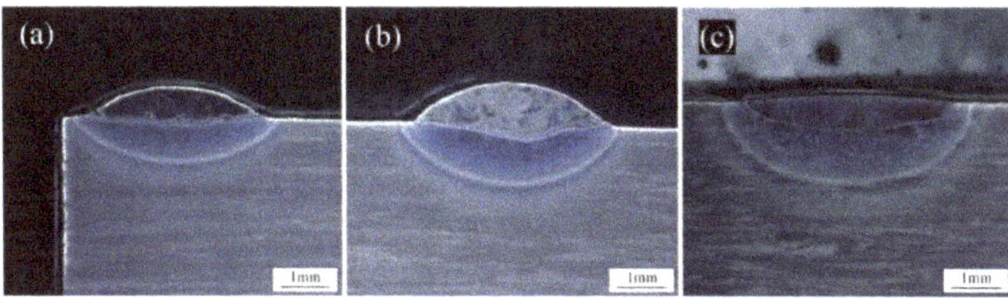

Figure 4. Transverse section of the high entropy alloy laser cladding layer under different specific energies. (**a**) 40.9 J/mm^2; (**b**) 58.3 J/mm^2; (**c**) 89.3 J/mm^2.

Table 3. The size of the cladding layer under different specific energy.

Case	E/(J/mm^2)	H/(mm)	W/(mm)	h/(mm)
1	40.9	0.58	3.69	0.15
2	58.3	0.8	4.04	0.35
3	89.3	0.28	4.18	0.55

3.2. Phase Transformation

X-ray Diffraction (XRD) is a commonly used phase analysis method [31,32]. The XRD patterns of the pre-placed powders and the FeCoCrNi coatings are shown in Figure 5, which shows that the HEA coatings are mainly composed of the FCC phase and BCC phase, precipitating a small amount of Fe-Cr phase and Laves phase, while the raw powders possess the single-phase FCC crystal structure only. The addition of the Ti element influences the valence electron concentration (VEC) inside the FeCoCrNi high entropy alloy, thus inhibiting the formation of the FCC phase in the cladding layer. Meanwhile, the atomic size of the Ti element is quite different from that of Fe, Co, Cr, Ni, which is beneficial to the formation of the BCC phase. It is also can be found from Figure 5 that the addition of the Ti element causes the change of the lattice constant of the FCC phase. This is also due to the above-mentioned atomic size. Compared with other elements, Ti has a larger atomic radius. Therefore, its addition led to severe lattice distortion, thereby increasing the lattice constant of the FCC phase.

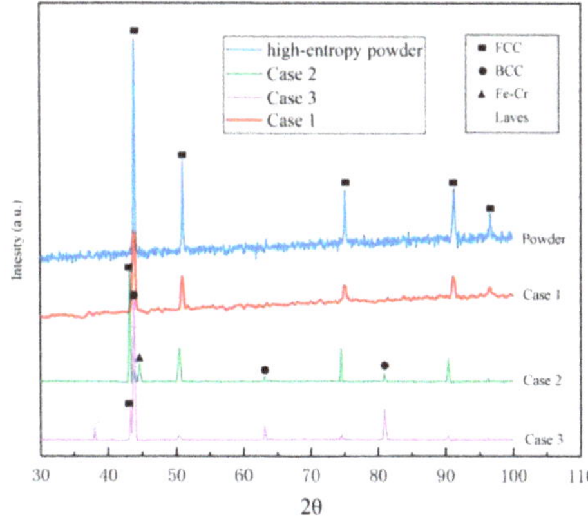

Figure 5. XRD patterns of as-alloyed powder and high entropy alloy (HEA) coatings.

Besides, Ti and Co\Cr elements have a low composite enthalpy, hence, the excessively high temperature inside the molten pool provides an opportunity for the reaction between the elements. There is a great possibility for the Ti element in Case 3 to be combined with the Co element and Ni element to form $(Co, Cr)_2Ti$ Laves phase (as shown in Figure 5). However, the Ti element content in Case 1 is extremely small, and the XRD test result is the phase on the top of the cladding layer, so no detection results are consistent with the detection results of the high entropy alloy powder, and no obvious BCC phase and Fe-Cr phase are detected. As the specific energy increases, the Ti element content gradually increases, so the BCC characteristic gradually becomes obvious, and Fe-Cr can be detected. When the specific energy is too large, the Laves phase inside the cladding layer is significantly precipitated.

3.3. Microstructure

Based on the substantial investigation on the macro morphology and phase transformation, the effect of specific energy on the crystal growth and microstructure is figured out in this research. Figure 6 presents the SEM morphology of the coatings obtained under the condition of different specific energy. Metallurgical bonding has been obtained between the base metal and the coatings of which the bonding region is mainly composed of columnar crystal and shrinkage cavities, as shown in Figure 6. There is a slight increase in the width of the bonding region from 25.5 µm to 60 µm at the function of specific energy varied from 40.9 to 89.3 J/mm^2. With regard to the thermal cycling process, the increase of specific energy leads to the decrease of actual cooling rate and thus provides an opportunity for the mixing of HEA alloy and TC4 substrate. As implied in the detailed image, the transformation from columnar crystal to the cavities is an important feature in this region. It also can be concluded from Figure 6g–i that there are obvious differences in the size of columnar crystals under different specific energy. There is a significant increase in the width of the columnar crystal zone from 3.2 µm to 12.1 µm at the function of specific energy varied from 40.9 to 89.3 J/mm^2. It is attributed that the higher heat provides more energy for the growth or coarsening of sub-grain. Thus, the columnar crystals grow larger slightly with the increase of specific energy.

The microstructure at the top of the HEA laser cladding layer under different specific energies is presented in Figure 7. Meanwhile, the chemical composition of different regions, which are shown in Figure 7, is exhibited in Table 4. It can be seen from Figure 7 and Table 4 that there is a slight difference between Case 1 and Case 2, while Case 3 shows an obvious difference. The microstructure of Case 1 is composed of the Fe-Cr phase (point A) and matrix. Compared to Case 1, the rich-Ti phase is formed with the higher specific energy which provides a promotion to the melting of the TC4 substrate. However, the extremely high specific energy results in sufficient reaction between powders and substrate. Thus, the (Ni, Co, Ti)-rich phase is formed because of the more negative ΔH_{mix}.

EDS analysis is then carried out along the path from the upper layer to the substrate as marked in Figure 8. Little fluctuation in the relative concentration curves of Cr, Co, Fe, Ti, and Ni elements is found in the upper and middle layers in the coatings. However, the content of all the elements increased and decreased sharply, respectively, in the bottom layers of the coatings, especially in the bonding region. As the content of Ti elements in the substrate is much higher than that of the coatings, the obvious dilution effect by the substrate upon the cladded coatings led to the increase of Ti elements and the formation of the bonding region. Besides, the element change of the sample with lower specific energy is strongly sharper than that with a higher specific energy. Among the experiments, increasing specific energy leads to an increase in the dilution ratio. More melted substrate metal is melted to give birth to a higher dilution rate and thus less obvious difference in the chemical composition between the HEA coating and substrate.

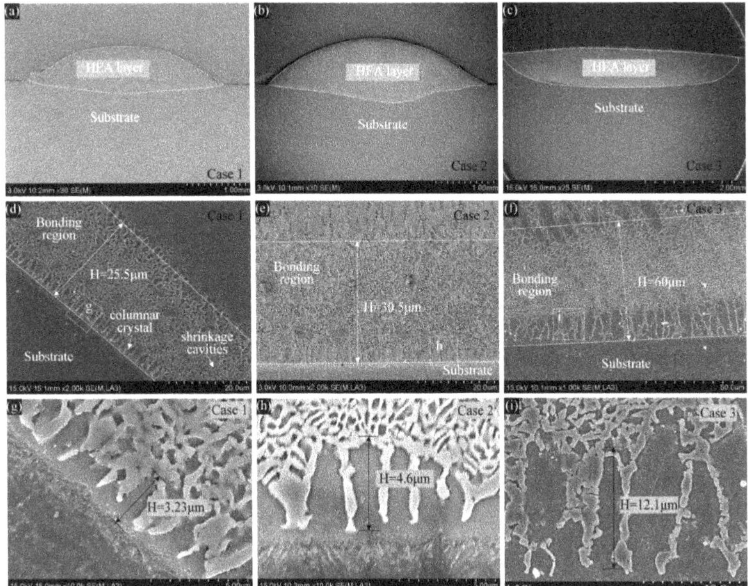

Figure 6. (a) SEM image of the transverse section of Case 1, (b) SEM image of the transverse section of Case 2, (c) SEM image of the transverse section of Case 3, (d) high-magnification SEM image of bonding region in Case 1, (e) high-magnification SEM image of bonding region in Case 2, (f) high-magnification SEM image of bonding region in Case 3, (g) high-magnification SEM image of the localized region marked in (d), (h) high-magnification SEM image of the localized region marked in (e), (i) high-magnification SEM image of the localized region marked in (f).

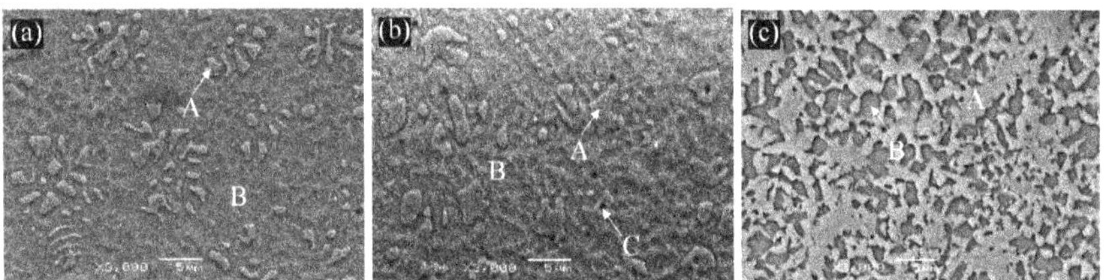

Figure 7. SEM images of laser cladded coatings in different region. (a) Case 1, (b) Case 2, (c) Case 3.

Table 4. The chemical composition of different points in Figure 7.

Case	Point	Al(at%)	Ti(at%)	V(at%)	Cr(at%)	Fe(at%)	Co(at%)	Ni(at%)
1	A	1.8	6.95	1.1	35.58	25.63	16.70	12.25
	B	1.75	17.06	00.47	20.78	21.43	19.92	18.59
2	A	2.20	7.99	0.83	36.49	24.53	15.04	12.92
	B	2.42	16.54	0.7	21.37	20.69	19.71	18.58
	C	2.85	47.08	0	17.70	12.78	10.65	8.93
3	A	09.31	40.34	02.29	09.95	07.58	15.56	14.97
	B	09.97	61.57	02.30	08.54	06.11	05.29	06.22

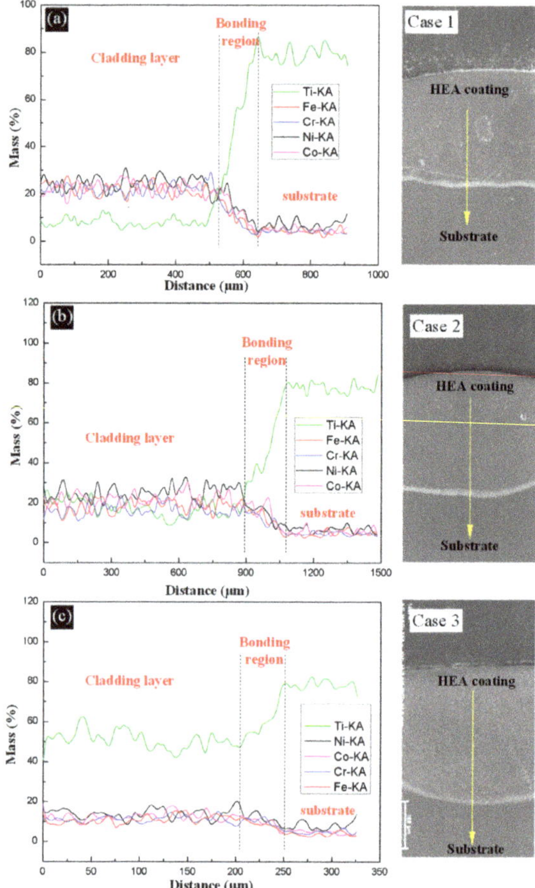

Figure 8. (**a**) EDS line scanning of Case 1, (**b**) EDS line scanning of Case 2, (**c**) EDS line scanning of Case 3.

3.4. Hardness

Figure 9 presents the hardness depth profile in the transverse cross-section of the laser clad coating. The hardness was tested with a load of 1000 g. The variation of hardness reflects the coating microstructure and element changes, which are responsible for the hardness variation. Due to the interaction between the multi-principal elements that will promote the internal fine-grain strengthening, dispersion strengthening, and solid solution strengthening of the alloy during the solidification process, the hardness of the HEA coating is much higher than that of the Ti6Al4V substrate. The hardness of FeCoCrNi coating exceeds 850 Hv when the parameters of Case 1 and Case 2 are selected. Results indicate that a better FeCoCrNi coating is obtained than that from the literature [28]. Meanwhile, Figure 9b shows that the average hardness of HAZ is slightly higher than the substrate. This increase is likely to be attributed to the localized metallurgical changes that occur in the HAZ associated with the rapid heating and cooling above the α-β phase-transition temperature of this specific alloy but below its melting point. It can also be seen that the microhardness of HEA coating in Case 3 is significantly lower than that in Case 1 and Case 2, which is related to the element content in coatings, as shown in Figure 8. The content of Ti is much higher than the other element, hence the balance of high entropy is broken.

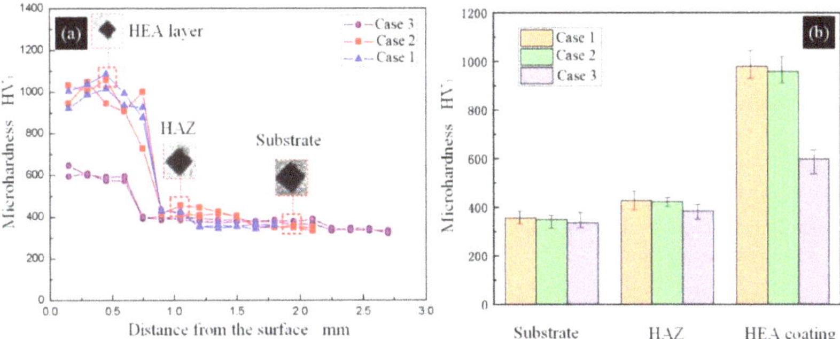

Figure 9. (**a**) Hardness depth profile in the transverse cross-section of the laser cladded coating, and (**b**) average microhardness on the cross-section of samples under different specific energy.

3.5. Wear Resistance

Figure 10 shows the wear morphology of the sample surface after the above-mentioned wear resistance test. It can be seen from the figure that the surface wear of the TC4 substrate is very serious, and there are pits and adhesive materials, which is attributed to the that under the action of plastic deformation and large pressure, the surface material of the substrate will adhere to the surface of the friction pair and transfer during the sliding process. Therefore, pits and adhesive materials are formed. In addition, deep furrows were also found. This is because metal particles caused by wear will adhere to the surface of the friction pair, thereby forming grooves on the surface of the substrate. Meanwhile, the main component of the cladding layer of Case 3 is the Ti element, so there is not much difference in the wear morphology. Compared with the TC4 matrix, it shows slight adhesion wear and abrasive wear. A large amount of Laves phase is distributed in the cladding layer, which increases the hardness of the cladding layer, so abrasive wear is effectively suppressed. XRD results show that a HEA solid solution phase is generated in the cladding layer, which improves the plasticity of the material, and the adhesion wear is also suppressed.

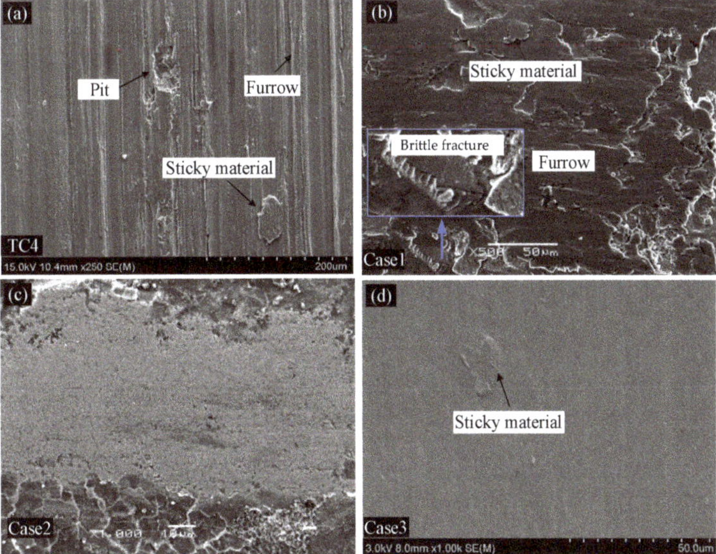

Figure 10. The wear morphology of TC4 and coatings. (**a**) TC4, (**b**) Case 1, (**c**) Case 2, (**d**) Case 3.

Figure 10b,c are the wear morphology of Case 1 and Case 2, respectively. A nearly straight crack with a smooth fracture appearance can be found on the wear morphology of Figure 10b, so brittle fracture will likely occur under the action of larger pressure, which will seriously affect the practical application of Case 1. Therefore, sufficient specific energy is essential during laser cladding of FeCoCrNi high entropy alloy on the Ti substrate. Case 2 only suffered slight scratches during the wear process. Besides, as can be seen from the figure, the wear resistance of Case 2 is significantly higher than that of the TC4 matrix. This is because Case 2 has an equal proportion of elements, solid solution strengthening, and second phase strengthening, which results in a significant improvement in the wear resistance of the cladding layer.

Due to the obvious brittle fracture of Case 1, it is difficult to define the size of the wear region. Therefore, only the wear size statistics of the TC4 substrate and Case 2 and Case 3 are performed. The results are shown in Table 5. It can be seen that the wearing depth and width of the HEA cladding layer is significantly lower than that of the TC4, and the wear cross-sectional area is calculated using the probe of the friction-wear machine, and its wear rate was further calculated, as shown in Table 5. The results show that the HEA cladding layer plays a significant role in improving the wear resistance of the TC4 surface, which can effectively protect the matrix material. Results indicate that the wear resistance of Case 2 is improved by 5 times relative to the TC4 matrix, while the wear resistance of Case 3 is improved by 2.84 times. The reason is that as the hardness increases, the wear resistance of the material will also increase. Therefore, the feasibility and extent of improve wear resistance of titanium alloy by laser cladding of HEA is revealed. It is worth noting that the appropriate process parameters are limited to a small range during laser cladding of FeCoCrNi high entropy alloy on the Ti substrate.

Table 5. The size of the wear region of the cladding layer and substrate under different specific energy.

	TC4	Case 3	Case 2
Wearing depth(μm)	116.3	68.6	27.3
Wearing width(μm)	954	569.8	366
Wearing area(μm^2)	6.11×10^4	2.15×10^4	9.9×10^3
Wearing volume(mm^3)	0.305	0.107	0.050
Wear resistance(mm^3/s)	5.08×10^{-4}	1.78×10^{-4}	8.33×10^{-5}

4. Conclusions

FeCoCrNi HEA coatings have been successfully deposited directly on Ti6Al4V alloy using the laser cladding method. The microstructure evolution and enhanced properties of laser cladded FeCoCrNi HEA on TC4 substrate have been studied and the main conclusions are as follows:

1. The HEA coating is mainly composed of the FCC phase and BCC phase, precipitating a small amount of Fe-Cr phase and Laves phase. With the increase of specific energy, there is an intensification in the transformation from the FCC phase to the BCC phase and precipitation of the Laves phase.
2. Specific energy has an important influence on the cladding morphology. The coating width will increase with higher specific energy. Meanwhile, the length and width of the columnar crystal at the bottom of the cladding layer increase significantly with higher specific energy.
3. Attributed to the various strengthening mechanism, the microhardness of HEA coating reaches 1098 HV, which is about 200% higher than that of the TC4 substrate.
4. The appropriate process parameters are limited to a small range during laser cladding of FeCoCrNi high entropy alloy on the TC4 substrate. The wear resistance is significantly improved by the HEA coating by using optimized process parameters. The effect of scanning path and strategy of the microstructure and properties of laser cladded FeCoCrNi high entropy alloy are worth deeply investigation in the future.

Author Contributions: Conceptualization, L.W. and X.Z.; methodology, Z.G.; formal analysis, M.W.; investigation, T.L.; writing—original draft preparation, L.W.; writing—review & editing, F.W.; project administration, X.Z.; funding acquisition, X.Z. All authors have read and agreed to the published version of the manuscript.

Funding: This work was supported by the National Natural Science Foundation of China (grant number 51975285) and the National Laboratory for Remanufacturing of Academy of Armored Forces Engineering (grant number 6142005200402).

Conflicts of Interest: The authors declare no conflict of interest.

References

1. Quazi, M.M.; Ishak, M.; Fazal, M.A.; Arslan, A.; Rubaiee, S.; Qaban, A.; Aiman, M.H.; Sultan, T.; Ali, M.M.; Manladan, S.M. Current research and development status of dissimilar materials laser welding of titanium and its alloys. *Opt. Laser Technol.* **2020**, *126*, 106090. [CrossRef]
2. Li, W.; Cao, C.; Yin, S. Solid-state cold spraying of Ti and its alloys: A literature review. *Prog. Mater. Sci.* **2020**, *110*, 100633. [CrossRef]
3. Chern, A.H.; Nandwana, P.; Yuan, T.; Kirka, M.M.; Dehoff, R.R.; Liaw, P.K.; Duty, C.E. A review on the fatigue behavior of Ti-6Al-4V fabricated by electron beam melting additive manufacturing. *Int. J. Fatigue* **2019**, *119*, 173–184. [CrossRef]
4. Gao, F.; Yu, W.; Song, D.; Gao, Q.; Guo, L.; Liao, Z. Fracture toughness of TA31 titanium alloy joints welded by electron beam welding under constrained condition. *Mater. Sci. Eng. A* **2020**, *772*, 138612. [CrossRef]
5. Qi, C.; Zhan, X.; Gao, Q.; Liu, L.; Song, Y.; Li, Y. The influence of the pre-placed powder layers on the morphology, microscopic characteristics and microhardness of Ti-6Al-4V/WC MMC coatings during laser cladding. *Opt. Laser Technol.* **2019**, *119*, 105572. [CrossRef]
6. Li, J.; Zhou, J.; Feng, A.; Huang, S.; Meng, X.; Sun, Y.; Sun, Y.; Tian, X.; Huang, Y. Investigation on mechanical properties and microstructural evolution of TC6 titanium alloy subjected to laser peening at cryogenic temperature. *Mater. Sci. Eng. A* **2018**, *734*, 291–298. [CrossRef]
7. Chew, Y.; Bi, G.J.; Zhu, Z.G.; Ng, F.L.; Weng, F.; Liu, S.B.; Nai, S.M.L.; Lee, B.Y. Microstructure and enhanced strength of laser aided additive manufactured CoCrFeNiMn high entropy alloy. *Mater. Sci. Eng. A* **2019**, *744*, 137–144. [CrossRef]
8. Lee, C.; Song, G.; Gao, M.C.; Feng, R.; Chen, P.; Brechtl, J.; Chen, Y.; An, K.; Guo, W.; Poplawsky, J.D.; et al. Lattice distortion in a strong and ductile refractory high-entropy alloy. *Acta Mater.* **2018**, *160*, 158–172. [CrossRef]
9. Li, R.; Niu, P.; Yuan, T.; Cao, P.; Chen, C.; Zhou, K. Selective laser melting of an equiatomic CoCrFeMnNi high-entropy alloy: Processability, non-equilibrium microstructure and mechanical property. *J. Alloys Compd.* **2018**, *746*, 125–134. [CrossRef]
10. Yeh, J.W.; Chen, S.K.; Lin, S.J.; Gan, J.Y.; Chin, T.S.; Shun, T.T.; Tsau, C.H.; Chang, S.Y. Nanostructured High-Entropy Alloys with Multiple Principal Elements: Novel Alloy Design Concepts and Outcomes. *Adv. Eng. Mater.* **2004**, *6*, 299–303. [CrossRef]
11. Liu, J.; Liu, H.; Chen, P.; Hao, J. Microstructural characterization and corrosion behaviour of AlCoCrFeNiTix high-entropy alloy coatings fabricated by laser cladding. *Surf. Coat. Technol.* **2019**, *361*, 63–74. [CrossRef]
12. Liu, H.; Liu, J.; Chen, P.; Yang, H.; Hao, J.; Tian, X. Microstructure and Properties of AlCoCrFeNiTi High-Entropy Alloy Coating on AISI1045 Steel Fabricated by Laser Cladding. *J. Mater. Eng. Perform.* **2019**, *28*, 1544–1552. [CrossRef]
13. Anupam, A.; Kottada, R.S.; Kashyap, S.; Meghwal, A.; Murty, B.S.; Berndt, C.C.; Ang, A.S.M. Understanding the microstructural evolution of high entropy alloy coatings manufactured by atmospheric plasma spray processing. *Appl. Surf. Sci.* **2020**, *505*, 144117. [CrossRef]
14. Wang, W.; Qi, W.; Xie, L.; Yang, X.; Li, J.; Zhang, Y. Microstructure and Corrosion Behavior of (CoCrFeNi)95Nb5 High-Entropy Alloy Coating Fabricated by Plasma Spraying. *Materials* **2019**, *12*, 694. [CrossRef]
15. Dou, B.; Zhang, H.; Zhu, J.H.; Xu, B.Q.; Zhou, Z.Y.; Wu, J.L. Uniformly Dispersed Carbide Reinforcements in the Medium-Entropy High-Speed Steel Coatings by Wide-Band Laser Cladding. *Acta Metall. Sin. Engl. Lett.* **2020**, *33*, 1145–1150. [CrossRef]

16. Gao, J.; Wu, C.; Hao, Y.; Xu, X.; Guo, L. Numerical simulation and experimental investigation on three-dimensional modelling of single-track geometry and temperature evolution by laser cladding. *Opt. Laser Technol.* **2020**, *129*, 106287. [CrossRef]

17. Savrai, R.A.; Soboleva, N.N.; Malygina, I.Y.; Osintseva, A.L. The structural characteristics and contact loading behavior of gas powder laser clad CoNiCrW coating. *Opt. Laser Technol.* **2020**, *126*, 106079. [CrossRef]

18. Liu, H.; Liu, J.; Chen, P.; Yang, H. Microstructure and high temperature wear behaviour of in-situ TiC reinforced AlCoCrFeNi-based high-entropy alloy composite coatings fabricated by laser cladding. *Opt. Laser Technol.* **2019**, *118*, 140–150. [CrossRef]

19. Gao, L.; Song, J.; Jiao, Z.; Liao, W.; Luan, J.; Surjadi, J.U.; Li, J.; Zhang, H.; Sun, D.; Liu, C.T.; et al. High-Entropy Alloy (HEA)-Coated Nanolattice Structures and Their Mechanical Properties. *Adv. Eng. Mater.* **2018**, *20*, 1700625. [CrossRef]

20. Li, J.; Huang, Y.; Meng, X.; Xie, Y. A Review on High Entropy Alloys Coatings: Fabrication Processes and Property Assessment. *Adv. Eng. Mater.* **2019**, *21*, 1900343. [CrossRef]

21. Tian, Y.; Shen, Y.; Lu, C.; Feng, X. Microstructures and oxidation behavior of Al-CrMnFeCoMoW composite coatings on Ti-6Al-4V alloy substrate via high-energy mechanical alloying method. *J. Alloys Compd.* **2019**, *779*, 456–465. [CrossRef]

22. Shi, C.; Lei, J.; Zhou, S.; Dai, X.; Zhang, L.-C. Microstructure and mechanical properties of carbon fibers strengthened Ni-based coatings by laser cladding: The effect of carbon fiber contents. *J. Alloys Compd.* **2018**, *744*, 146–155. [CrossRef]

23. Weng, F.; Yu, H.; Chen, C.; Liu, J.; Zhao, L.; Dai, J.; Zhao, Z. Effect of process parameters on the microstructure evolution and wear property of the laser cladding coatings on Ti-6Al-4V alloy. *J. Alloys Compd.* **2017**, *692*, 989–996. [CrossRef]

24. Huang, C.; Zhang, Y.; Shen, J.; Vilar, R. Thermal stability and oxidation resistance of laser clad TiVCrAlSi high entropy alloy coatings on Ti–6Al–4V alloy. *Surf. Coat. Technol.* **2011**, *206*, 1389–1395. [CrossRef]

25. Huang, C.; Zhang, Y.; Vilar, R.; Shen, J. Dry sliding wear behavior of laser clad TiVCrAlSi high entropy alloy coatings on Ti–6Al–4V substrate. *Mater. Des.* **2012**, *41*, 338–343. [CrossRef]

26. Joseph, J.; Jarvis, T.; Wu, X.; Stanford, N.; Hodgson, P.; Fabijanic, D.M. Comparative study of the microstructures and mechanical properties of direct laser fabricated and arc-melted AlxCoCrFeNi high entropy alloys. *Mater. Sci. Eng. A* **2015**, *633*, 184–193. [CrossRef]

27. Jiang, L.; Wu, W.; Cao, Z.; Deng, D.; Li, T. Microstructure Evolution and Wear Behavior of the Laser Cladded CoFeNi2V0.5Nb0.75 and CoFeNi2V0.5Nb High-Entropy Alloy Coatings. *J. Spray Technol.* **2016**, *25*, 806–814. [CrossRef]

28. Cai, Y.; Chen, Y.; Manladan, S.M.; Luo, Z.; Gao, F.; Li, L. Influence of dilution rate on the microstructure and properties of FeCrCoNi high-entropy alloy coating. *Mater. Des.* **2018**, *142*, 124–137. [CrossRef]

29. Shu, F.; Zhang, B.; Liu, T.; Sui, S.; Liu, Y.; He, P.; Liu, B.; Xu, B. Effects of laser power on microstructure and properties of laser cladded CoCrBFeNiSi high-entropy alloy amorphous coatings. *Surf. Coat. Technol.* **2019**, *358*, 667–675. [CrossRef]

30. Sui, X.; Lu, J.; Hu, J.; Zhang, W. Effect of specific energy on microstructure and properties of laser cladded TiN/Ti3AlN-Ti3Al composite coating. *Opt. Laser Technol.* **2020**, *131*, 106428. [CrossRef]

31. Kozlovskiy, A.L.; Kenzhina, I.E.; Zdorovets, M.V. FeCo– Fe2CoO4/Co3O4 nanocomposites: Phase transformations as a result of thermal annealing and practical application in catalysis. *Ceram. Int.* **2020**, *46*, 10262–10269. [CrossRef]

32. Zdorovets, M.V.; Kozlovskiy, A.L. Study of phase transformations in Co/CoCo2O4 nanowires. *J. Alloys Compd.* **2020**, *815*, 152450. [CrossRef]

Publisher's Note: MDPI stays neutral with regard to jurisdictional claims in published maps and institutional affiliations.

Article

Analysis of the Process Parameters, Post-Weld Heat Treatment and Peening Effects on Microstructure and Mechanical Performance of Ti–Al Dissimilar Laser Weldings

Paola Leo [1,*], **Sonia D'Ostuni** [1], **Riccardo Nobile** [1], **Claudio Mele** [1], **Andrea Tarantino** [1] **and Giuseppe Casalino** [2]

[1] Innovation Engineering Department, University of Salento, Via per Arnesano s.n., 73100 Lecce, Italy; sonia.dostuni@unisalento.it (S.D.); riccardo.nobile@unisalento.it (R.N.); claudio.mele@unisalento.it (C.M.); andrea.tarantino@studenti.unisalento.it (A.T.)

[2] Politecnico di Bari, DMMM, Via Orabona, 4, 70125 Bari, Italy; giuseppe.casalino@poliba.it

* Correspondence: paola.leo@unisalento.it

Abstract: Dissimilar Ti–Al laser weldings are very interesting due to their difficulties in being processed because of the different physical properties of the alloys and the crack formations during cooling and solidification. In this study, the effect of laser offset and defocusing on microstructure, geometry and mechanical properties response of 2 mm thick dissimilar AA6061/Ti-6Al-4V laser welds was analyzed. Moreover, in order to reduce residual stresses, the joints were both heat-treated and mechanically treated by ultrasonic peening. The welds microstructure was found to be martensitic in the Ti-6Al-4V fusion zone, columnar dendritic in the AA6061 fusion zone and partially martensitic in the Ti-6Al-4V heat-affected zone. Intermetallic compounds based on the Al–Ti system were detected at the AA6061/Ti-6Al-4V interface and in the aluminum fusion zone. Both negative defocusing and higher laser offset improved the tensile performance of the welds, mainly by reducing the amount of brittle intermetallic compounds. The stress relaxation heat treatment, leading to the aging of the martensite and the increasing of the size of the intermetallic compound, reduced the tensile strength and ductility of the joints. On the contrary, for dissimilar Al–Ti welds, mechanical treatment was effective in increasing joints ductility and, moreover, corrosion resistance.

Keywords: corrosion susceptibility; defocusing; hardness; microstructure; offset; stress relief heat treatment; tensile test; ultrasonic peening

Citation: Leo, P.; D'Ostuni, S.; Nobile, R.; Mele, C.; Tarantino, A.; Casalino, G. Analysis of the Process Parameters, Post-Weld Heat Treatment and Peening Effects on Microstructure and Mechanical Performance of Ti–Al Dissimilar Laser Weldings. *Metals* **2021**, *11*, 1257. https://doi.org/10.3390/met11081257

Academic Editor: Sergey N. Grigoriev

Received: 7 July 2021
Accepted: 2 August 2021
Published: 9 August 2021

Publisher's Note: MDPI stays neutral with regard to jurisdictional claims in published maps and institutional affiliations.

1. Introduction

Titanium (Ti) and aluminum (Al) alloys are characterized by low density (equal to 4.43 g/cm^3 for Ti and 2.80 g/cm^3 for Al) and good mechanical properties (the average ultimate tensile strength is equal to 300 MPa for Al Alloys and 1205 MPa for Ti alloys) [1]. Al–Ti dissimilar welds are of significant interest in aeronautical and automotive applications, where weight reduction, combined with high mechanical strength and corrosion resistance, is required. Moreover, cost reduction can also be obtained by using Aluminum alloy coupled with Titanium [2]. For example, the passenger seat track after AIRBUS conceptual design was manufactured with Ti6Al4V crown and aluminum alloy web by laser beam welding [3]. The benefits and the recent trend of laser welding of Ti/Al alloys in the academic sector and industry were explored by Quazy et al. [4]. Recent trends also enclose dissimilar Al–Ti laser weldings based on high-strength aluminum–lithium alloys for aviation applications [5,6].

However, joining of Al/Ti alloys remains a difficult technological challenge because of the different physical properties and the formation of intermetallic phases (IMC) that lead to crack formation during cooling and solidification [4,7,8]. The successful joining of aluminum to titanium requires the reduction of the thickness of IMC layer at the interface as well as of the number of IMC particles dispersed on the aluminum fused zone (FZ).

In fact, although intermetallic compounds based on Ti–Al system are attractive, being characterized by good strength at high temperature, the development of IMC leads to restrictions on the available deformation modes. So, the occurrence of those particles is associated with increased strength but reduced ductility and fracture toughness [5].

Both the convective mixing and diffusion phenomena that increases the occurrence of IMC can be limited by diffusion bonding [6] friction welding [7], laser welding [4] of these alloys. In laser welding, the mismatch in thermophysical properties of chosen dissimilar alloy can be adjusted by shifting the laser beam to one of the substrates. For example, according to [4], a shift of the laser beam to the aluminum side promoted the formation of a thin IMC layer and yielded a high tensile strength, whereas a shift to the titanium side promoted the formation of a thick TiAl3 phase that contained cracks and voids. On the contrary, according to [8], the thickness of IMC can be significantly reduced by shift the laser beam on Ti side.

In addition, defocusing the laser beam can modify the IMC thickness of the layer and IMC particles amount due to the change in the beam energy density and temperatures values [9]. The defocus distance is the position of the focal plane relative to the in focus beam plane. A negative defocusing distance occurs when the focal plane is placed under the "in focus" plane, whereas in positive defocusing, the focal plane is above it. The negative defocusing leads to a deeper melt pool due to its convergent nature, by comparing with in focus and positive defocusing. Particularly, when keeping constant laser power and scanning speed, the defocusing distance (positive and negative sense) can affect the microstructure and defects presence [9–11].

Post welding heat treatments (PWHT) are usually applied to the welds. The welds residual stresses may lead to brittle fracture and reduced fatigue life [9,10]. In addition, the increased dislocation density in fusion welds can reduce ductility of the joints [12]. As the PWHT temperature increases, residual stresses and dislocation density are reduced faster. In the case of dissimilar Al–Ti welds the different properties of the alloys, limit the feasibility of an effective stress relief heat treatment. For example, a significative control of the Ti-6Al-4V welds could require heat treatments at temperature higher than the liquidus temperature of Al Alloys [12]. Moreover, the highest temperature suitable must also avoid grain size and IMC coarsening resulting in decreasing of strength and ductility [13,14]. Many studies were devoted to defining the role of PWHT on microstructure and IMC growth for dissimilar welds. For example, for a laser welded AA5754/Ti-6Al-4 V joint, the microstructure of the IMC did not change after heat treatment at 350 °C for 336 h [13]. At 450 °C for 168 h, nucleation and growth of TiAl3 were observed with consequent decreasing of joint tensile strength [13]. In another study regarding forty-layered Ti–Al composites fabricated in a single-shot explosive welding process, the thickness of the TiAl3 layer was only 0.7 ± 0.1 μm after heat treatment at 500 °C for 100 h [15]. Solid-state diffusion between Ti and Al was investigated between 520–650 °C in multi-laminated Ti/Al [16]. The authors found that the thickness of the TiAl3 layer reached 10.50 ± 0.78 μm and 24.11 ± 5.85 μm after annealing at 520 °C-4 h and 650 °C-64 h [16].

As an alternative to the PWHT, mechanical treatments can be applied to reduce residual stress of welds. The ultrasonic peening treatment (UPT) aims to redistribute residual stresses by eliminating tensile stresses and introducing compressive stresses [17,18]. The principle is the same as hammer peening [19,20] and consists of subjecting the material surface to a plastic compression deformation through a hammer gun that vibrates. In the case of UPT, the hammer gun vibrates at high frequency due to a piezoelectric actuator [17,18]. It has been shown [21] that UPT induces a nanocrystalline surface layers with increased microhardness. The nanocrystalline surface layer also enhanced the corrosion resistance of the Zr-2.5%Nb alloy in saline solution [21]. Moreover, this technique allows micro-cracks and pores closure [18] that are well-known to improve the fatigue strength. The efficiency of the UPT treatment obviously depend on the geometry of the impactor, on the percussion intensity and on the duration of the treatment. These parameters need to be accurately regulated to avoid that the treatment is too mild and thus ineffective, but at the same time not too much severe, to not irremediably damage the material [22]. Most papers study the effects of UPT on fatigue behaviors showing that under correct conditions UPT produces high improvement in fatigue life [23–25] or wear and corrosion resistance [21,22]. However, few investigations aimed to evaluate the effects of UPT on tensile properties of laser welded dissimilar joints.

In this paper, Al-Mg -Si alloy was laser welded to Ti-6Al-4V alloy. The effect of both focused and negative defocused beam coupled with two different laser offset values on fusion zone size, heat-affected zone size, intermetallic compounds occurrence and amount, hardness and tensile performance were investigated. Moreover, the occurrence and distribution along the thickness of the intermetallic particles has also been characterized. Stress relaxation heat treatment effects at 530 °C for 2 h followed by air cooling on the joints mechanical properties has been analyzed in term of hardness measurements and tensile test. The stress relaxation heat treatment effects were compared with those induced by the mechanical treatment of the joints. Potentiodynamic polarization test in NaCl solution was used to study the corrosion resistance both of aluminum alloy and welds before and after mechanical treatment. The novelty of this study is that it provides a hitherto unexplored chance to overpass the main limit of the dissimilar joints that is the low ductility by the mean of a mechanical treatment.

2. Materials and Experimental Methods

The chemical compositions of as-received Al and Ti alloys are shown in Tables 1 and 2. Butt joints Al-Mg-Si/Ti-6Al-4V were processed on 2 mm plates. The laser beam was focused on the Titanium side (Figure 1), respectively, 0.1 mm and 0.3 mm away from the aluminum edge, which is the laser offset. For each fixed laser offset, two different laser beam focus positions were used as in the schemes of Figure 1: in focus beam named T as Top (focus at the top of the thickness joint) and negative laser defocusing (-2 mm) named B as Bottom (defocused beam at the bottom of the thickness joint), respectively.

Table 1. AA6061 alloy chemical composition.

Si	Fe	Cu	Mn	Mg	Cr	Zn	Ti	Al
0.40	0.40	0.10	0.50	2.6–3.6	0.3	0.20	<0.15	bal.

Table 2. Ti6Al4V alloy chemical composition.

C	Fe	N_2	O_2	Al	V	H_2	Ti	C
<0.08	<0.25	<0.05	<0.2	5.5	3.5	<0.03	bal.	<0.08

In Figure 1, the macrostructure of the welds cross section is also shown. So, four samples were processed. The welding speed was 2500 mm/min and the laser power 1500 W. An Ytterbium Fiber Laser System (IPG YLS-4000), with a maximum output power equal to 4 kW was used for the welding. The laser beam was delivered with a diameter equal 200 μm and has a beam parameter product (BPP) equal to 6.3 mm/rad. The laser beam has a wavelength of 1070.6 nm. Argon and helium were employed as shielding gas with volumetric flow rate equal to 10 L/min. Particularly, since helium has a specific weight less than atmospheric air, it was employed for the bottom surface, while the heavier Argon was used on top surface.

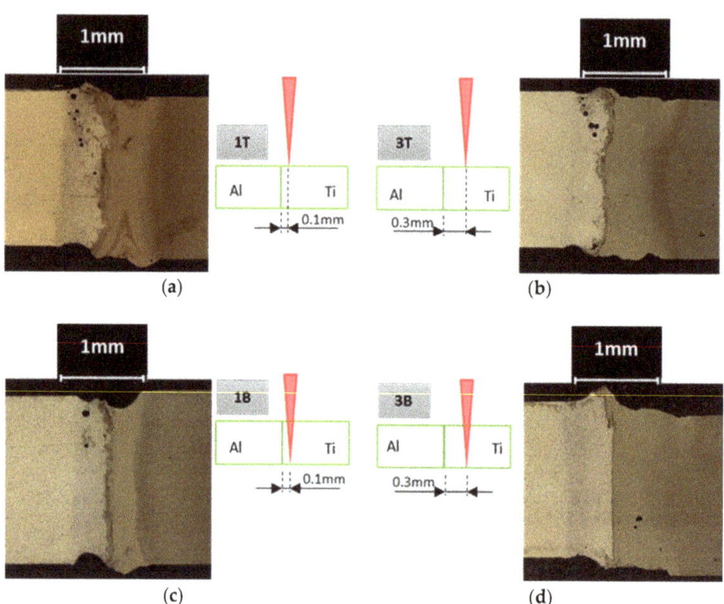

Figure 1. Cross section of the joints 1T (**a**), 3T (**b**), 1B (**c**) and 3B (**d**) coupled with the schemes of the process.

The welds were cut perpendicular to the welding direction to characterize the transverse section. They were analyzed by optical microscopy (OM; Nikon Epipkot 200) and scanning electron microscopy (SEM; Zeiss EVO 40) equipped with energy dispersive spectrometer (EDS Bruker). The cross sections of the samples were prepared using the standard metallographic grinding and polishing techniques and attached using Keller reagent (95 mL H_2O, 2.5 mL HNO_3, 1.5 mL HCl, HF 1 mL). The grain structure was revealed by electro-polishing (20% perchloric acid and 80% ethanol at 0 °C, electro-polishing parameters: 15 V and 60 s), anodic oxidation (Barker etching, anodizing parameters: 20 V and 80 s) and subsequent investigation under polarized light in OM. The area size of FZ, HAZ and intermetallic layer were evaluated using NIS-Element software for imaging analysis. In addition, the width of FZ at different position along the thickness of the joints were evaluated using NIS-Elements. NIS Element is a NIKON software supplied with Epiphot 200 OM. The software is tailored to facilitate image capture, object measurement and counting. The role of the energy density on hardness evolution was investigated by Vickers indentation in the cross section of the joints along a line perpendicular to joint interface. For Vickers microhardness profiles, a Vickers Affri Wiky 200JS2 microhardness tester was used, with load 0.3 kg and time of indentation 15 s (0.3/15 s). The distance between indentations was equal to 300 μm. For the joints T and B, the hardness measurements were done 200 and 1800 μm from upper surface. All the welded joints were subjected to tensile tests. Rectangular samples (length: 20 mm; width: 5 mm; thickness: 2 mm), with sample length perpendicular to the weld line, were cut for tensile test. Tensile tests were carried out on a servo-hydraulic testing machine having MTS810 having a load capacity of 100 kN. The samples were tested using strain rate equal to 0.2 mm/min. The joints 3T and 3B, characterized by having best mechanical performances, were further treated by stress relief heat treatment (SRHT) at 530 °C-2 h to relax welding thermal stress. After that, joints 3B and 3T were characterized in terms of hardness at a half of thickness and tensile tests. On the welded joints 3T and 3B, UPT was applied only on the Aluminum longitudinal surfaces for 1 min/side by an Ultrasonic Peening Sintes UP 600. The diameter of the peening needle was 3 mm and the vibration frequency was 20 kHz. The treated zone (about 6 mm from the Al/Ti interface) included both Al FZ, HAZ and BM.

Electrochemical measurements were performed with an AMEL 5000 potentiostat/galvanostat in an aqueous 3.5% NaCl solution. All experiments were conducted in naturally aerated, near neutral solutions at ambient temperature. A conventional three-electrode cell was employed, with a platinized titanium expanded mesh counter electrode, a Ag/AgCl reference electrode and a specimen of AA6061 aluminum alloy, before and after UTP, as a working electrode. All the potentials were referred to Ag/AgCl. The samples were insulated using teflon to mask their cut edges and back sides, in order to leave only a surface area of 0.5 cm^2 exposed to the electrolyte. The steady state potential was determined after 60 min of immersion in the solution at open circuit potential (OCP). Following the determination of the steady state OCP, potentiodynamic polarisation measurements were performed from -0.15 V vs. OCP to $+1.5$ V vs. Ag/AgCl at a scan rate of 1 mV s^{-1}. Each corrosion test was repeated on two samples.

3. Results and Discussion

3.1. Analysis of Welds Macrostructure and Microstructure

The macrostructure of the joints is shown in Figure 1 together with the scheme of the processing. The chemical etchant clearly reveals the different zones of the joints consisting of unaffected base material (BM), FZ and HAZ. In addition, highlighting the intermetallic layer (IMCL) at the interface. Some small round pores can be observed in the FZ, above all on the Al side.

The BM AA6061, supplied in the annealed condition, is a precipitation-hardenable aluminum alloy, containing magnesium and silicon as its major alloying elements. Figure 2a highlights the annealed grains of the alloy. The Vickers micro-hardness (HV0.3/15) is 52.5 ± 1.15. The Titanium base material is supplied in mill annealed condition and the microstructure is characterized BCC (body-centered cubic) beta phase (darker zone in Figure 2b) that outlines the HCP (hexagonal closest packed) alfa phase (lighter in Figure 2b). The Vickers micro-hardness (HV0.3/15) of the alloy is equal to 290.3 ± 10.2.

All the joints exhibit an aluminum fusion zone with columnar grains that grow from the HAZ/FZ interface towards the Ti side (Figure 2c). The growth of columnar grains, promoted by steep heat gradient, occurred in the opposite direction of the heat flow (perpendicular to the FZ/HAZ interface). In the HAZ of the aluminum alloy, coarsening or dissolution of soluble Mg_2Si particles, developed due to annealing of BM, can happens [25–31]. The particles dissolution is more significant as the distance from fused zone is reduced since, being closer to the fusion zone, the HAZ thermal cycle is characterized by higher peak temperature [9,10]. Figure 3a shows the evolution of the 3B joint Al side microstructure in the different zones (FZ, HAZ) with respect to the BM.

In the titanium side, FZ, the microstructure is martensitic (α'), as shown in Figure 2d [5,27,30]. In fact, as reported in the literature [5,8–10,13,31], the microstructure of the laser fusion zone of Ti-6Al-4V alloy is due to the complete evolution of the beta phase microstructure to martensitic one due to the high cooling rate. The HAZ is a mixture of martensitic and primary α grain [5,8,13,31]. In the HAZ, during the welding heat thermal cycle, an increase of β phase occurs upon heating but, due to the lower peak of temperature, this transformation is never complete and some alpha phases remain unchanged, while the transformed beta grains develop the martensitic microstructure upon cooling. Figure 3a shows the evolution of the Ti side microstructure in the 3B joint in the different zones (FZ, HAZ) with respect to the BM.

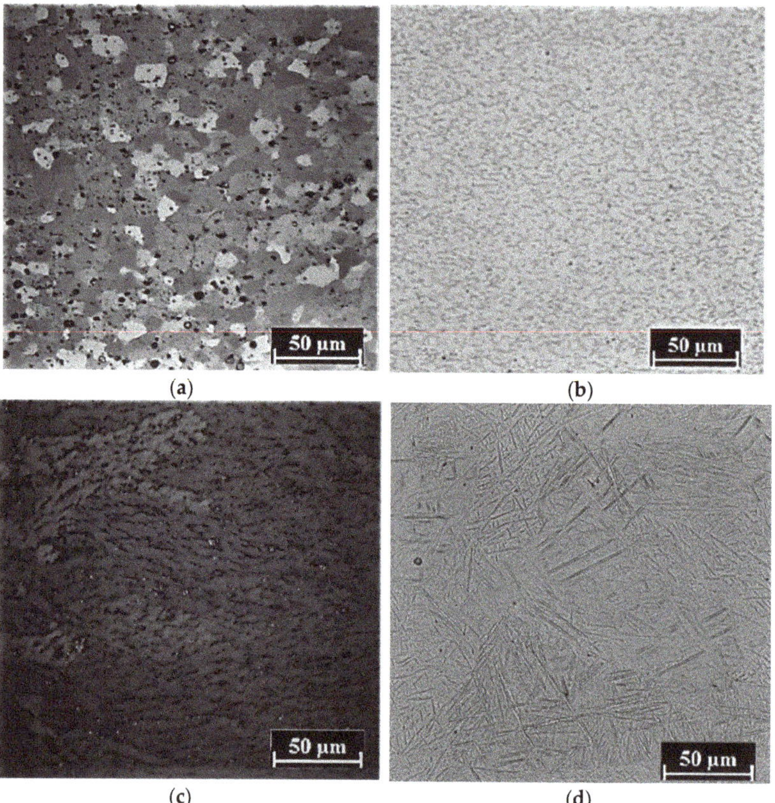

Figure 2. Optical micrographs of the AA6061 BM (**a**), Ti-6Al-4V BM (**b**), AA6061 FZ (**c**) and Ti-6Al-4V FZ (**d**).

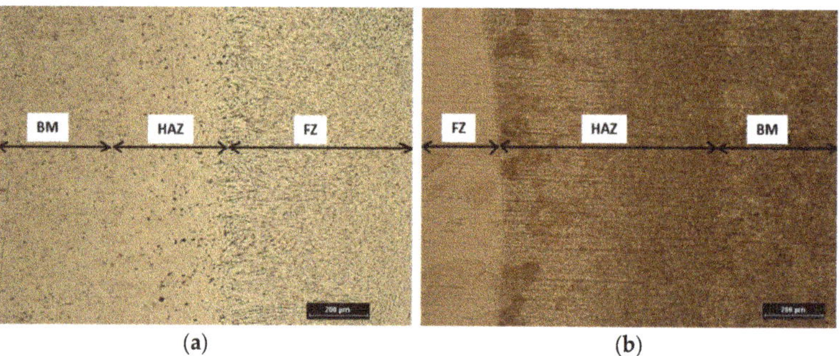

Figure 3. Optical micrographs showing microstructural evolution in the Al side (**a**) and Ti side (**b**).

3.2. Intermetallic Layer and Geometry of Welds Fusion Zones

The IMCL generated during welding at the Al/Ti fused zone interface is due to the melting and mixing of the different alloys. The shape of the intermetallic layer change with process parameters. At fixed defocusing, the interlayer was more curved when the offset was small (Figure 1). At fixed offset, the shape of IMCL become flatter with decreasing defocusing (Figure 1). In fact, both in focus beam and low laser offset promote higher temperature close to the interface favoring convective mixing of the melted alloys, as well as atomic diffusion [8,13]. Consequently, the thickness of the interlayer was also lower with negative defocusing (samples B exhibit lower size of IMCL with respect to the samples T in Figure 4) and increased laser offset (samples 3 exhibit lower size of IMCL with respect to the samples 1 in Figure 4).

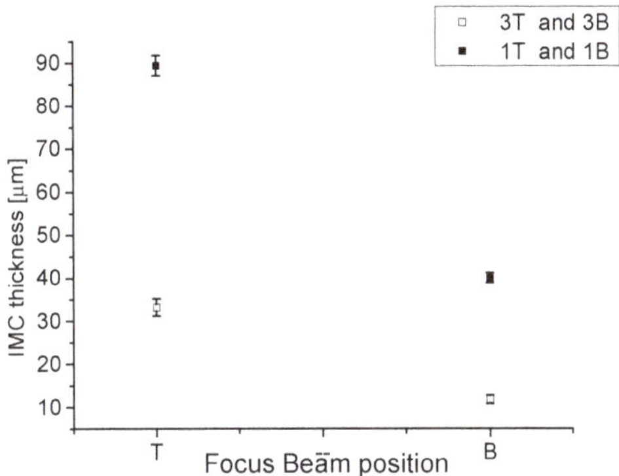

Figure 4. IMCL thickness size for joints 1T and 1B and joints 3T and 3B as function of the focus beam position.

For the same previously discussed role of energy density on the mixing of two alloys, acicular IMC particles could nucleate and grow mainly in the welds processed with focused beam (Figure 5a,b) thanks to the augmented titanium atoms concentration next to the interlayer (Figure 5).

The microstructure of the FZ and HAZ of all the joints does not change with process parameters. Nevertheless, the size and geometry of the different zones exhibit some differences, as shown in Figures 6 and 7. Regarding the FZ and HAZ sizes, the joints 1T and 1B are characterized by a lower FZ and HAZ with respect to the joints 3T and 3B (Figure 6a). Moreover, for the Ti side of joints 1T and 1B, those sizes slightly decreased with defocusing. In the joints 3T and 3B, due to the higher distance of laser beam from the Al alloy, the heat input is better confined in the Ti side. The lower heat conductivity of this alloy [5,9,10,27,29] leads to higher time in temperature and promotes higher size for the FZ and HAZ (Figure 6b). This effect is strengthened by the lowering of energy density with negative defocusing, as better shown below in Figure 7, that leads to lower temperature gradient [9,10].

For the joints 1T and 1B, the heat input better involves the aluminum side, which easily conducts away the heat due to its higher heat conductivity [5,9,10,27,29]. Therefore, the size of the FZ are reduced with respect to the joints 3T and 3B (Figure 6). The high thermal conductivity of aluminum alloy also compensates the effect of the negative defocusing, so the Al side fused zones do not change significantly with defocusing, while Ti FZ and HAZ are decreased.

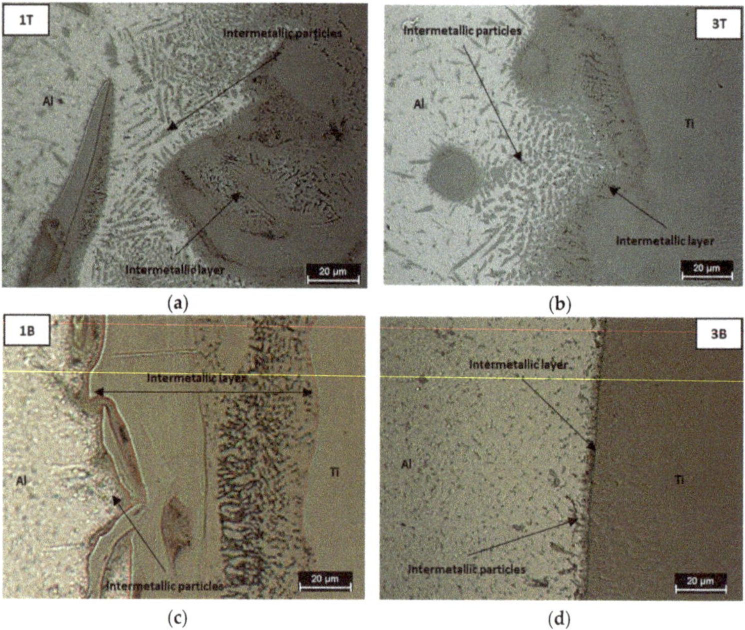

Figure 5. Intermetallic layer at a half of thickness in the welds cross sections of joints 1T (**a**), 1B (**c**), 3T (**b**) and 3B (**d**).

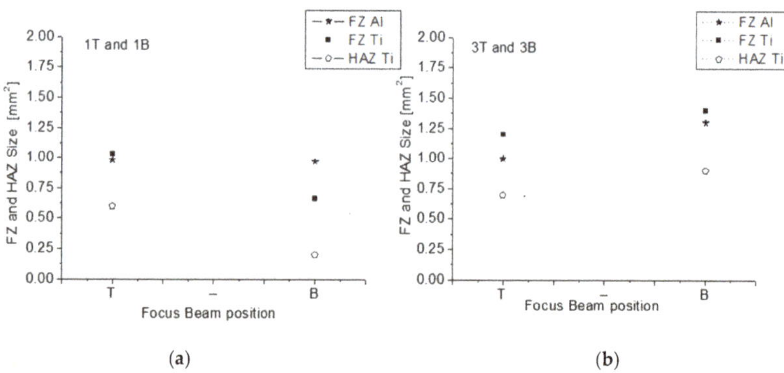

Figure 6. FZ and HAZ size for joints 1T and 1B (**a**) and joints 3T and 3B (**b**). Considering both focused (T) and defocused beam (B), the joints 1T and 1B (**a**) are characterized by a lower size of the FZ with respect to joints 3T and 3B (**b**).

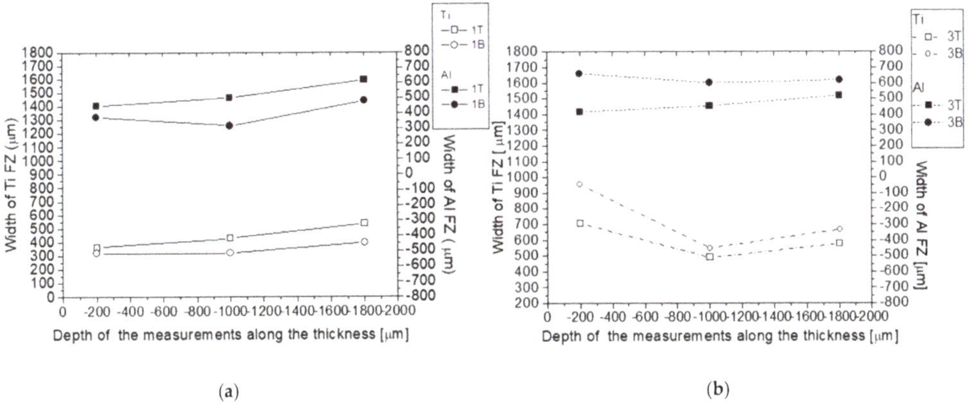

(a)   (b)

Figure 7. Width of the FZ side along the thickness for joints 1T and 1B (**a**) and joints 3T and 3B (**b**): the size of joint 3B along the thickness is always higher than the size of 3T joint (**b**), while the opposite is true for joints 1T and 1B (**a**).

The geometry of the joints FZ at different process parameters was evaluated by measuring the width of FZ along the joints thickness. Three measurements were made for each welds, respectively, at 200 μm, 1000 μm and 1800 μm from the top surface, as shown in Figure 7. The defocused beam strengthens the effect of Ti and Al heat conductivities on the size of FZ; therefore, the size of joint 3B along the thickness is always higher than the size of 3T joint (Figure 7b), while the opposite happens for joints 1T and 1B (Figure 7a).

For the joints processed with the laser offset equal to 0.3 mm, the highest difference in size with defocusing width was found to be 200 μm from the top (in the Ti side, it was equal to 700 micron for 3T vs. 960 micron for 3B). Even for the Al side, the highest increasing of width is registered at 200 μm from the top surface, but the width of Al FZ is more uniform along the joint thickness.

To summarize the results in Figures 6 and 7, the FZ size is decreased (or increased) with decreasing (or increasing) laser offset. The negative defocusing strengthens the effect of laser offset by increasing (decreasing) the size of FZ at the high (low) laser offset. Therefore, the highest sizes of FZ and HAZ were observed for the 3B joint (Figures 6 and 7). The most significant change in the geometry of the FZ size was observed for the Ti side of joint 3, where the width of FZ decreases moving from the top toward the bottom of the joints. The opposite happens for 1T/1B joints, but less significantly.

3.3. Intermetallic Particles: Count and Distribution

The effect of the process parameter on the occurrence and distribution of intermetallic particles in aluminum matrix was also investigated. The number and the size of particles at the Al–Ti interface was evaluated and showed as number of particles for classes of area (Figure 8). For all the welds, the highest number of particles had an area of lower than or equal to 10 μm. The number of particles is reduced by defocusing and, at fixed defocusing, it increases for the welds processed with laser offset equal to 0.1 mm, confirming the role of the peak temperature on the amount of IMC particles. Moreover, it is interesting to note that the number of particles decreases by 64%, comparing the 1T to the 3T joint (as well as from 1T to 1B) and decreases by 78% from the 1B to 3B joint (as well as from 3T to 3B). Therefore, the shift of 0.2 mm in the laser offset position leads to the same effect of a 2 mm shift in defocus distance.

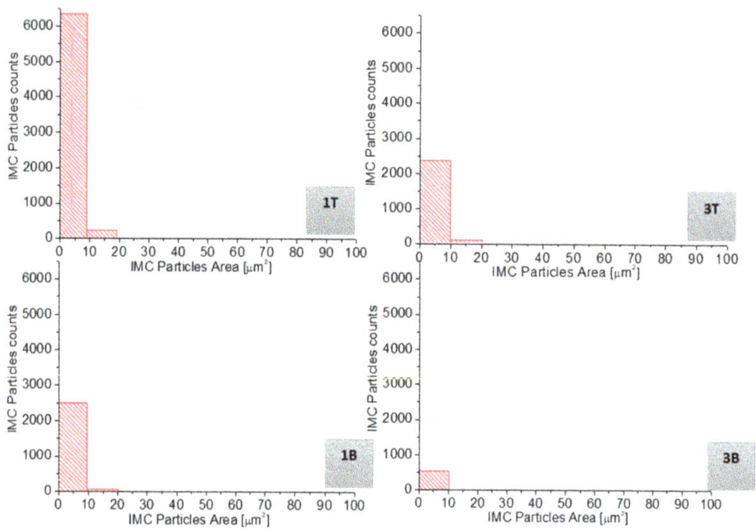

Figure 8. Distribution of IMC particles as function of particles area in 1T, 1B, 3T and 3B joints.

A more detailed analysis on IMC particles distribution was developed, considering the number and distribution of the IMC particles along the thickness. The FZ of the Al side was divided in three rectangular window 1000×600 µm. The middle window was centered at a half of the thickness in the aluminum FZ. The other two windows are, respectively, on the top side (top window) and the bottom side (bottom window) of the central window, as in the schemes of Figures 9 and 10. Inside each window, the number and size of the particles were counted. The analysis of the results shows that the amount of particles was always highest at the top of all the joints, even if, as showed previously, the maximum number decreases with negative defocusing and increased laser offset. The negative defocusing lead to energy loss and so lower average temperature at the interface; the same occurs when increasing the laser offset. In the T samples, the highest number of particles at the top of the joints is justified by the highest energy density and temperature, which favors a better mixing of the two alloys. For the negative defocusing, the increasing number of IMC particles at the top of the joints could be due to the divergent beam at the top, which imposes a longer time for cooling, favoring atomic diffusion [9,10,29].

The role of the energy density on hardness evolution was investigated by Vickers indentation (0.3/15 s) in the cross section of the joints along a line perpendicular to joint interface. For joints T and B, the hardness measurements were conducted, respectively, at 200 and 1800 µm away from the upper surface (Figure 11). These distances were chosen to investigate the role of energy density in the focus point (joints T) or defocused positions (joints B). By processing the welds using negative defocusing, the hardness values close to the AA6061/Ti interface were reduced. The beam defocusing reduces the hardness imposing lower cooling rate due to lower energy density and the consequent microstructure coarsening. Moreover, in 3T and 3B welds, the hardness values were closer to each other for the two different focus conditions (Figure 11b) in both the Al and Ti sides of the welds.

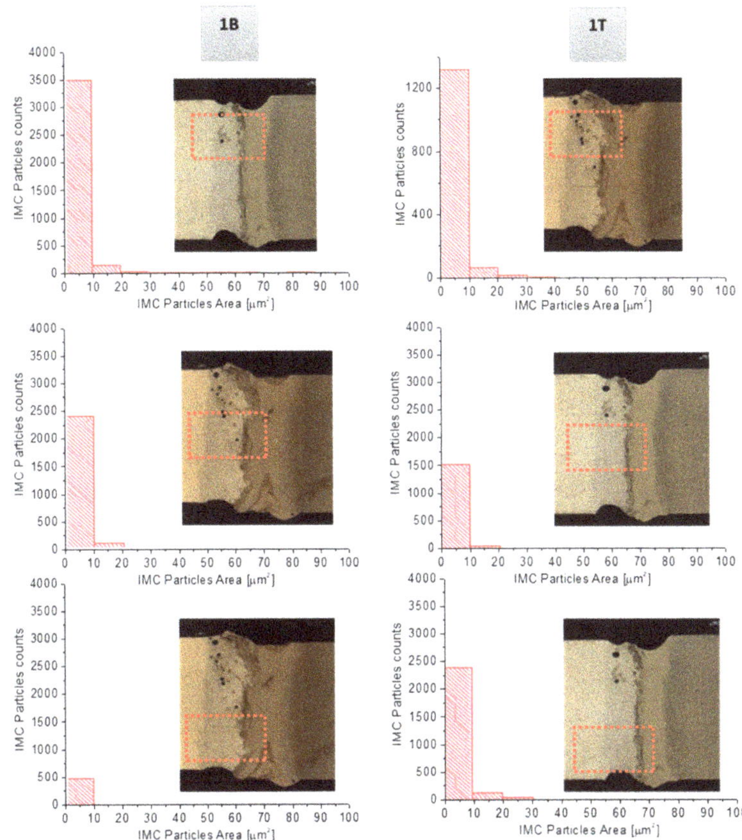

Figure 9. IMC particles distribution along the thickness of joints 1T and 1B.

The effect of the defocused beam on the hardness values is mitigated as laser offset increases, probably due to the prominence of the Ti heat conductivity effect on the cooling rate. With increasing laser offset, Ti heat conductivity dominates the cooling rate and the microstructure evolution and, therefore, the gap in hardness due to beam defocusing is less significant with respect to the joints processed with laser offset equal to 0.1 mm.

3.4. Mechanical Properties: Microhardness and Tensile Test of the as Welded Samples

Table 3 shows the average results after performing the tensile test on two specimens of each type. 3B and 3T welds exhibited the best tensile performances. As shown by the previous analysis, this behavior can be justified by the lowest thickness of IMCL, the lowest number of IMC particles and the lowest gap of hardness values at the interface (Figures 4, 8 and 11). Table 3 highlights the main limit of the joints that is the low value of ductility. Ductility values depend on microstructure [9,10] and intermetallic compounds developed in dissimilar Al–Ti welds that are very brittle [5,27,29,30], but also affected by residual stresses. Residual stresses are significant in the high energy density process (such as laser or electron beam welding) due to the severe temperature gradients and the high heating/cooling rate [9,10]. Residual stress leads to brittle fracture and reduced fatigue life and can be reduced both by heat treatments [9,10,12–16] and mechanical treatment [17–25].

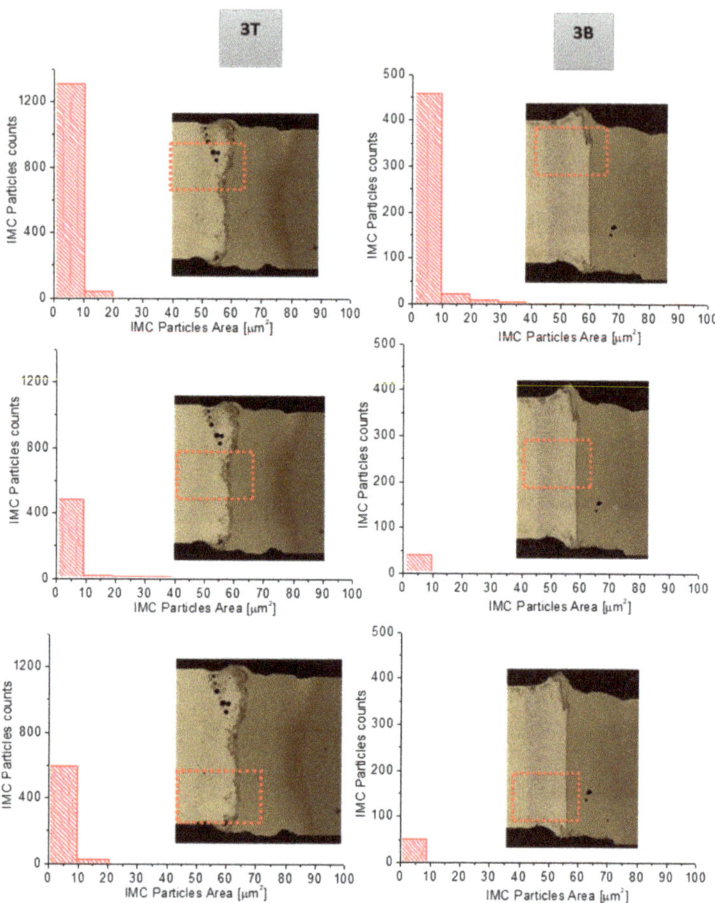

Figure 10. IMC particles distribution along the thickness of joints 3T and 3B.

3.5. Microstructural Evolution and Mechanical Properties due to SRHT

For the analyzed joints, a feasible PWHT should reduce the residual stress without hardening the Ti side or softening the Al side in order to not further limit the ductility and toughness of the welds. A significative control of the Ti side microstructure requires heat treatments at temperatures higher than the Beta Transus Temperature of Ti-6al-4V (900 °C). Those values of temperature should lead to a completely BCC beta microstructure that allows one to obtain the appropriate microstructure and hardness values upon cooling. Such a temperature is higher than the melting point of AA6061 Al Alloy [27,28,30]. Moreover, the highest heat treatment temperature and holding time suitable for Al alloy must also avoid grain size coarsening. Therefore, the different properties of the alloys limit the effectiveness of the stress relief heat treatment. Considering that the liquidus T of the AA6061 alloy is equal to 582 °C [32], the temperature of the SRHT was chosen to be 530 °C and the holding time equal to 2 h, followed by air cooling. The treatment was applied to joint 3, characterized by the best mechanical performance (both strength and ductility). Hardness measurements were conducted on the cross sections of the heat-treated joints, at a half of thickness. The results indicate that the HAZ and FZ of Ti6Al4V harden significantly after SRHT, while in the AA6061 side, the increasing of hardness is not significant (five point more or less). In the Ti side, the highest increases of hardness, obtained close to the weld interface, were equal to 60 HV in 3B joint and 110 HV in 3T Joint (Figure 12).

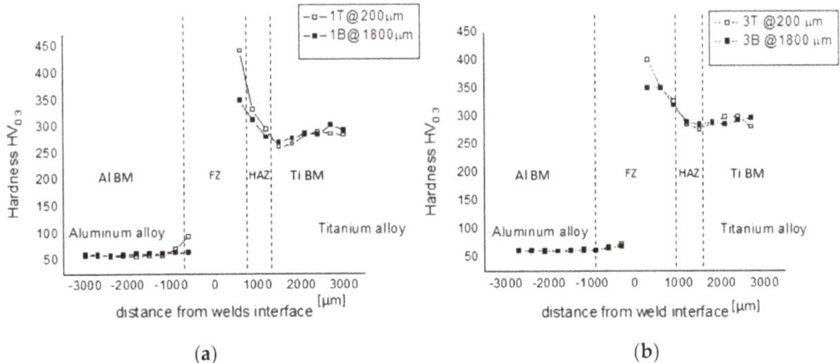

(a)

(b)

Figure 11. Micro hardness profile as a function of distance from the weld interface at 200 and 1800 μm from upper surface for 1T and 1B weld (**a**) and 3T and 3B welds (**b**).

Table 3. Average results coming from tensile test of two specimens for joints 1T, 1B, 3T, 3B.

Joints	Yield Stress [MPa]	Ultimate Tensile Stress [Mpa]	Strain [%]
1T	95 ± 3	110.2 ± 6	1.4 ± 0.3
1B	88 ± 4	100.0 ± 3	1.2 ± 0.2
3T	118 ± 5	158.1 ± 8	1.8 ± 0.4
3B	98 ± 4	172.7 ± 8	4.5 ± 0.9

The increase in the hardness of the Ti side is due to the aging of the martensite, which leads to the precipitation of hardening $\alpha + \beta$ particles [5,27,30,33]. In addition, some amount of the retained β phase could possibly exist in the FZ and HAZ welds due to rapid cooling. In this case, the effect of the heat treatment should lead to the microprecipitation of the α-phase platelet, it being the α phase most stable at low temperature. Therefore, from the transformation of the retained β, a further hardening effect is obtained, with the β-phase being comparatively soft and ductile relative to the α phase [5,27,30,33].

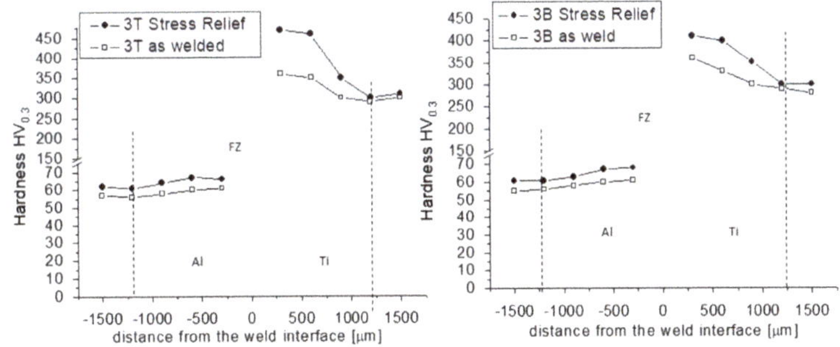

Figure 12. Micro hardness profile as a function of distance from the weld interface at a half of thickness for joints before and after SRHT.

Moreover, it was also observed that the IMC thickness changes with the SRHT and both IMCL and particles in the Al matrix close to the interface grow up upon the SRHT. The average growth of IMCL was equal to 2.60 ± 0.6 μm and 1.30 ± 0.3 μm, respectively, for the 3T and 3B joints. The growth was evaluated along the thickness and it is also easily observable due to its lighter color in the optical micrographs, as shown in Figure 13. SEM micrographs at high magnification (Figure 14) allow one to better observe the growth of the IMC particles due to SRHT. In the SEM micrographs, the new compounds appear darker. According to the EDS analysis on the new nucleated particles (particle 1 in Figure 14f, Table 4) and the previous study of P. Leo et al. [13], the composition of the new nucleated and grown compounds is Al_3Ti, being a thermodynamically favored compound at the highest percentage of aluminum [5,13,27,29].

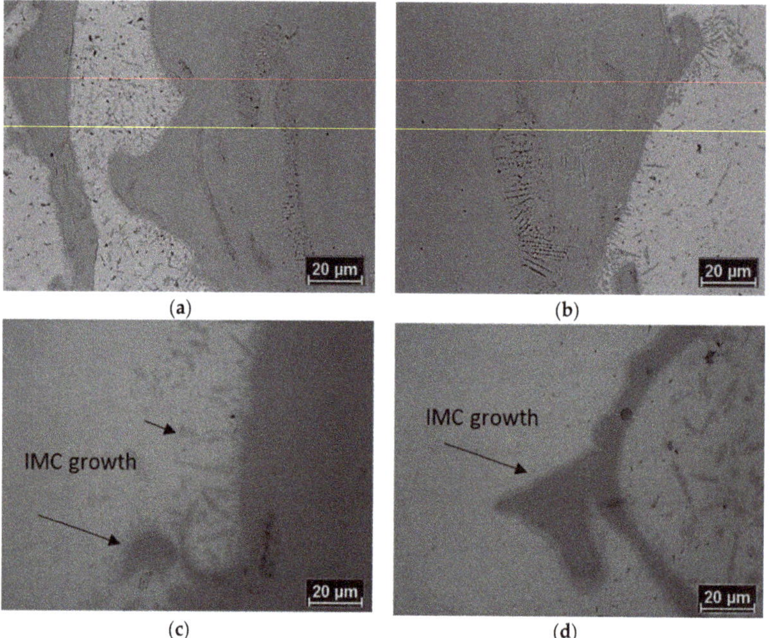

Figure 13. Optical micrographs before (**a,b**) and after (**c,d**) SRHT for welds 3 T (**a,c**) and 3B (**b,d**).

Due to the microstructure evolution induced by SRHT, both the ductility and strength of the welds were reduced, as shown in Figure 15.

It is noteworthy that a previous study [33] showed that aging heat treatments of Al alloy in dissimilar Al–Ti welds could increase the tensile strength, but only with a strongly reduction of the ductility. In that work, the aging treatment at 180 °C required, as a first step, a solution treatment at a temperature higher than 500 °C. The low temperature heat treatment increased the strength of the Al side by the precipitation of a small and uniformed dispersion needle shaped Mg_5Si_6 particles named β'' [14,27,29,32], but the strong decrease in weld ductility could be mainly due to the first high temperature step. That first step is responsible for the increases in IMCL thickness/IMC particle size (Figures 13 and 14) and for the strong increase in the hardness of the Ti side, as shown in Figure 12; as a consequence, the gap in the mechanical property at the weld interfaces results increased leading to reduced both fracture strength and ductility [33].

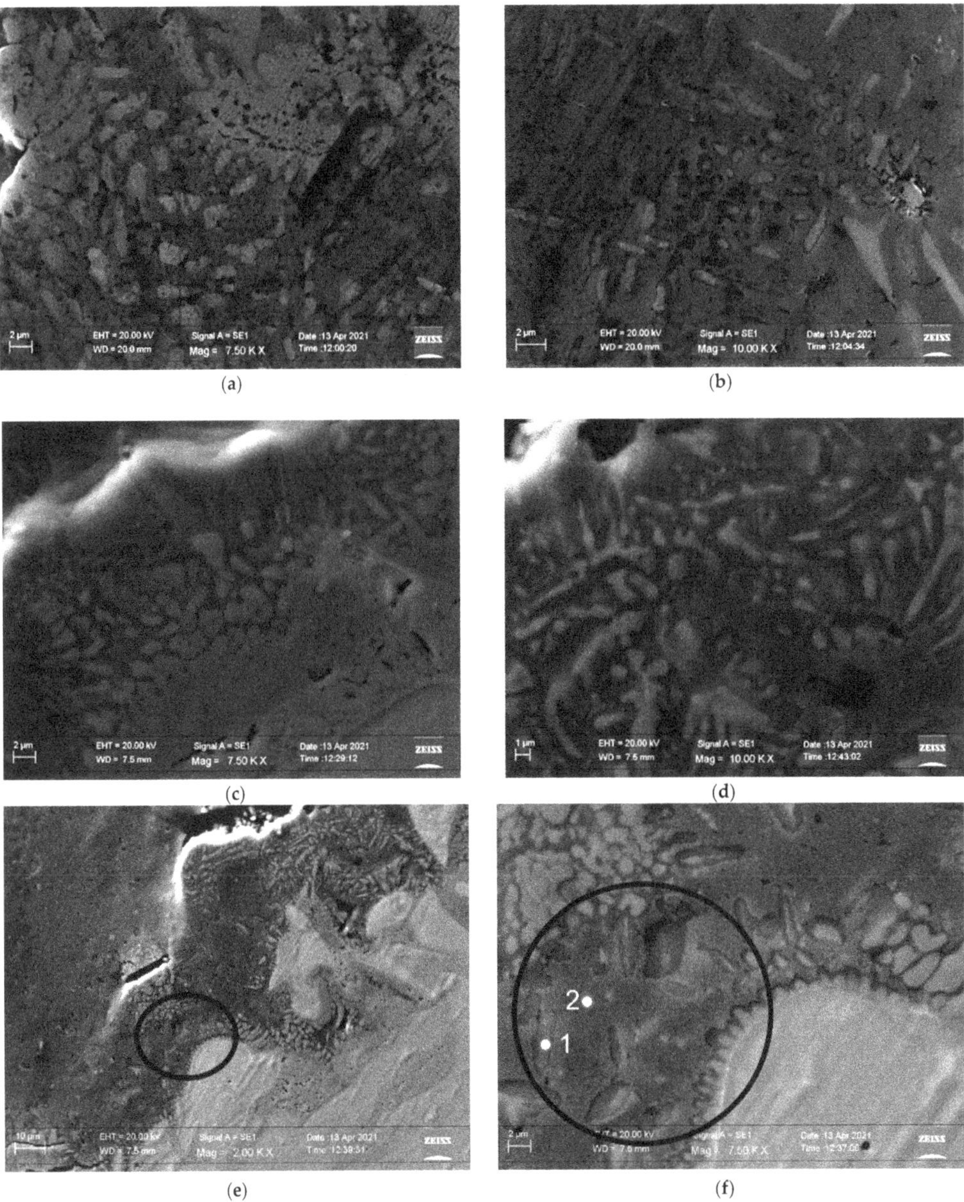

Figure 14. Scanning electron micrographs before (**a**,**b**) and after (**c**,**d**) SRHT for 3T weld at 7500× (**a**,**c**) and 10,000× (**b**,**d**). 3T sample microstructure after SRHT at 2000× and 7500× with circled the zone of EDS investigation (**e**,**f**), together with the points of analysis (**f**).

Table 4. EDS result on particle (point 1) and aluminum matrix (point 2) shown in Figure 14f.

Elements [at%]	Point 1 (Particle)	Point 2 (Matrix)
Al	72.2 ± 2.6	85.4 ± 3.0
Ti	26.9 ± 1.0	10.7 ± 0.8

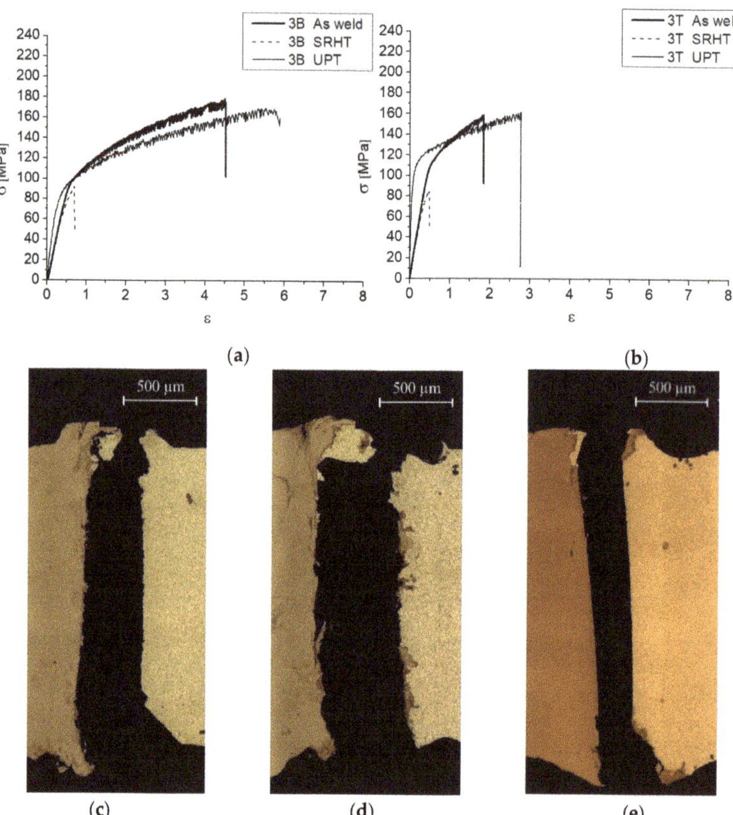

Figure 15. Tensile curves for joints 3B (**a**) and 3T (**b**) as welded, after SRHT and after UPT. Fractured cross section of B sample after tensile test in as welded state (**c**), after UPT (**d**) and after SRHT (**e**).

The main difficulty in providing good service performance of the dissimilar welds involves the low toughness and ductility. Therefore, the PWHT of the Al–Ti welds at temperatures higher than 500 °C is not the correct way to adjust tensile properties. Therefore, a possible application of UTP treatments is proposed in this study.

3.6. Role of UPT on Tensile Properties and Corrosion Resistance

The tensile curve of the 3B and 3T joints after UPT (Figure 15) are characterized by higher elastic modulus with respect the as welded ones. These results could be due to the reduction of porosity induced by UPT, as shown by Chawla et al. [34] and also verified by Tian et al. [18] for a AA6061 Al Alloy. The increases in fracture strain observed after UPT could be justified by the reduction of porosity together with the reduced strain hardening of the UPT samples (compare the tensile curves in Figure 15) [18,34]. In fact, even if the ultimate tensile strength of UPT is treated and welded samples do not differ significantly, the work hardening behavior is different (before and after UPT) because it is affected by the presence and the amount of residual stress [34,35]. To describe the mechanical behavior of the joints after UPT, residual stress analysis and work hardening characterization should be developed. As shown by Zhoua et al. [36], the reconstruction of the residual stress field associated with the work hardening is of great importance for an accurate assessment of the mechanical performances. Therefore, further detailed studies are necessary to define the relationship between mechanical properties and UPT treatment for Al/Ti dissimilar joints and possibly introduce further improvements. The data presented in this study are only intended to provide a hitherto unexplored chance to overpass the main limit of the dissimilar joints that is the low ductility.

In all the tensile samples, fracture cracks initiate in the Al FZ, at the interface with IMCL or IMC particles, and propagate mainly along the IMCL, as shown in Figure 15c–e by the micrographs of the tensile fracture cross sections of the 3B sample in a welded state, after UPT and after SRHT.

Figure 16a shows the variation of the OCP aluminum alloy before and after UPT, joint before and after UPT and Ti-6Al-4V specimens, in the near-neutral aqueous 3.5% NaCl solution, monitored for 60 min. As expected, the OCP value for the investigated AA6061 alloy is about -0.75 V vs. Ag/AgCl [37]. It is also evident that the OCP value of the treated sample becomes less negative during immersion until reaching a constant potential, probably due to the coarsening of the oxide film. A clear difference can be observed in the behavior of the two samples with an ennobling of the material obtained after the treatment. The significantly higher OCP values of titanium indicate its more noble behavior. The OCP values of the untreated and treated joints are similar to those of the untreated and treated Al6061 samples, respectively, which represent the least noble element of the joints. The potentiodynamic polarization behavior of the investigated samples is reported in Figure 16b. The cathodic branch is mainly controlled by oxygen diffusion, indicating that the corrosion mechanism of the samples is analogous. The anodic branch shows a continuous increase in current density, indicating susceptibility to pitting corrosion until pseudo-passivation region is reached, where the current density increases much more slowly as the potential increases. From the polarization curves, the values of corrosion potential (E_{corr}) and corrosion current density (i_{corr}) can be calculated using the Tafel extrapolation method and are summarized in Table 5. Polarization shows much lower E_{corr} and higher i_{corr} for the aluminum specimen, both alone and in the joint, indicating an enhanced corrosion resistance in the aggressive solution containing chloride ions after UPT treatment. The slightly more negative values of E_{corr} and slightly higher of the i_{corr}, evaluated in the joints compared to the aluminum specimens, may be due to a galvanic coupling effect with Ti-6Al-4V.

In the pictures of Figure 17, the corroded windows for the Al samples and 3B welds before and after UPT are shown. In the optical micrographs, the edge between the corrosion window and the Al alloy (shown in the circle of each picture) is highlighted. The dark zone in the micrographs is the corroded zone, while the lighter one is the Al alloy. The effect of the UPT is evident on the treated Al surfaces (Figure 17c,d) compared to the untreated ones (Figure 17a,b).

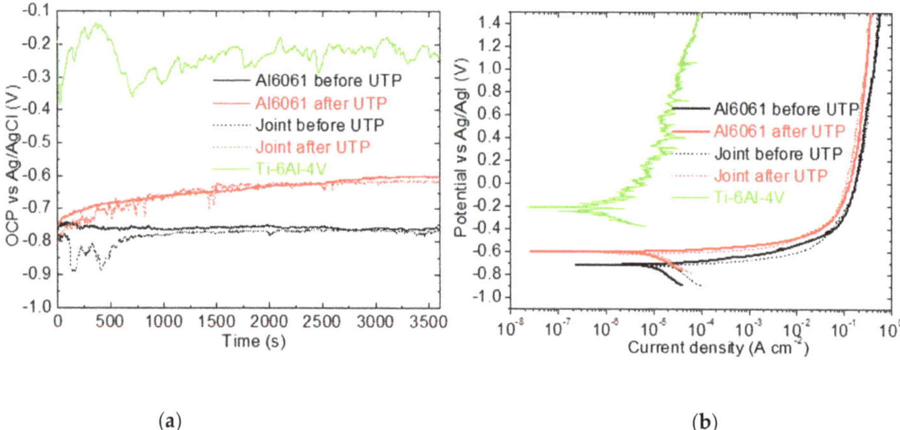

(a)   **(b)**

Figure 16. (**a**) Open circuit potential and (**b**) potentiodynamic polarization curves of AA6061 before and after UPT, 3B joint before and after UPT, Ti-6Al-4V samples tested in 3.5% NaCl solution.

Table 5. Corrosion potential and corrosion current density extracted from potentiodynamic polarization curves reported in Figure 16b.

Material	E_{corr}	i_{corr}
	V vs. Ag/AgCl	$\mu A \chi \mu^{-2}$
AA6061 before UPT	−0.719	11.05
AA6061 after UPT	−0.603	6.22
Ti-6Al-4V	−0.209	0.26
3B weld before UPT	−0.724	16.12
3B weld after UPT	−0.617	9.28

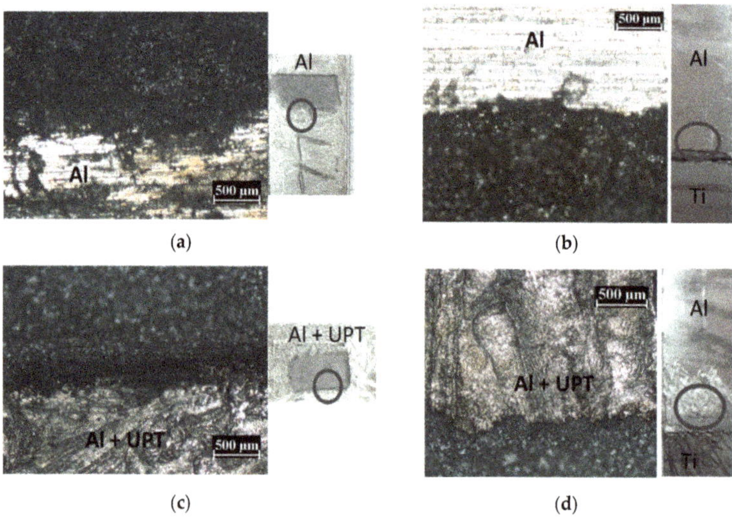

(a)   **(b)**

(c)   **(d)**

Figure 17. Pictures of corroded windows of the samples and optical micrographs of one edge of the corrosion window (circled in the pictures) for Al sample before UPT (**a**), Al sample after UPT (**c**), 3B weld before UPT (**b**) and 3B weld after UPT (**d**).

4. Conclusions

In this study, the effect of defocusing and laser offset on the microstructures of dissimilar AA6061/Ti-6Al-4V joints, intermetallic layer size and intermetallic particles size and distribution were analyzed. Moreover, the tensile performances with and without the stress relief heat treatment at 530 °C for 2 h and ultrasonic peening treatment were studied and compared. The aim of this research is to provide an opportunity for researchers and the welding industry to overcome the low ductility of Al–Ti welds using mechanical treatment, since, to date, the research based on post-weld heat treatments on dissimilar joints has not been effective.

The main conclusions are the following:

1. All the analyzed joints exhibit martensitic microstructure in the Ti fused zone and partially martensitic microstructure in Ti heat-affected zone. The Al fused zone was columnar dendritic. The size of those zones decreased with decreasing laser offset. The negative defocusing strengthens the effect of the laser offset by decreasing the size of the fused zones as the laser offset decreases and by increasing the size of the fused zones as the laser offset increases. Therefore, the highest size was observed for the joint 3B. Specifically, the fusion zone size for the 3B weld was close to 1.4 mm^2 for both Al and Ti size.

2. The thickness and shape of the intermetallic layer, as well as the number of the intermetallic particles, varied with both defocusing and the laser offset. The layer was flatter and less thick in the case of negative defocusing and laser offset increased due to the lower peak temperature at the interface and reduced mixing. Therefore, the lowest size, equal to 12 µm, was found in the 3B joint. For the same reason, the amount of particles, highest at the top of all the joints, also decreased with negative defocusing and increased with laser offset.

3. The hardness measurements executed in the focus (1T, 3T) and defocusing point (1B,3B) indicated that negative defocusing reduces the hardness, imposing a lower cooling rate (due to lower energy density) and consequent microstructure coarsening. With increasing laser offset, the effect of Ti heat conductivity mitigated the role of beam energy density. Consequently, the gap in hardness between joint 3T and 3B (equal to 50 HV) was less significant as compared to that of joint 1T and 1B (equal to 100 HV).

4. Joint 3B exhibited the best tensile performances, with an ultimate tensile stress close to 173 MPa and strain to fracture equal to 4.5%. This behavior is justified by the lowest amount of intermetallic compound and the lowest gap of hardness values at the Al–Ti interface, as shown in the previous results.

5. The stress relief heat treatment at 530 °C for 2 h significantly hardens the Ti side and induces the growth of the intermetallic compounds. The average growth of IMCL was equal to 2.60 ± 0.6 µm and 1.30 ± 0.3 µm, respectively, for the 3T and 3B joints. The microstructure evolution induced by the heat treatment leads to both ductility and strength reduction.

6. The UPT-treated welds are characterized by higher elastic modulus and fracture strain with respect to the welded ones, probably due to the reduction of porosity. The fracture strain of the 3B joint was equal to 4.5% in the weld state and 5.8% after UPT. The ultimate tensile strength of UPT-treated and welded samples do not differ significantly, but work hardening behavior is different. Further studies are necessary to define the relationship between mechanical properties and UPT treatment for Al/Ti dissimilar joints and, possibly, introduce further improvements.

7. The results of the electrochemical measurements indicate that the corrosion susceptibility of the welds is enhanced by means of UPT treatment.

Author Contributions: Conceptualization, P.L.; methodology, P.L.; validation, P.L.; formal analysis, P.L.; investigation, P.L., S.D., R.N., C.M., A.T.; resources, G.C.; data curation, P.L.; writing—original draft preparation, P.L.; writing—review and editing, P.L., R.N., C.M., S.D.; visualization, P.L.; supervision, P.L. All authors have read and agreed to the published version of the manuscript.

Funding: This research received no external funding.

Institutional Review Board Statement: Not applicable.

Informed Consent Statement: Not applicable.

Data Availability Statement: The data presented in this study are available on request from the corresponding author.

Conflicts of Interest: The authors declare no conflict of interest.

References

1. Jiang, P.; Chen, R. Research on interfacial layer of laser-welded aluminum to titanium. *Mater. Charact.* **2019**, *154*, 264–268. [CrossRef]
2. Martinsen, K.; Hu, S.J.; Carlsond, B.E. Joining of dissimilar materials. *CIRP Annals* **2015**, *64*, 679–699. [CrossRef]
3. Vaidya, W.V.; Horstmann, M.; Ventzke, V.; Petrovski, B.; Koçak, M.; Kocik, R. Improving interfacial properties of a laser beam welded dissimilar joint of aluminum AA6056 and titanium Ti6Al4V for aeronautical applications. *J. Mater. Sci.* **2010**, *45*, 6242–6254. [CrossRef]
4. Quazia, M.M.; Ishaka, M.; Fazalb, M.A.; Arslanc, A.; Rubaieeb, S.; Qabane, A.; Aimana, M.H.; Sultanf, T.; Alig, M.M.; Manladanh, S.M. Current research and development status of dissimilar materials laser welding of titanium and its alloys. *Opt. Laser Technol.* **2020**, *126*, 106090. [CrossRef]
5. Malikova, A.; Vitoshkina, I.; Orishicha, A.; Filippova, A.; Karpova, E. Effect of the aluminum alloy composition (Al-Cu-Li or Al-Mg-Li) on structure and mechanical properties of dissimilar laser welds with the Ti-Al-V alloy. *Opt. Laser Technol.* **2020**, *126*, 106135. [CrossRef]
6. Malikova, A.; Vitoshkina, I.; Orishicha, A.; Filippova, A.; Karpova, E. Microstructure and mechanical properties of laser welded joints of Al-Cu-Li and Ti-Al-V alloys. *J. Manuf. Process.* **2020**, *53*, 201–212. [CrossRef]
7. Xue, X.; Wu, X.; Liao, J. Hot-cracking susceptibility and shear fracture behavior of dissimilar Ti6Al4V/AA6060 alloys in pulsed Nd:YAG laser welding. *Chin. J. Aeronaut.* **2020**, *34*, 375–386. [CrossRef]
8. Wang, Z.; Shen, J.; Hu, S.; Wang, T.; Bu, X. Investigation of welding crack in laser welding-brazing welded TC4/6061nd TC4/2024 dissimilar butt joints. *J. Manuf. Process.* **2020**, *60*, 54–60. [CrossRef]
9. Messler, R.W., Jr. *Principles of Welding: Processes, Physics, Chemistry, and Metallurgy*; Wiley: Hoboken, NJ, USA, 1999.
10. Kou, S. *Welding Metallurgy*, 2nd ed.; Wiley: Hoboken, NJ, USA, 2003.
11. Metelkova, J.; Kinds, Y.; Kempen, K.; de Formanoir, C.; Witvrouw, A.; Hooreweder, B.V. On the influence of laser defocusing in Selective Laser Melting of 316L. *Addit. Manuf.* **2018**, *23*, 161–169. [CrossRef]
12. Ferro, P.; Berto, F.; Bonollo, F.; Romanina, L.; Salemic, G. Post welding heat treatment improving mechanical properties on Ti-6Al-4V. *Procedia Struct. Integr.* **2020**, *26*, 11–19. [CrossRef]
13. Leo, P.; D'Ostuni, S.; Casalino, G. Low temperature heat treatments of AA5754-Ti6Al4V dissimilar laser welds: Microstructure evolution and mechanical properties. *Opt. Laser Technol.* **2018**, *100*, 109–118. [CrossRef]
14. Li, P.; Lei, Z.; Zhang, X.; Chen, Y. Effects of a post-weld heat treatment on the microstructure and mechanical properties of dual-spot laser welded-brazed Ti/Al butt joints. *J. Manuf. Process.* **2021**, *61*, 492–506. [CrossRef]
15. Lazurenko, D.V.; Bataev, I.A.; Mali, V.I.; Bataev, A.A.; Maliutina, I.N.; Lozhkin, V.S.; Esikov, A.M.; Jorge, J. Explosively welded multilayer Ti-Al composites: Structure and transformation during heat treatment. *Mat. Des.* **2016**, *102*, 122–130. [CrossRef]
16. Xu, L.; Cui, Y.Y.; Hao, Y.L.; Yang, R. Growth of intermetallic layer in multi-laminated Ti/Al diffusion couples. *Mat. Sci. Eng. A* **2006**, *435*, 638–647. [CrossRef]
17. Malaki, M.; Ding, H. A review of ultrasonic peening treatment. *Mat. Des.* **2015**, *87*, 1072–1086. [CrossRef]
18. Tian, Y.; Shen, J.; Hu, S.; Liang, Y.; Bai, P. Effects of ultrasonic peening treatment on surface quality of CMT-welds of Al alloys. *J. Mat. Proc. Tech.* **2018**, *254*, 193–200. [CrossRef]
19. Hacini, L.; Van Lê, N.; Bocher, P. Evaluation of Residual Stresses Induced by Robotized Hammer Peening by the Contour. *Method. Exp. Mech.* **2009**, *49*, 775–783. [CrossRef]
20. Hacini, L.; Van Lê, N.; Bocher, P. Effect of impact energy on residual stresses induced by hammer peening of 304 L plates. *J. Mater. Process. Technol.* **2008**, *208*, 542–548. [CrossRef]
21. Mordyuk, B.N.; Karasevskaya, O.P.; Prokopenko, G.I. Structurally induced enhancement in corrosion resistance of Zr–2.5%Nb alloy in saline solution by applying ultrasonic impact peening. *Mater. Sci. Eng.* **2013**, *559*, 453–461. [CrossRef]
22. Yanga, B.; Tana, C.; Zhaob, Y.; Wua, L.; Chena, B.; Songa, X.; Zhaoa, H.; Fenga, J. Influence of ultrasonic peening on microstructure and surface performance of laser-arc hybrid welded 5A06 aluminum alloy joint. *J. Mater. Res. Technol.* **2020**, *9*, 9576–9587. [CrossRef]
23. Statnikov, E.S.; Muktepavel, V.O.; Blomqvist, A. Comparison of Ultrasonic Impact Treatment (UIT) and Other Fatigue Life Improvement Methods. *Weld. World.* **2002**, *46*, 20–32. [CrossRef]
24. Ooi, S.W.; Garnham, J.E.; Ramjaun, T.I. Review: Low transformation temperature weld filler for tensile residual stress reduction. *Mater. Des.* **2014**, *56*, 773–781. [CrossRef]

25. Lixing, H.; Dongpo, W.; Wenxian, W.; Yufeng, Z. Ultrasonic Peening and Low Transformation Temperature Electrodes used for Improving the Fatigue Strengthof Welded Joints. *Weld. World.* **2004**, *48*, 34–39. [CrossRef]
26. Mordyuk, B.N.; Prokopenko, G.I.; Milman, Y.V.; Iefimov, M.O.; Sameljuk, A.V. Enhanced fatigue durability of Al–6Mg alloy by applying ultrasonic impact peening: Effects of surface hardening and reinforcement with AlCuFe quasicrystalline particles. *Mater. Sci. Eng. A* **2013**, *563*, 138–146. [CrossRef]
27. Polmear, I.J. *Light Alloys: Metallurgy of the Light Metals*; Butterworth Heinemann: Oxford, UK, 1995.
28. Mondolfo, L.F. *Aluminium Alloys: Structure and Properties*; Butterworth: Oxford, UK, 1976.
29. Porter, D.A.; Easterling, K.E. *Phase Transformations in Metals and Alloys*; Chapman & Hall: London, UK, 1992.
30. Smith, W.F. *Structure and Properties of Engineering Alloys*; McGraw-Hill: New York, NY, USA, 1992.
31. D'Ostuni, S.; Leo, P.; Casalino, G. FEM Simulation of Dissimilar Aluminum Titanium Fiber Laser Welding Using 2D and 3D Gaussian Heat Sources. *Metals* **2017**, *7*, 307. [CrossRef]
32. Maisonnette, D.; Suery, M.; Nelias, D.; Chaudet, P.; Epicier, T. Effects of heat treatments on the microstructure and mechanical properties of a AA6061 aluminium alloy. *Mat. Sci. Eng.* **2011**, *528.6*, 2718–2724. [CrossRef]
33. Kabir, A.S.H.; Cao, X.; Gholipour, J.; Wanjara, P.; Cuddy, J. Effect of Postweld Heat Treatment on Microstructure, Hardness, and Tensile Properties of Laser-Welded Ti-6Al-4V. *Met. Mat. Trans. A* **2012**, *43*, 4171–4184. [CrossRef]
34. Chawla, N.; Deng, X. Microstructure and mechanical behavior of porous sintered steels. *Mater. Sci. Eng. A* **2005**, *390*, 98–112. [CrossRef]
35. Zhoua, J.; Suna, Z.; Kanoutea, P.; Retrainta, D. Reconstruction of residual stress and work hardening and their effects on the mechanical behaviour of a shot peened structure. *Mech. Mater.* **2018**, *127*, 100–111. [CrossRef]
36. Muna, K.; Abbass, K.S.H.; Abbas, S.A. Study of Corrosion Resistance of Aluminum Alloy AA6061/SiC Composites in 3.5% NaCl Solution. *Int. J. Mater. Mech. Manuf.* **2015**, *3*, 31–35.
37. Fahimpour, V.; Sadrnezhaad, S.K.; Karimzadeh, F. Corrosion behavior of aluminum AA6061 alloy joined by friction stir welding and gas tungsten arc welding methods. *Mater. Des.* **2012**, *39*, 329–333. [CrossRef]

Article

Laser Micro Polishing of Tool Steel 1.2379 (AISI D2): Influence of Intensity Distribution, Laser Beam Size, and Fluence on Surface Roughness and Area Rate

André Temmler [1,*], Magdalena Cortina [2], Ingo Ross [2], Moritz E. Küpper [3] and Silja-Katharina Rittinghaus [2]

[1] Fraunhofer Institute for Applied Optics and Precision Engineering (IOF), Albert-Einstein-Straße 7, 07745 Jena, Germany
[2] Fraunhofer Institute for Lasertechnology (ILT), Steinbachstraße 15, 52074 Aachen, Germany; magdalenacortina@gmail.com (M.C.); ingo.ross@rwth-aachen.de (I.R.); siljakatharina.rittinghaus@ilt.fraunhofer.de (S.-K.R.)
[3] Chair for Lasertechnology, RWTH Aachen University, Steinbachstraße 15, 52074 Aachen, Germany; moritz.kuepper@rwth-aachen.de
* Correspondence: andre.temmler@iof.fraunhofer.de or andre.temmler@rwth-aachen.de

Citation: Temmler, A.; Cortina, M.; Ross, I.; Küpper, M.E.; Rittinghaus, S.-K. Laser Micro Polishing of Tool Steel 1.2379 (AISI D2): Influence of Intensity Distribution, Laser Beam Size, and Fluence on Surface Roughness and Area Rate. *Metals* **2021**, *11*, 1445. https://doi.org/10.3390/met11091445

Academic Editors: Sergey N. Grigoriev, Aleksander Lisiecki, Marina A. Volosova and Anna A. Okunkova

Received: 5 July 2021
Accepted: 7 September 2021
Published: 13 September 2021

Publisher's Note: MDPI stays neutral with regard to jurisdictional claims in published maps and institutional affiliations.

Abstract: Within the scope of this study, basic research was carried out on laser micro polishing of the tool steel 1.2379 (AISI D2) using a square, top-hat shaped intensity distribution. The influence of three different quadratic laser beam sizes (100 µm, 200 µm, 400 µm side length) and fluences up to 12 J/cm^2 on the resulting surface topography and roughness were investigated. Surface topography was analyzed by microscopy, white light interferometry, spectral roughness analysis, and 1D fast Fourier transformation. Scanning electron microscopy and electrical discharge analyses indicate that chromium carbides are the source of undesired surface features such as craters and dimples, which were generated inherently to the remelting process. Particularly for high laser fluences, a noticeable stripe structure was observed, which is typically a characteristic of a continuous remelting process. Although the micro-roughness was significantly reduced, often, the macro-roughness was increased. The results show that smaller laser polishing fluences are required for larger laser beam dimensions. Additionally, the same or even a lower surface roughness and less undesired surface features were created for larger laser beam dimensions. This shows a potential path for industrial applications of laser micro polishing, where area rates of up to several m^2/min might be achievable with commercially available laser beam sources.

Keywords: laser melting; surface roughness; laser polishing; quadratic laser spot; tool steel 1.2379; area rate

1. Introduction

Manual polishing of metallic surfaces is still a widespread and common manufacturing process, which primarily aims to significantly reduce the surface roughness. However, the polishing result often depends heavily on the condition and skill of the person who conducts the manual polishing. Additionally, the manual polishing process gets increasingly cost intensive and time consuming when the surface to be polished is not a plane but an arbitrarily formed surface with small surface features and small curvature radii. Laser polishing is a new manufacturing process with the potential to at least partially replace manual polishing for small and complex shaped surfaces [1]. This is underscored e.g., by the proven potential for full automation, by the flexible use of wear-free "polishing tools", and by the high spatial resolution of the laser polishing process [2]. Therefore, it is understandable that an increasing number of researchers and studies aim to strengthen the fundamental understanding of the laser polishing process.

The special research focus of this study lies on the surface roughness evolution after laser micro polishing (LµP) of cold working steel 1.2379 (DIN X153CrMoV12, AISI D2)

using square intensity distributions of different sizes. D2 is typically used for deep drawing tools since it contains high amounts of carbon and chromium. Adapted heat treatment cycles usually lead to the precipitation of $M_{23}C_6$ and M_7C_3 carbides within the martensitic steel matrix [3]. In conventional forming operations, this results in a high resistance against the adhesive and abrasive wear of deep drawing tools made from this material (Sing et al. [3]). Within a DFG priority program of the German Research Association SPP 1676 ("Dry metal forming"), the cold working steel was selected as one 'standard material'. Therefore, laser processing and its effects on surface topography, mechanical properties, and wear received broad attention and were investigated, e.g., for rotary swaging [4,5], cold massive forming [6], cold extrusion [7,8], or deep drawing [9–12].

Since laser polishing is a highly localized process including extraordinary temperature gradients of up to 10^9 K s^{-1}, an intense interaction of radiation and material is typical for this energy beam-based process [13]. An introduction to the topic of energy beams for surface modifications including laser polishing was given by Deng et al. [14]. The review article of Krishnan and Fang [15] focuses more on the specifics of the laser-polishing process and presents a decent collection of relevant studies and works on laser polishing. Although already a few years old, Bordatchev et al. [16] also delivers a systematic evaluation of achieved laser polishing results for a wide range of metals. Bhaduri et al. [17] lists some key publications on laser polishing (using cw and pulsed laser radiation) for various materials with high relevance in industrial applications. Concerning specifically LµP, the studies of Temmler et al. [18,19] include a more detailed introduction to the specifics and characteristics of LµP for metals. A more general introduction on the interaction of pulsed laser radiation and material was recently compiled by Li and Guan [20] and gives a good overview on theoretical fundamentals, which are also partially relevant for pulsed laser remelting. However, LµP typically aims not to work in the ablation regime, so that a reduction of surface roughness is achieved by the redistribution of molten material [21]. Thus, the melt duration is decisive for the effective reduction of spatial frequencies or wavelengths in LµP [22]. Kuisat et al. [8] found even for the direct laser interference patterning (DLIP on Ti64 and Scalmalloy©) process using ns laser pulses that this smoothing effect is inherent to the remelting process and occurs simultaneously to the DLIP process. This is similar to the WaveShape process, which achieves structuring and polishing in one process step [23] or even utilizes a spatially adapted laser power modulation for the reduction of waviness on a surface (Oreshkin et al. [24]). Overall, the laser polishing process can be seen as a spatial low-pass filtering of a surface [25,26], resulting in an effective reduction of surface roughness. Therefore, the reduction of surface roughness in LµP is specifically pronounced for the micro-roughness, while the waviness of the surface stays typically unaffected [27]. Thus, special tools for surface roughness analysis are usually required such as spatial frequency [28] or spatial wavelength analysis [27] based on an adapted fast Fourier transformation. The longest spatial wavelength or smallest spatial frequency that is effectively reduced by LµP is referred to as critical wavelength [29] and critical frequency [30], respectively. Nonetheless, a fundamental and complete understanding of the specific roughness evolution for an arbitrary material is still not in sight.

In addition to surface roughness, specifically, the mechanical properties after LµP have been the focus of many studies. A detailed and extensive study on material properties after laser remelting (continuous wave (cw) and pulsed) of tool steel was presented by Temmler et al. [27]. Guan et al. [31] investigated the effect of pulse duration and heat transfer for laser pulses in the microsecond domain on Mg alloy AZ91D and found that discrete laser remelting occurs with a characteristic, homogeneous microstructure. In addition to surface structuring, Ma et al. [32] showed for additively manufactured Ti alloys that laser polishing is an effective method to reduce surface roughness, increase gloss, and enhance microhardness. Morrow and Pfefferkorn [13] found local hardness variations due to a heterogenic microstructure after pulsed laser remelting on tool steel S7. In the context of surface hardening in pulsed surface treatment, Maharjan et al. [33] compared the hardness of 50CrMo4 steel after laser treatment using fs, ps, ns, ms, and cw laser radiation and found that the most pronounced hardening effect was achieved for longer pulse duration and cw laser radiation. Furthermore, alone, the change in microstructure leads to characteristic surface features resulting from the track overlap of the scanning strategy. Both Morrow and Pfefferkorn [13] as well as Ma et al. [33] found a characteristic backtempering effect from overlapping laser pulses due to carbon diffusion [27]. Li et al. [34] found through a thermal history analysis of the laser-polishing process that martensitic phase formation in the remelted layer leads to significant effects on fatigue, residual stress, and strength. Temmler et al. [27] found particularly high residual stresses after LμP of tool steel H11 in an Argon environment, while Bhaduri et al. [35] found high tensile stresses after LμP of AlSi10Mg parts in an atmospheric environment. Liang et al. [36] found for laser polishing of additively manufactured Ti6Al4V an increase in hardness due to reduced porosity, an improved cycle fatigue life, and an improved cell biocompatibility.

Although different approaches for theoretical and FEM models of the LμP process exist [37] (e.g., by Mai et al. [38], Ukar et al. [39], Chow et al. [40], Vadali et al. [25], Ma et al. [41], or Richter et al. [42]), a precise prediction of surface roughness is only possible for a few selected materials and a narrow range of process parameters [25]. A reason for this is the complex interactions between laser beam and material based on their multiple properties, which makes the integration of all aspects into a complete and coherent model a very challenging task [37]. Particularly, process-inherent surface structure formation, resulting from specific material properties or from the selected process parameters and processing strategy as described by Nüsser et al. [43], is not considered in any model. However, these process-inherent surface features often remain on the surface and lead to a considerable surface roughness after laser polishing.

Additionally, the specific shape of the intensity distribution (ID) of a laser beam might be important for a laser-based process. The effects of the laser beam intensity distribution for laser welding were discussed by e.g., Kaplan [44]. If possible, the ID should be specifically tailored to the requirements of the laser process, the material, and its application, so that an adapted time-dependent temperature profile can be created (Völl et al. [45]). With regard to squared IDs in laser-based processing, Khare et al. [46] came to the conclusion that squared IDs enable higher process speeds and reduce the risk of centerline solidification cracking. Furthermore, in comparison to circular laser beams, the squared ID leads to changes in the microstructural growth, which is dominantly axial instead of columnar (Kou [47]). More specifically regarding laser polishing, Nüsser et al. [48] found that a squared ID might be beneficial for LμP of Ti6Al4V.

In this context, this study contributes to expand and strengthen the empirical data basis for laser micro polishing of high-carbon steels. This study provides an in-depth analysis of an LμP polished surfaces of 1.2379+ tool steel and demonstrates the transition from a discrete, pulsed to continuous remelting process for high fluences. Additionally, it provides a systematical analysis by directional 1D FFT analyses and shows different smoothing behaviors in the x and y directions. Furthermore, this study investigates a correlation between carbides and the formation of craters in pulsed laser processing. Finally, this study contains results and discussion regarding achievable area rates in LμP, and regarding a correlation of micro-hardness evolution and process characteristics in laser polishing (LμP or cw laser polishing). Overall, the material 1.2379+ poses multiple challenges for laser polishing, which were systematically identified, and potential mechanisms were discussed. Therefore, this work joins the steadily increasing list of experimental studies on laser polishing over the past decade.

2. Materials and Methods

2.1. Opto-Mechanical Set-Up

The "POLAR" laser polishing machine (Figure 1a) is a prototype and was set up by Fraunhofer ILT (Aachen, Germany) and Maschinenfabrik Arnold (Ravensburg, Germany) as part of a publicly funded project. The mechanical basis for this machine was a 5-axis milling center C600U from Hermle (Gosheim, Germany), in which an optical set-up for laser materials processing was integrated (Figure 1b).

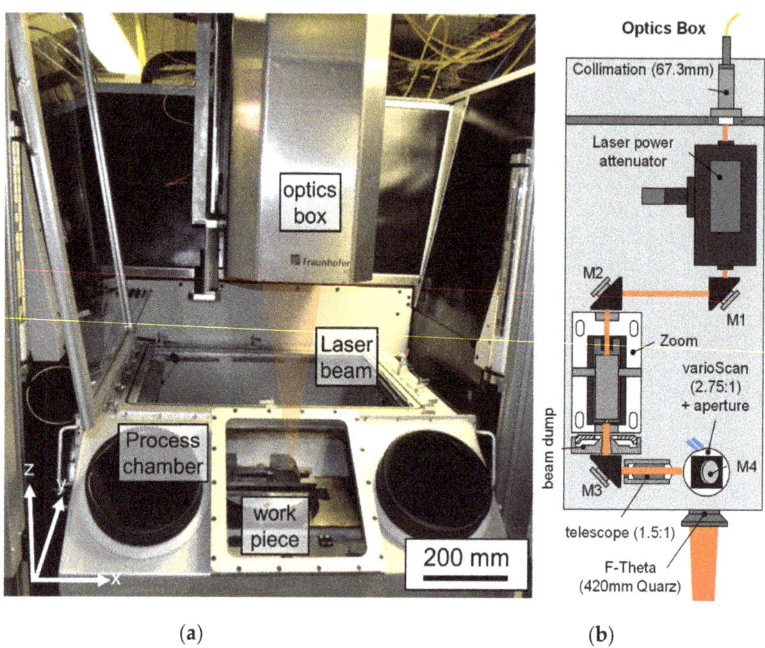

Figure 1. (**a**) Photo of processing chamber and optics box of the POLAR machine, and (**b**) schematic of the optical set-up and beam path.

A diode-pumped Yb:YAG disk laser (TruMicro 7051, Trumpf GmbH, Ditzingen, Germany) was used in the POLAR machine for the experimental investigation. The maximum average laser power of the TruMicro 7051 is approximately $P_{L,max,TM}$ = 550 W at λ_{em} = 1030 nm for pulse repetitions rates between f_P = 5 and 20 kHz. In the pulsed mode, Q-Switch controlled laser pulses were generated with pulse durations ranging from t_P = 1 µs to approximately 3 µs, which strongly depends on laser power. The square laser beam results from an optical transport fiber with a square, step-index fiber core and edge lengths of d_{fiber} = 100 µm, which was projected onto the work piece. Different laser beam dimensions were achieved by means of a zoom telescope, which offers a 20 mm beam aperture and enables continuous magnification in the range from approximately 0.2–1.8. A laser scanning system from Scanlab AG (Puchheim, Germany) consisting of a VarioScan 30, a HurryScan 25, and an f-theta objective (focal length 420 mm) was used for fast laser beam deflection. Additionally, a laser power attenuator was part of the optical set-up, so that the TruMicro7051 was operated at maximum laser power and the laser power attenuator was used to control the laser power on the workpiece. This was done to maximize the pulse stability of the laser beam source and achieve a fixed pulse duration of approximately t_P = 1.2 µs for all laser powers and laser beam sizes. Laser polishing took place in an approximately 200 L process chamber using a closed-loop control to adjust and stabilize the residual oxygen content.

2.2. Intensity Distribution and Laser Beam Characteristics

An optical delivery fiber with a square step-index fiber core (100 µm × 100 µm) was used for beam guidance, delivery, and shaping of the intensity distribution. Using a continuous zoom telescope, three different dimensions of the intensity distribution were realized and measured by a so-called "MicroSpotMonitor" (Primes GmbH, Pfungstadt,

Germany). The "LaserDiagnoseSoftware v2.98" (Primes GmbH, Pfungstadt, Germany) (LDS) was used to analyze beam caustic and particularly intensity distribution in the focal plane (Figure 2).

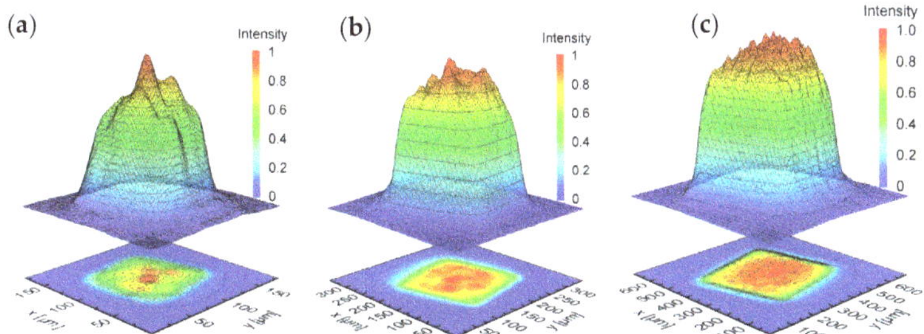

Figure 2. Normalized intensity distribution of square laser beams in their focal plane with side lengths of (**a**) 100 μm (Q100), (**b**) 200 μm (Q200), and (**c**) 400 μm (Q400).

Figure 2 shows intensity distributions for three laser beams with a square basis and side lengths of 100 μm, 200 μm, and 400 μm, respectively. In the following, these will be referred to as Q100, Q200, and Q400. Based on 86% energy inclusion, the LDS analysis delivers the following laser beam characteristics for Q100, Q200, and Q400 (Table 1). Furthermore, a square asymmetric super Gaussian distribution was fitted to the measured intensity distribution Equation (1) [49].

$$I(x,y) = I_0 \cdot e^{-2\left(\left(\frac{x}{a}\right)^M + \left(\frac{y}{a}\right)^M\right)^N} = I_0 \cdot e^{-2\left(\left(\left(\frac{x}{a}\right)^M + \left(\frac{y}{a}\right)^M\right)^{\frac{1}{M}}\right)^{NM}}. \tag{1}$$

Table 1. Tabular overview of laser beam characteristics.

Material/Element	M^2	z_R [mm]	Φ [mmrad]	DoF [mm]	M	$N \cdot M$
Q100	5.90	3.16	50.27	1.0	4.50	4.61
Q200	9.31	4.15	55.12	1.3	6.89	6.93
Q400	16.4	10.7	45.44	3.4	8.23	8.44

The main characteristics are the beam diameter a and the numerical parameters M and $N \cdot M$. A larger parameter M represents an intensity distribution closer to a square shape, while a larger product of $N \cdot M$ represents an intensity distribution closer to a top-hat shape.

2.3. Material and Sample Preparation

Particularly in the fields of solid forming, as well as cutting and punching tools, tool steel 1.2379 (DIN X153CrMoV12, AISI D2) is widely used, since it combines dimensional stability, high wear resistance, and toughness. Furthermore, 1.2379 (D2) is a recommended material for e.g., punches, ejectors, and tool dies [13]. Dörrenberg Edelstahl GmbH (Engelskirchen-Ründeroth, Germany) provided a special variant of 1.2379, which was powder metallurgically (PM) produced and used for all experiments. A special characteristic of the PM variant is a low content of impurities such as sulfur and phosphorus, a high homogeneity in the distribution of chemical elements, and a segregation-free microstructure. In terms of suitability for laser polishing, Ross et al. [8] found that powder metallurgically remelted 1.2379+ is preferable to the standard variant since chromium carbides are more homogenously distributed within the bulk material, on average smaller, and mostly spherically formed.

An overview of the chemical composition (wt %) of D2 is shown in Table 2.

Table 2. Tabular overview of chemical composition for AISI D2 (in wt %).

Material/Element [1]	C	Si	Cr	Mo	Mn	V	Fe
AISI D2	1.56	0.4	11.86	0.83	0.38	0.84	Bal.
Dev.	±0.1	±0.1	±0.45	±0.2	±0.1	±0.1	-

[1] Based on supplier information.

The dimensions of the samples were approximately $50 \times 75 \times 15$ mm^3, while the initial heat treatment state was soft annealed. The surface of the samples was ground and showed an initial roughness of $Ra = 0.33 \pm 0.02$ μm. Both sides of the flat samples were prepared in the same way and used for the experimental investigations.

2.4. Process Principle, Scan Strategy and Process Parameters

Laser remelting of a surface shows a high degree of similarity to a micro-welding process with the exception that no additional material is required in the remelting process. Laser remelting not only requires no additional materials, but it is also non-subtractive, which means that no material is lost during the process. In contrast to additive and subtractive processes, laser remelting redistributes the material at the material's surface while it is molten. Typically, capillary, thermo-capillary, and gravitational forces lead to a smoothing of the surface. Particularly, the surface tension leads to a smooth melt pool surface and effective damping of capillary surface waves. Laser micro polishing (LμP) is a specific variant of laser remelting in which pulsed laser radiation is used to remelt a surface. (Figure 3a). The combination of pulse frequency and scan speed determines the pulse distance on the work piece. LμP is typically characterized as a discrete remelting process, which means that the melt duration is shorter than the temporal distance between laser pulses (reciprocal pulse frequency). This usually results in melt pool dimensions that show a high ratio of width to depth [21,22]. Areal processing is typically achieved by a meandering scanning strategy with a defined track offset between antiparallel remelting tracks (Figure 3b).

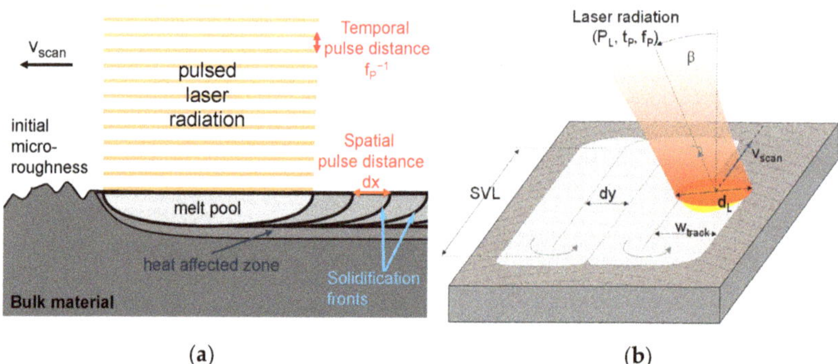

Figure 3. (a) Schematic of the process principle of laser micro polishing (LμP), and (b) schematic of a standard scanning pattern for areal laser processing.

Laser micro polishing utilizes pulsed laser radiation and was conducted using a laser beam source from Trumpf (TruMicro 7051, Ditzingen, Germany). Three similar investigations were conducted using different square laser beam sizes Q100, Q200, and Q400 and compared among each other. The experimental approach was based on an investigation on the influence of laser fluence on surface topography and surface roughness. An overview, a short description of the investigated process parameters, and the range of investigation is given in Table 3.

Table 3. Overview of relevant process parameters and range of investigation.

Process Parameter	Q100	Q200	Q400
Repetition rate f_{rep}	20 kHz	20 kHz	20 kHz
Side length of laser beam d_L [µm]	100	200	400
Scanning velocity v_{scan} [mm/s]	200	400	800
Spot offset dx [µm]	10	20	40
Track offset dy [µm]	10	20	40
Fluence F [J/cm^2]	4–12	4–12	4–12
Pulse energy E_P [mJ]	0.4–1.2	1.6–4.8	6.4–19.2
Laser power P_L [W]	8–24	32–96	128–384
Area rate [cm^2/min]	1.2	4.8	19.2
Inclination angle β [°]	0	0	0
Shielding gas	Ar + O_2	Ar + O_2	Ar + O_2
Residual oxygen $c(O_2)$ [ppm]	1000	1000	1000

The process parameters for the different laser beam dimensions were chosen based on the following considerations. The pulse frequency of 20 kHz was the maximum available at the laser beam source. Scanning velocity was adapted to the laser beam sizes and repetitions rate, so that the spot overlap (d_L/dx) in the scanning direction was 90% for all laser beams. Track offset was determined in the same way, so that track overlap was also 90%. This results in approximately 100 remelting cycles per irradiated area and strongly different area rates (polished area per unit time). Fluence is typically a decisive process parameter in LµP and was investigated in the same range for all laser beam dimensions. Therefore, the investigated range of pulse energy (and average laser power) was adapted to the laser beam dimensions. For all experiments, the inclination angle was $\beta = 0°$, and the shielding gas atmosphere was a mixture of Argon and 1000 ppm residual oxygen from the ambient air in the process chamber.

2.5. Surface Analysis

Surface analysis was done by micrographs of the surface using a stereo microscope M205 C by Leica Microsystems. An objective with a numerical aperture of 0.35 and a maximum optical resolution of up to 1050 LP/mm (line pairs per millimeter) was used.

Since stylus roughness measurements are often not meaningful to determine the effect of LµP on surface roughness [27], systematic white light interferometry (WLI) measurements of laser-polished areas were conducted. Surface analysis was done by an analysis of the spectral composition of surface roughness, which is typically presented as a roughness spectrum (Sa spectrum) [18,27,50,51]. The WLI measurements for these Sa spectra were conducted with a "Newview 7300" (Ametek-Zygo, Berwyn, PA, USA). The maximum vertical resolution was approximately 0.1 nm, while the spatial resolution ranged between 0.36 and 9.50 µm. The control and analysis software Zygo MetroPro (V10.3) was used for all measurements.

3. Results

3.1. Initial Surface and Material Analysis

The initial surface was measured by WLI, and characteristic sections were analyzed by fast Fourier transformation (FFT). The surfaces exhibit two kinds of marks, which are a result of the mechanical preparation process. The directions of the grinding marks are strongly heterogeneous but exhibit a clearly dominant orientation. Therefore, the profile analysis shows significant differences in the spatial frequency spectrum for horizontal or vertical sections through the initial surface. While Figure 4 focuses primarily on the macro-roughness of the surface at spatial frequencies smaller than 50 mm^{-1}, Figure 5 primarily highlights the micro-roughness up to spatial frequencies of 200 mm^{-1}. Figure 4 gives a representative impression of the initial surface, while the results displayed in Figure 5 will be used for an in-depth analysis of changes in micro-roughness and their corresponding spatial frequencies and wavelengths, respectively.

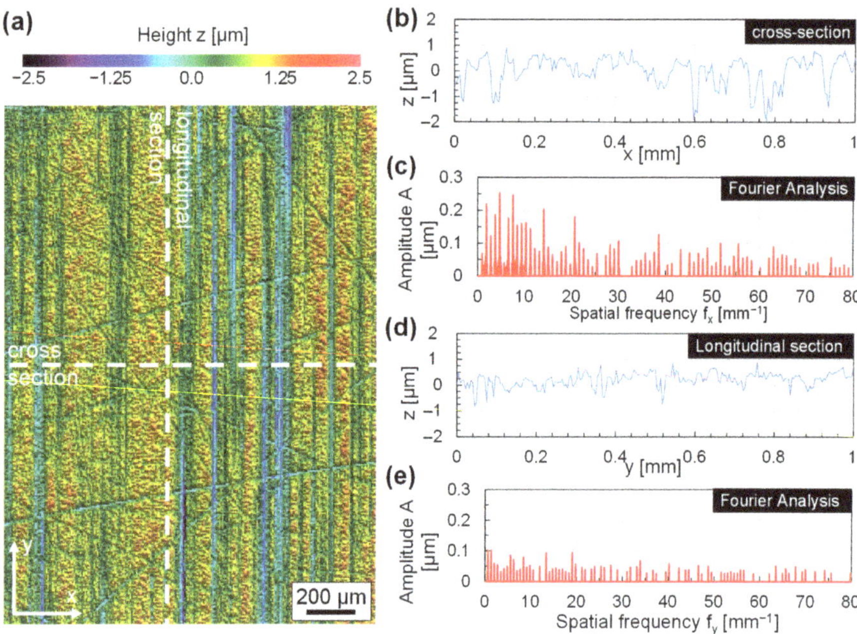

Figure 4. (**a**) WLI image of initial surface roughness; (**b**) Profile of cross-section along horizontal white dashed line; (**c**) One-dimensional (1D) Fourier analysis of profile along cross-section; (**d**) Profile of longitudinal section along vertical white dashed line; (**e**) One-dimensional (1D) Fourier analysis of longitudinal section.

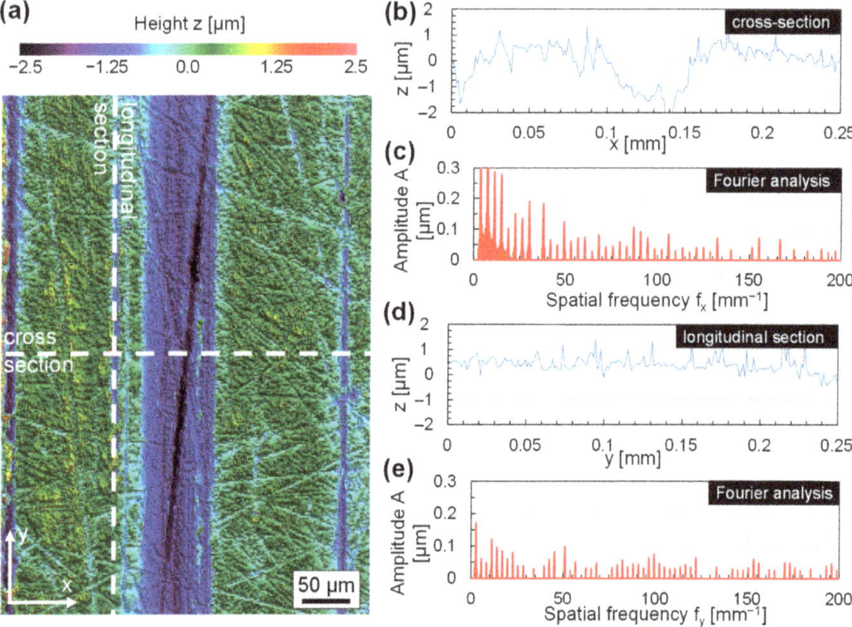

Figure 5. (**a**) WLI image of initial micro-roughness; (**b**) Profile of cross-section along horizontal white dashed line; (**c**) One-dimensional (1D) Fourier analysis of profile along cross-section; (**d**) Profile of longitudinal section along vertical white dashed line; (**e**) One-dimensional (1D) Fourier analysis of longitudinal section.

In addition to surface roughness, the initial microstructure and distribution of chemical elements at the surface was investigated by scanning electron microscopy (SEM) and energy-dispersive X-ray spectroscopy (EDX). In contrast to the experimental investigation, the surface was manually polished before SEM and EDX to get a clearer picture and less errors in measurement.

Figure 6 shows SEM images in different magnifications. While Figure 6a gives an overview of the overall distribution of chromium carbides (dark gray) in the steel matrix (light gray), Figure 6b provides a close-up of chromium carbides regarding sizes and geometrical shapes.

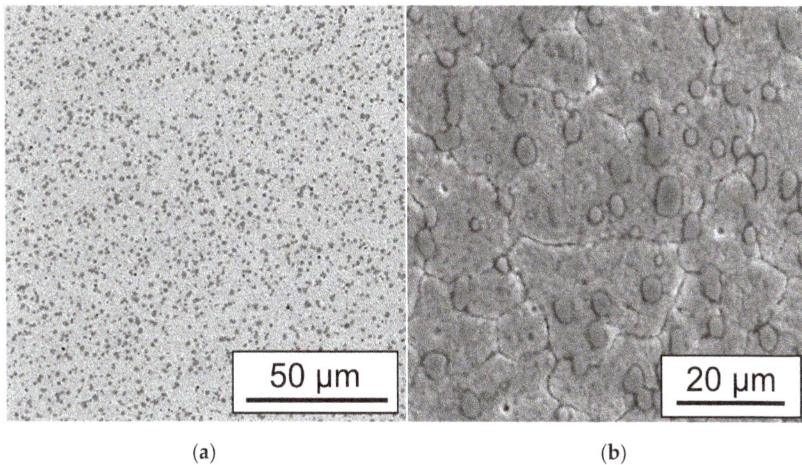

(a)   (b)

Figure 6. (a,b) SEM images of manually polished surface in two different magnifications to visualize size and distribution of chromium carbides (dark gray spots).

Firstly, the dark and light gray phases were characterized by EDX, and it was confirmed that the dark gray spots were chromium-enriched carbides (C39.55; Si 0; V3.69; Cr 31.83; Mn 0.39; Mo 1.04; Fe Bal.; in weight %), while the lighter gray areas were the steel matrix of AISI D2 containing less chromium (C 14.37; Si; 0.91; V 0.35; Cr 7.06; Mn 0.18; Mo 0.44; Fe Bal. in weight %). The size of the chromium carbides was primarily in the range from one to three micrometers, while the shape was predominantly ellipsoid and often spherical (Figure 6b). However, also, agglomerations of multiple chromium carbides as well as areas with a smaller density of chromium carbides were observed (Figure 6a). The size, shape, and distribution of chromium carbides are a characteristic feature of the powder metallurgical production process of 1.2379 (AISI D2). Therefore, this enhanced version of the material is typically referred to as 1.2379+ in comparison to 1.2379 (standard).

3.2. Surface Topography after Pulsed Laser Remelting

The surface topography and thus surface roughness after laser polishing were investigated in a full factorial design for three different laser beam dimensions—Q100, Q200, and Q400—and ten different laser fluences ranging from 3 to 12 J/cm^2 in equidistant steps of 1 J/cm^2. Figure 7 shows micrographs in a matrix form for selected laser fluences and for the laser beams Q100, Q200, and Q400. Figure 7a represents the initial surface topography since a fluence of 4 J/cm^2 at Q100 was not sufficient to introduce any topographical change in the initial surface.

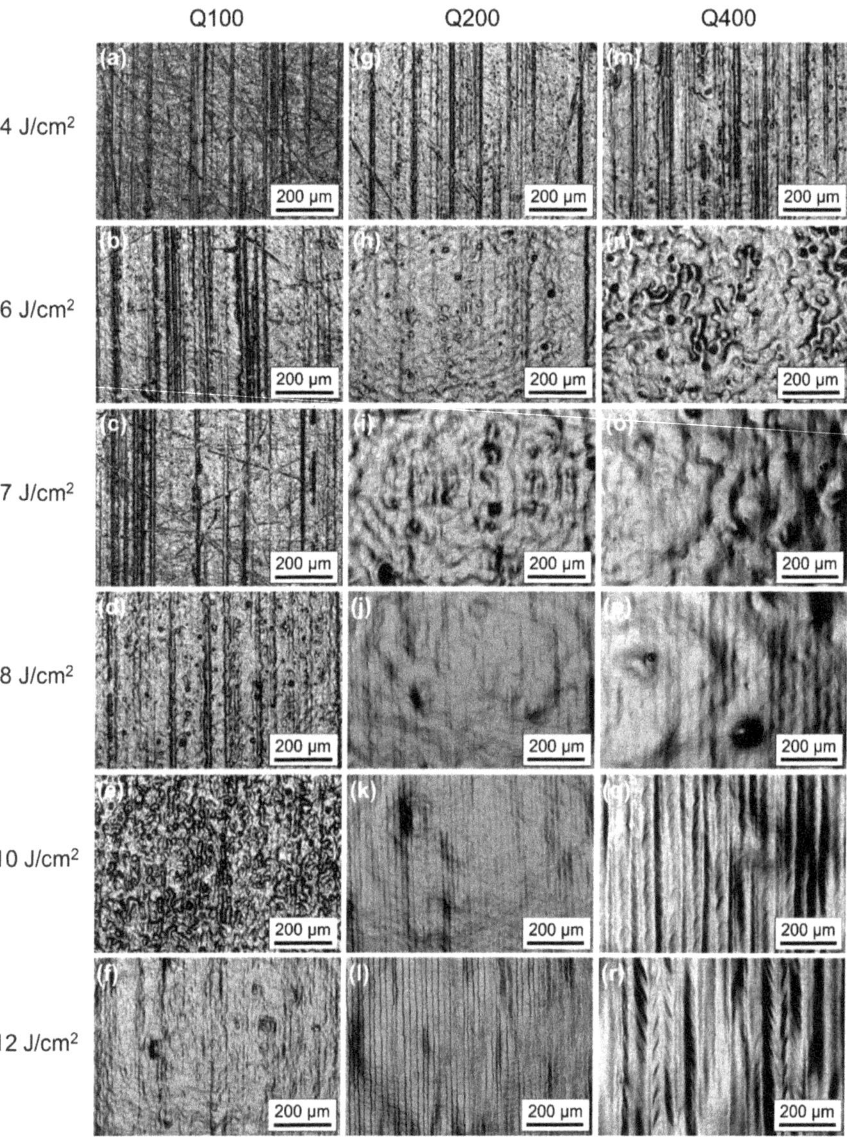

Figure 7. Micrographs of surfaces after laser micro polishing using tools Q100 (**a–f**), Q200 (**g–l**), and Q400 (**m–r**) for laser fluences ranging from 4 to 12 J/cm².

The smallest laser fluence investigated was $F = 3$ J/cm², but this fluence was not sufficient for any of the laser beam dimensions to remelt the surface at least partially. Remelting starts at a fluence of approximately $F = 4$ J/cm² for Q200 and Q400, while a fluence of approximately $F = 6$ J/cm² is required for Q100. The micrographs for these fluences show that various small, circular craters were formed during the remelting process. In general, it is observed that surface topography is significantly different for the same fluence at different laser beam dimensions (e.g., Figure 7b,h,n). A purely topographical description shows that small craters are preferably observed for smaller laser beams and fluences (Figure 7b,g,m). Larger laser beams and higher fluences, on the other hand, tend to produce dents, i.e., comparatively long-wavelength dimples or depressions (Figure 7j,o,p), which clearly differ in their topology from the craters at small beam sizes. At small fluences, basically only comparatively small, circular craters are visible in the laser-polished surface. A very large crater density was observed particularly at Q100 and 10 J/cm^2 (Figure 7e). Using larger beam sizes, surfaces with wavy depressions were formed. In particular, at large laser beam sizes and high fluences, surface ripples are formed in addition to the characteristic stripe texture (Figure 7q,r).

Additionally, visual process observation enables identifying certain characteristic process regimes and points toward different mechanisms of surface structure formation in the remelting process. Particularly, material ablation is a significant effect that is visible for all laser beam sizes at high fluences. However, the fluence required for material ablation strongly depends on the laser beam dimensions. Significant evaporation is visible for Q100 at approximately 12 J/cm^2, in comparison to approximately 10 J/cm^2 at Q200, and approximately 9 J/cm^2 at Q400. Laser ablation often coincides with clearly visible, parallel stripes on the surface (Figure 7k,l,q,r). The distance between the stripes corresponds to the track offset and is typically an indication for a continuous remelting process. These characteristic stripes get only visible for Q100 at the highest fluence of 12 J/cm^2 (Figure 7f). For Q200, these stripes are already hardly observable at 8 J/cm^2 (Figure 7j) but get clearly visible at 10 J/cm^2 (Figure 7k). Qualitatively, the same is true for Q400, but for lower fluences. Remelting stripes get hardly visible at approximately 7 J/cm^2 (Figure 7o) but get more pronounced for fluences of approximately 8–10 J/cm^2 (Figure 7p,q).

In particular, the surfaces remelted at high laser fluence appear to radiate a greater gloss (Figure 7q,r), i.e., to be significantly smoother at the micro-roughness level. The observed evaporation might lead to a comparatively small micro-roughness at the same time.

3.3. Micro-Roughness

In the following, the evolution of the micro-roughness of the laser remelted surfaces is presented in detail based on WLI images. The image size is 640 × 480 pixels, which corresponds to a spatial resolution of approximately 0.1 μm. This resolution and image size were chosen, because micro polishing particularly achieves or intends to achieve a reduction of micro-roughness in this range. Remelting has already started at 4 J/cm^2 for Q200 and Q400 (Figure 8g,m), which is evident from the rounded shape of the visible surface structures. In the case of Q100, effective remelting begins at fluences greater than 4 J/cm^2. At a fluence of 6 J/cm^2 (Figure 8b), a rounding of surface features can already be seen. However, in addition, crater-like depressions are observed with increasing frequency (Figure 8c,d,g,h). These tend to become deeper and wider with increasing fluence (Figure 8e,i,j). Furthermore, significant micro-porosity is visible at different laser beam dimensions and fluences (Figure 8i,j,o,p).

A comparatively small micro-roughness is observed at fluences of more than 8 J/cm^2 for large beam dimensions Q200 and Q400 (Figure 8j,k,o,p,q). At the same time, as the fluence increases, the stripe structure at the distance of the track offset also becomes more pronounced at increasing fluences (Figure 8k,l,q,r). In all cases, a certain waviness remains on the surface or cannot be smoothed by micro polishing.

In addition to a general overview of the resulting micro-roughness (Figure 8), indications for the origin of crater-like features were found at selected process parameters. Figure 9a shows a WLI image of the surface topography after laser remelting (Q200, $F = 4$ J/cm^2) in comparison to a representative SEM image of the surface near chromium carbide distribution in the initial bulk material (Figure 9b).

Even though an unambiguous correlation is difficult, Figure 9 reasonably suggests that the size, agglomeration, and distribution of chromium carbides are most likely the cause and starting point for randomly distributed craters of different sizes after laser remelting.

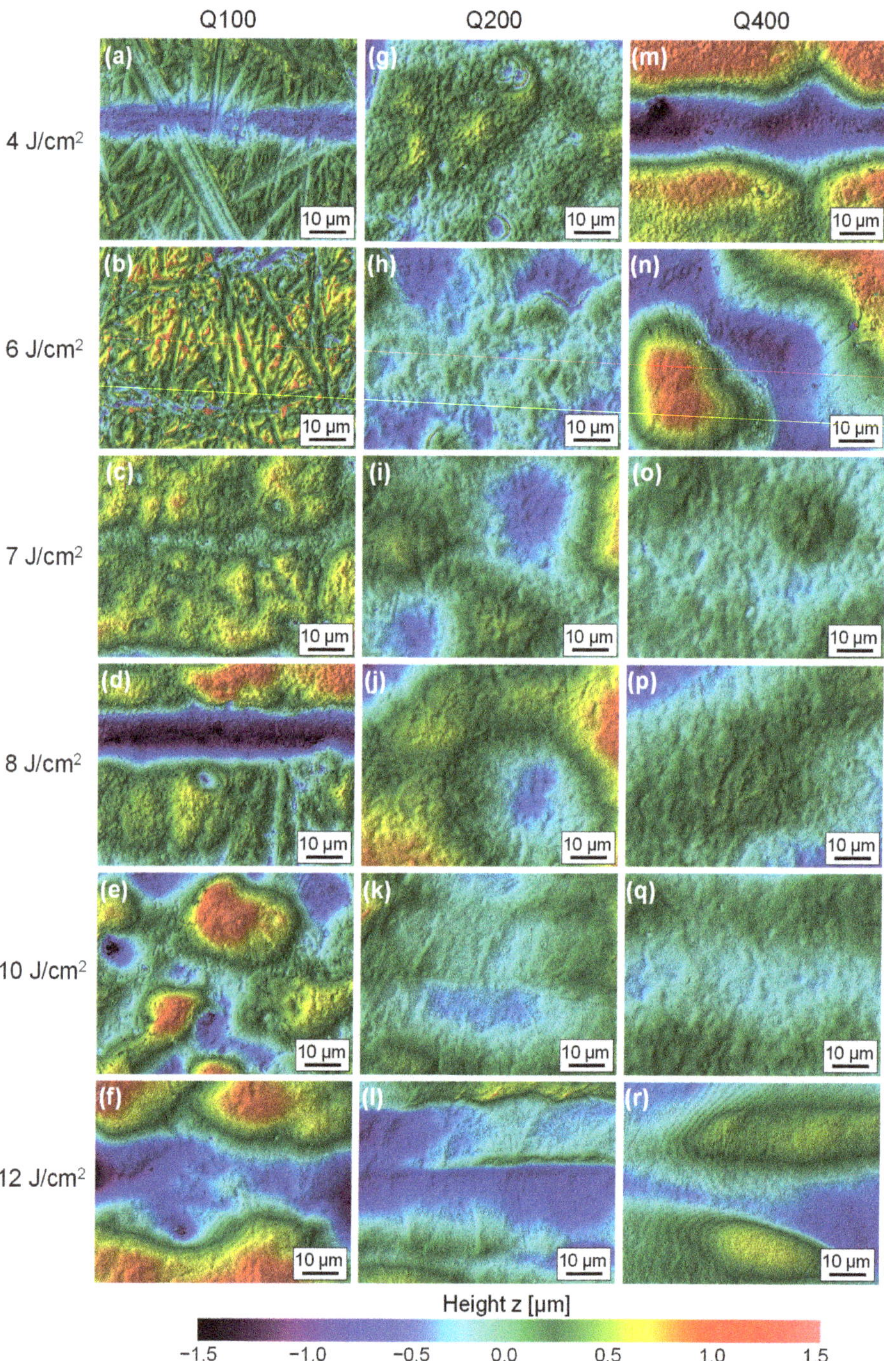

Figure 8. (a–r) WLI images of representative surface topographies as a function of laser beam dimension (Q100, Q200, Q400) and six selected laser fluence used for LµP (4, 6, 7, 8, 9, 10, 12 J/cm²).

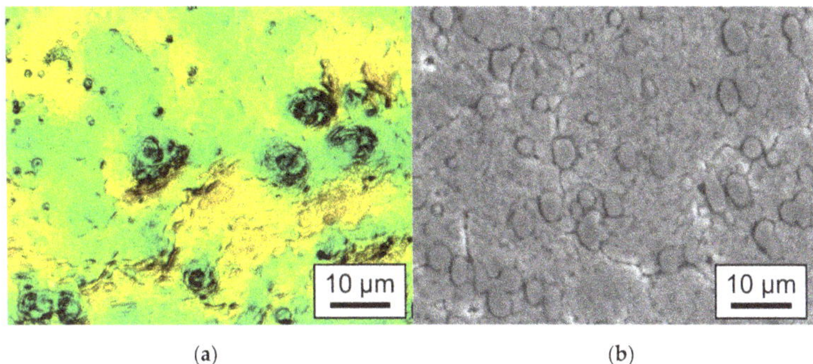

(a)   (b)

Figure 9. (**a**) WLI image of surface after laser remelting (Q200; $F = 4\,\text{J/cm}^2$) and (**b**) SEM images of size and distribution of chromium carbides of initial surface.

3.4. Surface Roughness

A detailed examination of the resulting surface roughness is carried out based on a systematic analysis of WLI data for each set of process parameters investigated. Figure 10a shows the surface roughness Sa as a function of the spatial wavelength (Sa spectrum) for the laser fluences investigated. Several observations are directly apparent for the smallest laser beam size Q100. The high standard deviation is due to the large number of craters on the surface (Figure 7b–e), so that the evaluation strongly depends on the number and the geometrical dimensions of the craters. Since the WLI data were measured at five randomly selected positions, this very high standard deviation results from large differences in the measured surface topographies. Furthermore, for fluences up to about 10–11 J/cm², only a small reduction in micro-roughness up to a spatial wavelength of approximately 10 μm was obtained. In contrast to the other fluences, a high number of craters were formed for these fluences, which even led to a significant increase in macro-roughness. A significant reduction in micro-roughness up to a critical wavelength of approximately 80 μm was achieved only for the highest fluence of $F = 12\,\text{J/cm}^2$. However, this laser remelting process was accompanied by a significant evaporation of material. Furthermore, a characteristic stripe structure is visible in the micrograph (Figure 7f), which indicates that the remelting process had changed from a discrete pulsed process to a continuous remelting process. Nonetheless, the lowest surface roughness was achieved for $F_{pol} = 12\,\text{J/cm}^2$, so that this was determined as laser polishing fluence F_{pol}.

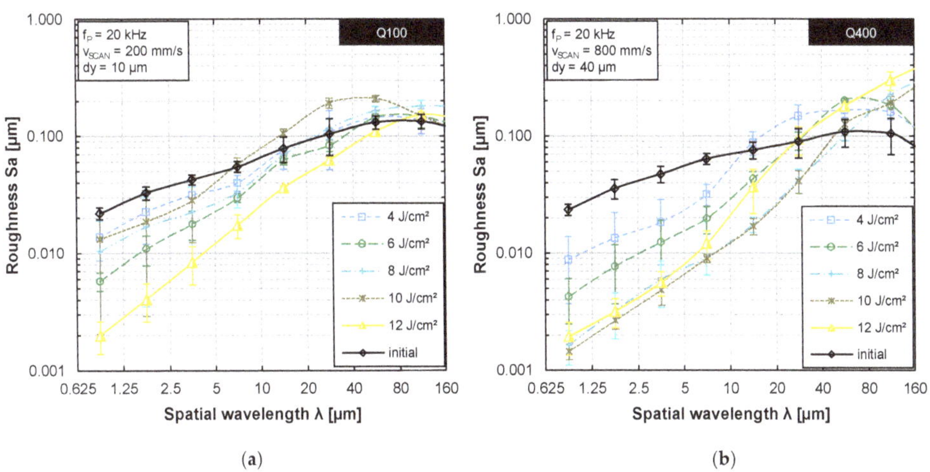

(a)   (b)

Figure 10. (**a**) Sa spectrum for Q100 and (**b**) Sa spectrum for Q400 each for fluences ranging from 4 to 12 J/cm².

The Sa spectrum for Q400 is significantly different in comparison (Figure 10b). At a fluence of approximately 4 J/cm^2, remelting of the surface started, and a reduction of micro-roughness was achieved for spatial wavelengths up to 10 μm. The reduction in micro-roughness and the critical wavelength were continuously increased for larger fluences. The critical wavelength was increased up to approximately 80 μm for fluences in the range of approximately 8–10 J/cm^2. However, at 10 J/cm^2, the visible evaporation of material already accompanies the remelting process, and characteristic stripes in the distance of the track offset are visible again on the remelted surface (Figure 7q). Laser polishing fluence at which the minimal roughness was achieved is approximately at $F_{pol} = 8$ J/cm^2

The series of Sa spectra is completed by Q200 (not shown) and exhibits qualitatively the same interdependencies as the Sa spectrum for the larger laser beam Q400. For Q200, the minimal roughness was achieved at $F_{pol} = 9$ J/cm^2. To conclude the analysis by Sa spectra, the lowest surface roughness for their respective beam dimensions and laser polishing fluence are compared in Figure 11a, while Figure 11b shows roughness evolution as a function of area rate for different spatial wavelength intervals. These comparisons reveal several insights. Firstly, it is revealed that the critical wavelength ($\lambda_{cr} = 80$ μm) is almost the same for all laser beam dimensions investigated. Secondly, laser polishing fluence increases for decreasing laser beam dimensions. Thirdly, for the larger laser beam size, a greater reduction particularly in micro-roughness was achieved. Fourthly, the achieved surface roughness for Q200 and Q400 are almost the same. Finally, macro-roughness was not reduced by any combination of process parameters, but process-inherent surface features were generated, which led even to an increase in macro-roughness. Furthermore, Figure 11b demonstrates explicitly that the minimal roughness achieved does not depend on area rate. This result is principally the same for each specific spatial wavelength interval up to 40 μm. Therefore, the results indicate that the area rate can be increased without an increase in resulting surface roughness after LμP. In this study, the area rate was increased by a factor of 16 from 1.2 cm^2/min to 19.2 cm^2/min (Figure 11b), while the required laser polishing fluence was reduced from approximately 12 to 8 J/cm^2 at the same time (Figure 11a). This led to an overall reduction of energy input by approximately 33%.

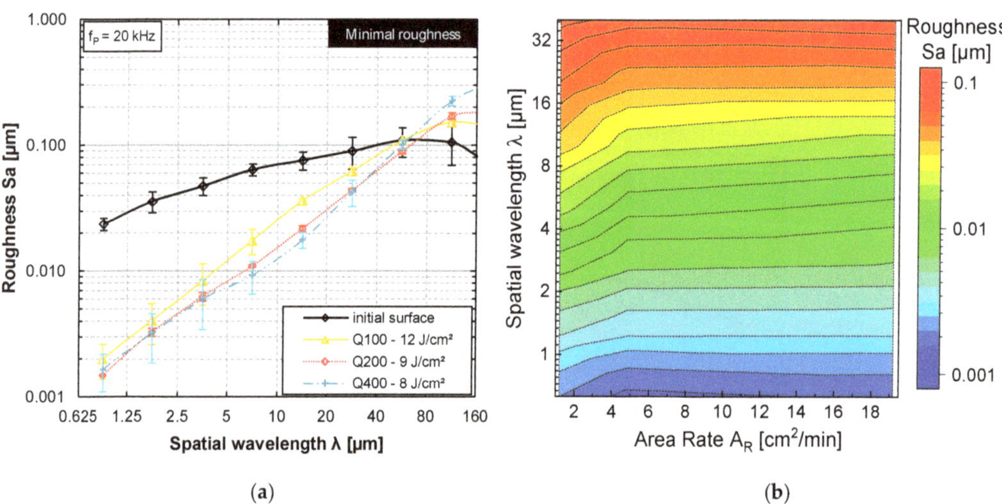

(**a**)   (**b**)

Figure 11. (**a**) Sa spectrum for the lowest surface roughness and laser polishing fluences achieved with each tool Q100, Q200, and Q400; (**b**) roughness Sa as a function of area rate and spatial wavelength at laser polishing fluence.

Sa spectra analyze the areal surface roughness but do not allow for any differentiation regarding a reduction in roughness depending on the orientation of surface features, such as grinding or milling grooves. The initial surface showed a strong heterogeneity regarding the directional roughness, since the milling grooves had a dominant direction, so that the roughness perpendicular to the grooves was significantly higher than parallel to them (cf. Figures 4 and 5). Therefore, a representative WLI measurement of the surface showing the lowest roughness after laser polishing was analyzed by 1D FFT analysis along two perpendicular sections (Figure 12a). The 1D FFT analysis provides a frequency spectrum for each section, which was then compared to a similar analysis of the initial surface (Figure 5).

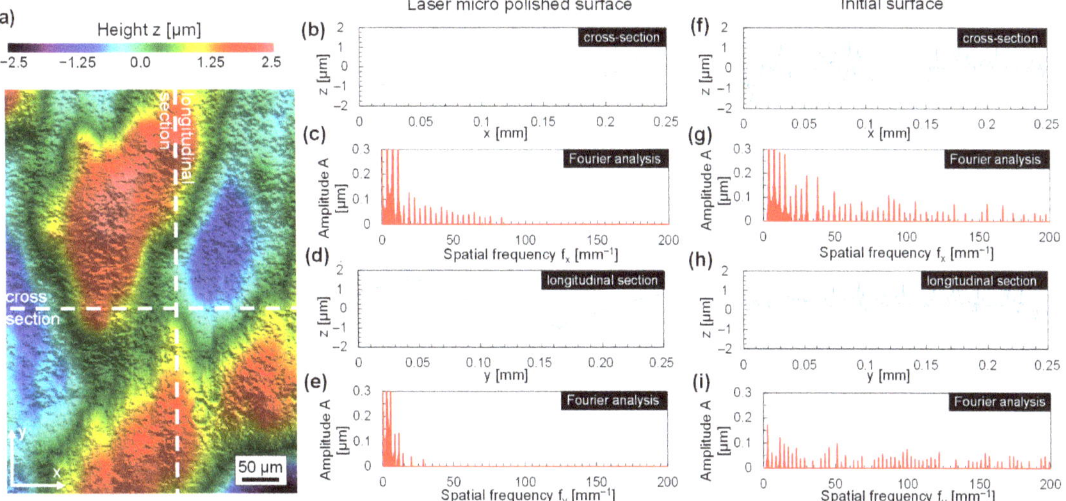

Figure 12. (**a**) WLI image of LµP surface exhibiting the lowest surface roughness (Q400, $F = 8$ J/cm^2); (**b**) cross-section along the marked line in WLI image and corresponding (**c**) Fourier analysis; (**d**) longitudinal section along marked line in WLI image and corresponding (**e**) Fourier analysis; in direct comparison to a representative (**f**) cross-section, (**h**) longitudinal, and their corresponding Fourier analyses (**g**,**i**).

Figure 12a shows a WLI image of the surface with the lowest micro-roughness (Q400; $F = 8$ J/cm^2). A cross-section along the white dashed line is shown in Figure 12b, while the profile of a representative longitudinal section is shown in Figure 12d. The corresponding frequency spectra are displayed in Figure 12c,d, respectively. A similar analysis is shown in Figure 12f,g for a representative cut-out of the initial surface (Figure 5).

A comparison of sections and their corresponding FFT analyses clearly visualizes the effect of LµP on surface roughness. Micro-roughness is significantly reduced along both sections down to spatial frequencies of approximately $f_y \approx 83$ mm^{-1} ($\lambda_{cr,y} \approx 12$ µm) for the cross-section (Figure 12c) and even down to $f_y \approx 33$ mm^{-1} ($\lambda_{cr,y} \approx 30$ µm) for the longitudinal section (Figure 12e). The comparison of an LµP surface (Figure 12b) and initial surface (Figure 12f) along the cross-section perpendicular to the milling grooves shows that these could not be completely smoothed out. This is also visualized and quantified in the corresponding FFT analyses (Figure 12c,g), which show that the amplitudes of characteristic spatial frequencies in the range of approximately 1–10 mm^{-1} could not be reduced. The profiles and FFT analyses for the comparison along the longitudinal section show that these low spatial frequencies were not only not reduced but significantly increased (Figure 12d,e,h,i). Therefore, roughness at low spatial frequencies (long spatial wavelengths) undergoes a directional homogenization, so that the amplitudes for the corresponding spatial frequencies in the x and y direction are almost equalized after LµP (Figure 12c,e).

3.5. Micro-Hardness

The influence of fluence on micro-hardness was exemplarily investigated for Q400. In this case, the micro-hardness HV0.1 (20s) was investigated by means of nano-indentation (Picodentor HM500, Helmut Fischer GmbH, Sindelfingen, Germany) for fluences ranging from 5 to 12 J/cm^2 (Figure 13). Ten hardness measuring points were examined on each laser-polished surface. The hardness measurements were sorted from the smallest to largest and are shown in Figure 13a. The initial hardness H0 of the initial surface was measured the same way, and a micro-hardness of 420 ($\pm$210) HV0.1 was determined (red, dashed line in Figure 13b). The standard deviation on this measurement was very large, depending on whether chromium carbides were completely or partially hit during the measurement (cf. Figure 6b). The maximum micro-hardness on the initial surface was approximately 827 HV0.1 (chromium carbides), while the minimum hardness was only 101 HV0.1 (steel matrix; not shown in Figure 13a).

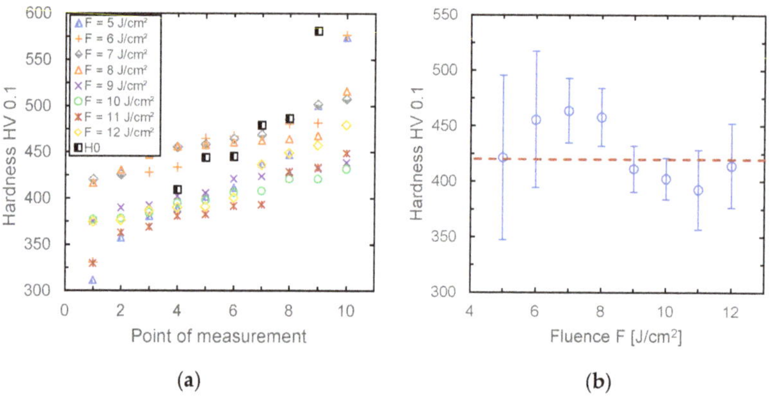

(a)   (b)

Figure 13. (a) Micro-hardness HV0.1 for ten individual measurements, and (b) average micro-hardness HV0.1 of laser polished surfaces for fluences ranging from 5 to 12 J/cm^2 (Q400).

Figure 13b shows the average micro-hardness and standard deviation as a function of laser fluence. At fluences of up to approximately 8 J/cm^2, the hardness was thereby increased, and the standard deviation was significantly reduced to approximately $\pm$29 HV0.1. For all laser remelted surfaces, a micro-hardness between approximately 382–464 HV was achieved. The maximum was reached at 464 ($\pm$29) HV0.1 (F = 8 J/cm^2), the minimum average micro-hardness was approximately 382($\pm$36) HV0.1 (F = 8 J/cm^2). There is a tendency for greater hardness to be achieved on the discretely remelted surfaces (F < 9 J/cm^2) compared to the presumably continuously remelted surfaces, on which the characteristic remelting stripes were also observed. This might be a result of the different characteristics of the remelting process and higher cooling rates associated with the discrete, pulsed LμP process.

4. Discussion

4.1. Surface Roughness Evolution

In general, surface roughness reduction is achieved, particularly a reduction of micro-roughness. This reduction depends among others on the initial roughness, which is clearly visible if the spatial frequency profile along a longitudinal and a cross-section are compared to each other (cf. Figure 12). Due to the mechanical preparation of the surfaces, characteristic grinding grooves were created on the surface (cf. Figures 4 and 5). The roughness perpendicular to these grooves (cross-section) was considerably higher than that parallel to these grooves (longitudinal section). Accordingly, the smoothing effect is different for these two directions. Although roughness spectra enable an evaluation of surface roughness, directional differences are not visible in these spectra. The spectral analysis along a longitudinal section shows that almost all spatial frequencies larger than approximately $f_y \approx 33$ mm^{-1} ($\lambda_{cr,y} \approx 30$ µm) are completely reduced to almost zero. In contrast, only spatial frequencies larger than approximately $f_y \approx 83$ mm^{-1} ($\lambda_{cr,y} \approx 12$ µm) were effectively smoothed out along the cross-section. The main reason for this is that the initial surface features are much more pronounced perpendicular to the milling grooves than parallel to them. Furthermore, the scanning strategy creates surface features perpendicular to the scanning direction, such as a stripe-like structure due to the track offset (Figure 7l,q). Therefore, although it seems that spatial frequencies smaller than approximately $f_y \approx 83$ mm^{-1} and $f_y \approx 33$ mm^{-1} were not completely smoothed, it is more likely that the initial roughness within this range was actually smoothed, but additional process-inherent surface features were created. This would explain the heterogeneity in the micro-roughness for these specific spatial frequencies.

4.2. Formation of Craters and Dimples

The formation of crater-like features in the surface was observed for many sets of process parameters but was particularly pronounced for Q100. As the initial analysis of microstructure regarding the size and distribution of chromium carbides indicates, it can be assumed that these chromium carbides are the source or at least the initiation side for these craters. The number of craters seems to correlate approximately with the carbides near the surface (cf. Figure 6). The formation of crater-like features is not unknown in LµP, but it was intensely investigated by Liebing [52] for bearing steel 100Cr6. However, Liebing [52] primarily identified oxides and sulfides as sources for melt pool disturbances or evaporation that lead to pronounced crater formation. According to the combined works of Tolochko et al. [53] and Boley et al. [54], the conclusion can be drawn that the absorption coefficient of chromium carbides at the laser wavelength is significantly larger than that of the surrounding steel matrix. Additionally, the intensity distribution for Q100 shows a significant intensity peak in the center of the distribution (cf. Figure 2a). Both effects in combination are assumed to result in the increased number of craters after laser remelting. This demonstrates that a top-hat-shaped intensity distribution or at least an intensity distribution without pronounced peaks is preferable for laser micro polishing. Additionally, it can be assumed that the chromium carbides also influence the surface topography evolution for the larger laser beam sizes. However, instead of pronounced craters, preferably long-wave dimples or depressions were created. Due to larger laser beam dimensions and thus larger melt pools, the influence of an evaporating chromium carbide of a certain size is relatively smaller than for a smaller melt pool. Additionally, larger melt pools lead to longer melt durations and increase the available time for the effective damping of capillary surface waves [28]. Capillary surface waves may result from disturbances of the melt pool such as localized evaporation [55]. On the other hand, larger melt pools lead to the remelting of more chromium carbides at the same time. However, an evolution mechanism of similar features was discussed by Nüsser et al. [43], who assume that these dimples result from macro ripple formation and the repeated remelting of these ripples. An additional effect might result from a low pulse stability of the laser beam source, which might lead to additional, undesired surface features with the approximate dimensions of the laser beam [19]. Furthermore, Spranger and Hilgenberg [56] found that a significant dissolution of carbides was a result of pulsed laser remelting of AISI D2. In sum, we assume that all these effects play a role in the formation of long-wave dimples. The partial evaporation and partial dissolution of chromium carbides are assumed to be the key reasons for disturbances of the melt pool volume, changes in melt pool dynamics, and deformation of the melt pool surface. These lead to the formation of craters at small laser beam dimensions and to long wave dimples when remelting with larger laser beams. Multiple remelting cycles (approximately 100 times per spot) lead to a partial smoothing and directional homogenization of the specific spatial roughness frequencies, as it is shown in the roughness spectra (Figures 10–12).

4.3. Transition to Continous Remelting Process and Micro-Hardness

This stripe-like structure with continuous boundary lines between the individual tracks (Figure 7l,q) is typically a feature of a continuous remelting process [27]. This is to be expected for long pulse duration, for large track and pulse overlap, and for high laser fluences. This effect is essentially due to heat accumulation (Weber et al. [57,58]), so that the molten pool no longer solidifies before the following laser pulse impinges on the surface. This is an effect that was observed in LμP particularly by Temmler et al. [18] at high pulse repetition frequencies for ns laser pulses. The authors defined an empirical formula for an estimation of a threshold scan speed $v_{scan,th}$ as a function of laser beam diameter d_L, pulse duration t_P, and pulse repetition frequency f_{rep}, at which the discrete pulsed remelting process changes to a continuous one (Equation (2)).

$$v_{scan,th} \leq d_L \cdot t_P \cdot f_{rep}^2 \tag{2}$$

Although the criterion from Equation (2) is not fulfilled for the process parameters used, the threshold scan speed ($v_{scan,\,th}$ = 192 mm/s; t_P = 1.2 μs) is just smaller than the used scan speed by a factor of approximately 4. Additionally, it must be added that the introduced criterion was established on an empirical basis and should be valid for a quasi-continuous remelting process at laser polishing fluence without material ablation. However, as already determined from the microscope images and visual observations during the process, a continuous remelting process occurs only at a fluence level, where significant material evaporation is already clearly observable without optical aids. Therefore, it is reasonable to assume that the stripes are a clear indication for a continuous remelting process. This would particularly help to explain that almost no craters were visible anymore for a small laser beam size Q100 and very high fluences (F = 12 J/cm^2).

The effect of surface ripple formation (cf. Figure 7q,r) is probably due on the one hand to the formation of a continuous melt pool and on the other hand to the irradiation of the surface in reoccurring, discrete time intervals of 50 μs (20 kHz). This leads to periodic fluctuations of the melt pool volume, which results in a periodic structuring of the surface. This principle is typically known from the WaveShape process, where it is specifically used to generate mostly periodic surface structures [7,59,60]. Furthermore, a low pulse stability of the laser beam source probably causes the generated ripples to fluctuate in shape and structure. A similar effect was specifically used by Pfefferkorn and Morrow [61] to achieve surface structuring by the spatial and temporal control of laser fluence.

The transition to the continuous remelting process also tends to be reflected in the results of the micro-hardness measurements (Figure 13). A dissolution and/or partial evaporation of chromium carbides is assumed to lead to a significant reduction of standard deviation in the micro-hardness measurements. This is presumably because only fewer, smaller, or no chromium carbides were hit in the micro-hardness measurements. Thus, the largest micro-hardness measured was approximately 550 HV instead of approximately 830 HV0.1 on the initial surface. However, at the same time, no areas of the comparatively soft steel matrix were hit any more ($\approx$100 HV 0.1), but the minimal micro-hardness in the laser-polished fields was approximately 320 HV0.1. This indicates that carbon from the chromium carbides was increasingly dissolved in the steel matrix. It is assumed that the entire surface boundary layer was hardened through the formation of martensite and achieved an average micro-hardness of up to approximately 464 HV0.1 with significantly smaller standard deviations (partially < 30 HV0.1). Additionally, the micro-hardness of the discretely, remelted LμP surface tends to be higher than for the continuously remelted surface. This is presumably due to the higher cooling rates in the LμP process in comparison to the continuous remelting process. A similar effect has been observed in earlier works on other steels e.g., H11 (Preußner et al. [62]), or S7 (Morrow and Pfefferkorn [13]). However, Maharjan et al. [33] found for 50CrMo4 steel that the most pronounced hardening effect was achieved for longer pulse duration and cw laser radiation, i.e., in principle for smaller cooling rates. The reason for these contradictory results is unclear, but it is most likely a consequence of the different chemical composition of these steels or the specifics of the temperature evolution during the laser treatment.

4.4. Critical Wavelength/Frequency

The critical wavelength or critical frequency was almost the same for all laser beam sizes. This underscores the importance of the pulse duration and that the laser beam size is of secondary importance. However, the laser beam dimensions are assumed to be at least twice as large as the critical wavelength. The critical wavelength for Q100 and $F = 12 \, J/cm^2$ is of special interest since a change to a continuous process is assumed. In a continuous remelting process, melt duration is typically approximated by the interaction time of the laser beam and material [39], which is typically calculated by $t_{int} = d_L \cdot v_{scan}^{-1}$ and is approximately $t_{int} = 0.5$ ms for Q100. This is two orders of magnitude larger than the pulse duration ($t_P = 1.2 \, \mu s$), but the critical wavelength was not significantly increased to more than 80 μm (in comparison to Q200 and Q400). This is a little bit surprising, but in this case, the critical wavelengths are presumably strictly limited by the laser beam dimensions, since no roughness with larger dimensions than the laser beam can effectively be smoothed in cw laser polishing. Additionally, melt pool disturbances due to the partial evaporation and dissolution of chromium carbides counteract a more effective smoothing. This is particularly imminent for Q200 and Q400, where also a continuous remelting process is assumed for high fluences of $F > 10 \, J/cm^2$. This critical wavelength is not only further increased but even reduced, since the process-inherent formation of surface structures, e.g., ripples or boundary lines at the edges of the laser tracks, counteracts the smoothing effect of surface remelting.

4.5. Laser Polishing Fluence and Area Rate

Another important result of this study is that the laser polishing fluence decreases for larger laser beam dimensions (Figure 11a). A possible explanation results from an energetic point of view. At fixed laser fluence, a larger laser-irradiated area absorbs more laser energy in total. An increase in absorbed energy not only results in larger melt pool dimensions but also in longer melt durations. Additionally, temperature gradients in larger melt pools are less steep, and thus, cooling to ambient temperature takes longer. Particularly for multi-pulse processing, this has several consequences. Firstly, if the temporal distance between subsequent laser pulses is the same as for smaller laser beam dimensions, heat accumulation is more pronounced due to larger energy input per pulse and smaller temperature gradients in the melt pool. Heat accumulation not only leads to a reduction in the threshold fluence for laser ablation [57] but also to a reduction in laser polishing fluence [18]. Secondly, particularly noteworthy at this point is that the overall energy input for a larger laser beam size is reduced. This prevents the global heat accumulation of the whole sample and should reduce thermal effects such as distortion and deformation. Since the laser polishing fluence is reduced and local heat accumulation is increased for larger laser beams, locally induced thermal stresses should also be reduced [63]. Based on the insights of Spranger at al. [64], one could also assume that the very high cooling rates lead to a very fine microstructure in each case; however, this was not particularly investigated in this study.

The results already show that larger laser beam sizes lead to higher area rates without increasing the resulting surface roughness if the number of remelting cycles is kept constant (Figure 11b). In principle, the influence of laser beam dimensions s_L^2, repetition rate f_{rep}, and number of remelting cycles n on an achievable area rate A_R can be discussed based on Equation (3).

$$A_R = \frac{s_L^2 \cdot f_{rep}}{n};$$

$$with \; n = \frac{s_L^2}{dx \cdot dy}; \; s_{L,max}^2 = \frac{E_{P,max}}{F_{pol}}; \; P_{L,max} = E_{P,max} \cdot f_{rep,max}$$

(3)

The maximum area rate A_R is the product of the remelted area per pulse ($A = s_L{}^2$) and the pulse repetition rate f_{rep} divided by the number of remelting cycles per spot n (effectively resulting from pulse and track overlap). Thus, for example, the side length s_L of a square laser spot has a quadratic effect on the area rate A_R. However, the required pulse energy E_P scales almost linearly with the remelted area A ($E_P \sim A$). Increasing the pulse frequency leads to a linear increase in the area rate. Therefore, preferably, the dimensions of the focused laser beam should be maximized if the pulse repetition rate is crucially limited. The maximum laser beam dimensions depend on the available maximum pulse energy $E_{P,max}$ and on the laser polishing fluence F_{pol}. In turn, the laser polishing fluence F_{pol} tends to decrease with increasing dimensions of the laser beam (Figure 11a). Furthermore, the maximum pulse frequency depends on the maximum available laser power $P_{L,max}$ at maximum pulse energy $E_{P,max}$. Applied specifically to LµP, and taking into account this and other studies on LµP, a reasonably achievable area rate can be estimated for a commercially available laser system. In this work, a spot on the surface was remelted on average approximately $n = 100$ times. However, typically, only approximately $n = 20$ remelting cycles are required to achieve a minimal surface roughness in LµP of steels [18,27]. Using the example of a commercially available laser system (e.g., IPG YLPN-HP $t_P = 120$ ns, $P_{L,max} = 5$ kW; $E_{P,max} = 100$ mJ; $f_{rep,max} = 50$ kHz), a realistic area rate can be estimated. The selected laser beam source already provides a square fiber cross-section. The maximum pulse energy of approximately $E_{P,max} = 100$ mJ enables a laser focus with an area of $s_L{}^2 = 20$ mm^2 at a required laser polishing fluence of approximately $F_{pol} = 5$ J/cm^2. The maximum pulse frequency of the laser system is 50 kHz, so that a maximum laser power of $P_{L,max} = 5$ kW is available. Thus, considering $n = 20$ remelting cycles per spot, area rates of up to $A_R = 3$ m^2/min might be achievable. Area rates of this order of magnitude are already remarkably interesting for industrial applications, even for large-area polishing of flat components.

Larger laser beam dimensions lead to larger area rates with reduced energy input, since the required laser polishing fluence decreases (Figure 11). If high spatial resolution is of secondary importance in LµP and large pulse energies are available, then large area rates are easily achievable. However, a similar effect of decreasing laser polishing fluence was observed for high pulse frequencies [18]. Thus, an increase in pulse frequency also leads to a reduced laser polishing fluence and to a reduction of the overall irradiated laser energy. Which effect outweighs the other cannot be said without further studies, but both ways—increasing single pulse energy and increasing pulse repetition frequency, respectively—are viable ways to significantly increase the area rate in LµP. Since the product of single pulse energy E_P and pulse frequency f_{rep} equals the average laser power P_L, Equation (3) can be boiled down to a simple dependency $A_R \sim P_{L,max}$, which is valid for many laser-based processes. The area rate in LµP basically scales linearly with the average laser power of the laser beam source. Nonetheless, there are limits to consider regarding the maximum single pulse energy (regarding damage thresholds of optical components etc.), pulse duration (maximum pulse peak power), and maximum pulse repetition frequency (maximum deflection speed of laser scanning systems, etc.).

As a visual conclusion of the discussion, Figure 14 shows the laser-polished surface with the lowest roughness achieved in this study (Figure 14b) compared to the initial surface roughness (Figure 14a). A considerable reduction in surface roughness and a particular increase in gloss are the obvious results displayed in these micrographs.

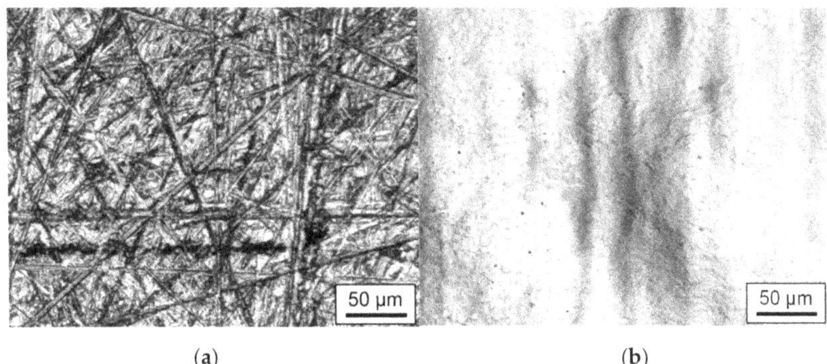

Figure 14. Micrographs of (**a**) initial surface roughness in comparison to (**b**) laser-polished surface with lowest surface roughness (Q400, F = 8 J/cm^2).

5. Conclusions

This work investigated laser micro polishing (LµP) of tool steel 1.2379 (AISI D2) using a square intensity distribution in three different sizes (Q100, Q200, Q400) and laser fluences ranging from 3 to 12 J/cm^2. A laser fluence of 3 J/cm^2 was below the melting threshold, while 12 J/cm^2 exceeded the evaporation threshold and led to significant material evaporation for all laser beam sizes. Additionally, the experimental results for LµP of 1.2379 led to some noteworthy insights:

- LµP significantly decreases micro-roughness up to a critical wavelength of approximately 80 µm and increases the gloss of the surface. (e.g., by a factor of ten for Q400 from Sa = 32 nm to Sa = 3 nm in the spatial wavelength regime from 1.25 to 2.5 µm).
- An increase in laser beam dimensions, while adapting pulse overlap and track offset accordingly, leads to a significant increase in area rate from 1.2 to 19.2 cm^2/min without increasing the resulting surface roughness.
- An increase in laser beam dimensions from Q100 to Q400 leads to a decrease in fluence from 12 J/cm^2 to approximately 8 J/cm^2 required for laser polishing, which results in an overall reduced energy input (approximately 33%).
- A significant increase in laser fluence above the polishing laser fluence leads to a change of the discrete, pulsed remelting process to a continuous remelting process. Particularly in the continuous remelting process, macro-roughness was partially increased, and surface rippling was observed.
- Near-surface chromium carbides are assumed to be partially evaporated during LµP. This presumably leads to melt pool disturbances and the formation of undesired surface features e.g., craters, which increase surface roughness.
- The dissolution of chromium carbides during the remelting process presumably leads to a homogenization of micro-hardness in the surface boundary layer of approximately 382–464 HV0.1.
- Intensity distributions with significant peaks are to be avoided since high peak intensities and local heterogeneities in material absorption increase the risk of uncontrolled material evaporation and the formation of undesired surface features.

Overall, it is concluded that an increase in laser beam size enables the use of higher pulse energies and larger average laser power, and it also leads to significantly higher area rates and less overall energy input. Therefore, besides an increase in pulse frequency, larger pulse energies and larger laser beam sizes are a viable way to significantly increase the area rate in LµP. For industrial applications, multi-kW laser systems with suitable pulse energies and pulse frequencies are widely available, which potentially enable area rates of several m^2 per minute in LµP.

Author Contributions: Conceptualization, A.T.; methodology, A.T., I.R.; software, A.T.; validation, A.T., M.E.K.; formal analysis, A.T., M.C.; investigation, M.C., M.E.K., S.-K.R.; resources, A.T.; data curation, M.C., A.T.; writing—original draft preparation, M.C., A.T.; writing—review and editing, A.T., M.C., S.-K.R.; visualization, A.T., I.R., M.E.K.; project administration, I.R., A.T.; funding acquisition, A.T., I.R.; supervision, A.T. All authors have read and agreed to the published version of the manuscript.

Funding: This research was partially funded by the Deutsche Forschungsgemeinschaft (DFG) in the collaborative research project "Lubricant-free tribology concepts for cold extrusion by interaction-minimized surface layers and structures", under grant number PO 591/37-3 within the priority program SPP 1676 "Dry Metal Forming".

Institutional Review Board Statement: Not applicable.

Informed Consent Statement: Not applicable.

Data Availability Statement: The data presented in this study are available on reasonable request from the corresponding author.

Acknowledgments: The authors would like to personally thank Konrad Wissenbach and Edgar Willenborg from the Fraunhofer Institute for Lasertechnology ILT (Aachen, Germany) for fruitful discussions and acknowledge their passionate drive to push laser polishing to new boundaries. Finally, we thank the anonymous referees for their effort and valuable comments.

Conflicts of Interest: The authors declare no conflict of interest. The funders had no role in the design of the study; in the collection, analyses, or interpretation of data; in the writing of the manuscript, or in the decision to publish the results.

References

1. Temmler, A.; Graichen, K.; Donath, J. Laser Polishing in Medical Engineering. *LTJ* **2010**, *7*, 53–57. [CrossRef]
2. Flemmer, J. CAM-NC-Datenkette für die Laserbearbeitung von Freiformflächen mit Simultaner Bewegung Mechanischer Achsen und Galvanometrischem Laserscanner. Ph.D. Thesis, RWTH Aachen University, 1. Edition, Apprimus. Aachen, Germany, 2018.
3. Singh, K.; Khatirkar, R.K.; Sapate, S.G. Microstructure evolution and abrasive wear behavior of D2 steel. *Wear* **2015**, *328–329*, 206–216. [CrossRef]
4. Hasselbruch, H.; Lu, Y.; Messaoudi, H.; Mehner, A.; Vollertsen, F. Tribological Properties of Multi-Layer a-C:H:W/a-C:H PVD-Coatings Micro-Structured by Picosecond Laser Ablation. *KEM* **2019**, *809*, 439–444. [CrossRef]
5. Henn, M.; Reichardt, G.; Weber, R.; Graf, T.; Liewald, M. Dry Metal Forming Using Volatile Lubricants Injected into the Forming Tool Through Flow-Optimized, Laser-Drilled Microholes. *JOM* **2020**, *1*, 59. [CrossRef]
6. Förster, D.J.; Weber, R.; Holder, D.; Graf, T. Estimation of the depth limit for percussion drilling with picosecond laser pulses. *Opt. Express* **2018**, *26*, 11546–11552. [CrossRef]
7. Temmler, A.; Comiotto, M.; Ross, I.; Kuepper, M.; Liu, D.M.; Poprawe, R. Surface structuring by laser remelting of 1.2379 (D2) for cold forging tools in automotive applications. *J. Laser Appl.* **2019**, *31*, 22017. [CrossRef]
8. Ross, I.; Temmler, A.; Küpper, M.; Prünte, S.; Teller, M.; Schneider, J.M.; Poprawe, R. Laser Polishing of Cold Work Steel AISI D2 for Dry Metal Forming Tools: Surface Homogenization, Refinement and Preparation for Self-Assembled Monolayers. *KEM* **2018**, *767*, 69–76. [CrossRef]
9. Mousavi, A.; Kunze, T.; Roch, T.; Lasagni, A.; Brosius, A. Deep drawing process without lubrication—An adapted tool for a stable, economic and environmentally friendly process. *Procedia Eng.* **2017**, *207*, 48–53. [CrossRef]
10. Brosius, A.; Mousavi, A. Lubricant free deep drawing process by macro structured tools. *CIRP Ann.* **2016**, *65*, 253–256. [CrossRef]
11. Häfner, T.; Heberle, J.; Hautmann, H.; Zhao, R.; Tenner, J.; Tremmel, S.; Merklein, M.; Schmidt, M. Effect of Picosecond Laser Based Modifications of Amorphous Carbon Coatings on Lubricant-free Tribological Systems. *JLMN J. Laser Micro/Nanoeng.* **2017**, *12*, 132–140. [CrossRef]
12. Jähnig, T.; Mousavi, A.; Steinhorst, M.; Roch, T.; Brosius, A.; Lasagni, A.F. Friction reduction in dry forming by using tetrahedral amorphous carbon coatings and laser micro-structuring. *Dry Met. Form.* **2019**, *5*, 25–30.
13. Morrow, J.D.; Pfefferkorn, F.E. Local Microstructure and Hardness Variation After Pulsed Laser Micromelting on S7 Tool Steel. *J. Micro Nano-Manuf.* **2016**, *4*, 209. [CrossRef]
14. Deng, T.; Li, J.; Zheng, Z. Fundamental aspects and recent developments in metal surface polishing with energy beam irradiation. *Int. J. Mach. Tools Manuf.* **2020**, *148*, 103472. [CrossRef]
15. Krishnan, A.; Fang, F. Review on mechanism and process of surface polishing using lasers. *Front. Mech. Eng.* **2019**, *14*, 299–319. [CrossRef]
16. Bordatchev, E.V.; Hafiz, A.M.K.; Tutunea-Fatan, O.R. Performance of laser polishing in finishing of metallic surfaces. *Int. J. Adv. Manuf. Technol.* **2014**, *73*, 35–52. [CrossRef]

17. Bhaduri, D.; Penchev, P.; Batal, A.; Dimov, S.; Soo, S.L.; Sten, S.; Harrysson, U.; Zhang, Z.; Dong, H. Laser polishing of 3D printed mesoscale components. *Appl. Surf. Sci.* **2017**, *405*, 29–46. [CrossRef]

18. Temmler, A.; Liu, D.; Luo, J.; Poprawe, R. Influence of pulse duration and pulse frequency on micro-roughness for laser micro polishing (LμP) of stainless steel AISI 410. *Appl. Surf. Sci.* **2020**, *510*, 145272. [CrossRef]

19. Temmler, A.; Ross, I.; Luo, J.; Jacobs, G.; Schleifenbaum, J.H. Influence of global and local process gas shielding on surface topography in laser micro polishing (LμP) of stainless steel 410. *Surf. Coat. Technol.* **2020**, *403*, 126401. [CrossRef]

20. Li, X.; Guan, Y. Theoretical fundamentals of short pulse laser–metal interaction: A review. *Nanotechnol. Precis. Eng.* **2020**, *3*, 105–125. [CrossRef]

21. Willenborg, E.; Wissenbach, K.; Poprawe, R. Polishing by Laser Radiation. In *Proceedings of the 2nd International WLT Conference Lasers in Manufacturing LIM 2003, Munich, Germany, 24–26 June 2003*; WLT: Munich, Germany, 2003; pp. 297–300.

22. Perry, T.L.; Werschmoeller, D.; Li, X.; Pfefferkorn, F.E.; Duffie, N.A. The Effect of Laser Pulse Duration and Feed Rate on Pulsed Laser Polishing of Microfabricated Nickel Samples. *J. Manuf. Sci. Eng.* **2009**, *131*, 31002. [CrossRef]

23. Temmler, A.; Wei, D.; Schmickler, T.; Küpper, M.E.; Häfner, C.L. Experimental investigation on surface structuring by laser remelting (WaveShape) on Inconel 718 using varying laser beam diameters and scan speeds. *Appl. Surf. Sci.* **2020**, 147814. [CrossRef]

24. Oreshkin, O.; Küpper, M.; Temmler, A.; Willenborg, E. Active reduction of waviness through processing with modulated laser power. *J. Laser Appl.* **2015**, *27*, 22004. [CrossRef]

25. Vadali, M.; Ma, C.; Duffie, N.A.; Li, X.; Pfefferkorn, F.E. Pulsed laser micro polishing: Surface prediction model. *J. Manuf. Process.* **2012**, *14*, 307–315. [CrossRef]

26. Chow, M.T.C.; Hafiz, A.M.K.; Tutunea-Fatan, O.R.; Knopf, G.K.; Bordatchev, E.V. Experimental Statistical Analysis of Laser Micropolishing Process. In *Proceedings of the ISOT 2010 International Symposium on Optomechatronic Technologies, Toronto, ON, Canada, 25–27 October 2010*; IEEE: Toronto, ON, Canada, 2010; pp. 1–6. ISBN 9781424476848.

27. Temmler, A.; Liu, D.; Preußner, J.; Oeser, S.; Luo, J.; Poprawe, R.; Schleifenbaum, J.H. Influence of laser polishing on surface roughness and microstructural properties of the remelted surface boundary layer of tool steel H11. *Mater. Des.* **2020**, *192*, 108689. [CrossRef]

28. Perry, T.L.; Werschmoeller, D.; Duffie, N.A.; Li, X.; Pfefferkorn, F.E. Examination of Selective Pulsed Laser Micropolishing on Microfabricated Nickel Samples Using Spatial Frequency Analysis. *J. Manuf. Sci. Eng.* **2009**, *131*, 21002. [CrossRef]

29. Nüsser, C.; Sändker, H.; Willenborg, E. Pulsed Laser Micro Polishing of Metals using Dual-beam Technology. *Phys. Procedia* **2013**, *41*, 346–355. [CrossRef]

30. Perry, T.L. A Numerical and Experimental Study of Pulsed Laser Micro Polishing of Metals at the Meso/Micro Scale. Ph.D. Thesis, University of Wisconsin-Madison, Madison, WI, USA, 2009.

31. Guan, Y.; Zhou, W.; Zheng, H.; Hong, M.; Zhu, Y.; Qi, B. Effect of pulse duration on heat transfer and solidification development in laser-melt magnesium alloy. *Appl. Phys. A* **2015**, *119*, 437–442. [CrossRef]

32. Ma, C.P.; Guan, Y.C.; Zhou, W. Laser polishing of additive manufactured Ti alloys. *Opt. Lasers Eng.* **2017**, *93*, 171–177. [CrossRef]

33. Maharjan, N.; Zhou, W.; Zhou, Y.; Guan, Y.; Wu, N. Comparative study of laser surface hardening of 50CrMo4 steel using continuous-wave laser and pulsed lasers with ms, ns, ps and fs pulse duration. *Surf. Coat. Technol.* **2019**, *366*, 311–320. [CrossRef]

34. Li, Y.-H.; Wang, B.; Ma, C.-P.; Fang, Z.-H.; Chen, L.-F.; Guan, Y. C.; Yang, S.-F. Material Characterization, Thermal Analysis, and Mechanical Performance of a Laser-Polished Ti Alloy Prepared by Selective Laser Melting. *Metals* **2019**, *9*, 112. [CrossRef]

35. Bhaduri, D.; Ghara, T.; Penchev, P.; Paul, S.; Pruncu, C.I.; Dimov, S.; Morgan, D. Pulsed laser polishing of selective laser melted aluminium alloy parts. *Appl. Surf. Sci.* **2021**, *558*, 149887. [CrossRef]

36. Liang, C.; Hu, Y.; Liu, N.; Zou, X.; Wang, H.; Zhang, X.; Fu, Y.; Hu, J. Laser Polishing of Ti6Al4V Fabricated by Selective Laser Melting. *Metals* **2020**, *10*, 191. [CrossRef]

37. Purushothaman, S.; Ravi Sankar, M. State of the art on atomistic modelling of laser polishing. *Mater. Today Proc.* **2021**, *44*, 689–695. [CrossRef]

38. Mai, T.A.; Lim, G.C. Micromelting and its effects on surface topography and properties in laser polishing of stainless steel. *J. Laser Appl.* **2004**, *16*, 221–228. [CrossRef]

39. Ukar, E.; Lamikiz, A.; Martínez, S.; Tabernero, I.; Lacalle, L.L.d. Roughness prediction on laser polished surfaces. *J. Mater. Process. Technol.* **2012**, *212*, 1305–1313. [CrossRef]

40. Chow, M.; Bordatchev, E.V.; Knopf, G.K. Impact of initial surface parameters on the final quality of laser micro-polished surfaces. In *Micromachining and Microfabrication Process Technology XVII*; Maher, M.A., Resnick, P.J., Eds.; SPIE: San Francisco, CA, USA, 2012; p. 824809.

41. Ma, C.; Vadali, M.; Li, X.; Duffie, N.A.; Pfefferkorn, F.E. Analytical and Experimental Investigation of Thermocapillary Flow in Pulsed Laser Micropolishing. *J. Micro Nano-Manuf.* **2014**, *2*, 21010. [CrossRef]

42. Richter, B.; Chen, S.; Morrow, J.D.; Sridharan, K.; Eriten, M.; Pfefferkorn, F.E. Pulsed laser remelting of A384 aluminum, part II: Modeling of surface homogenization and topographical effects. *J. Manuf. Process.* **2018**, *32*, 230–240. [CrossRef]

43. Nüsser, C.; Kumstel, J.; Kiedrowski, T.; Diatlov, A.; Willenborg, E. Process- and Material-Induced Surface Structures During Laser Polishing. *Adv. Eng. Mater.* **2015**, *17*, 268–277. [CrossRef]

44. Kaplan, A.F.H. Influence of the beam profile formulation when modeling fiber-guided laser welding. *J. Laser Appl.* **2011**, *23*, 42005. [CrossRef]

45. Völl, A.; Vogt, S.; Wester, R.; Stollenwerk, J.; Loosen, P. Application specific intensity distributions for laser materials processing: Tailoring the induced temperature profile. *Opt. Laser Technol.* **2018**, *108*, 583–591. [CrossRef]

46. Khare, J.; Kaul, R.; Ganesh, P.; Kumar, H.; Jagdheesh, R.; Nath, A.K. Laser beam shaping for microstructural control during laser surface melting. *J. Laser Appl.* **2007**, *19*, 1–7. [CrossRef]

47. Kou, S. *Welding Metallurgy*; John Wiley & Sons, Inc.: Hoboken, NJ, USA, 2002; ISBN 0471434914.

48. Nüsser, C.; Wehrmann, I.; Willenborg, E. Influence of Intensity Distribution and Pulse Duration on Laser Micro Polishing. *Phys. Procedia* **2011**, *12*, 462–471. [CrossRef]

49. Pirch, N.; Linnenbrink, S.; Gasser, A.; Wissenbach, K.; Poprawe, R. Analysis of track formation during laser metal deposition. *J. Laser Appl.* **2017**, *29*, 22506. [CrossRef]

50. Willenborg, E. Polieren von Werkzeugstaehlen mit Laserstrahlung. Ph.D. Thesis, RWTH Aachen University, Shaker. Aachen, Germany, 2006.

51. Nüsser, C. Lasermikropolieren von Metallen: Laser Micro Polishing of Metals. Ph.D. Thesis, RWTH Aachen University, First Edition, Apprimus. Aachen, Germany, 2018.

52. Liebing, C. Beeinflussung Funktioneller Oberflächeneigenschaften von Stahlwerkstoffen durch Laserpolieren. Ph.D. Thesis, RWTH Aachen University, Aachen, Germany, 2010.

53. Tolochko, N.K.; Khlopkov, Y.V.; Mozzharov, S.E.; Ignatiev, M.B.; Laoui, T.; Titov, V.I. Absorptance of powder materials suitable for laser sintering. *Rapid Prototyp. J.* **2000**, *6*, 155–161. [CrossRef]

54. Boley, C.D.; Mitchell, S.C.; Rubenchik, A.M.; Wu, S.S.Q. Metal powder absorptivity: Modeling and experiment. *Appl. Opt.* **2016**, *55*, 6496–6500. [CrossRef] [PubMed]

55. Temmler, A.; Liu, D.; Drinck, S.; Luo, J.; Poprawe, R. Experimental investigation on a new hybrid laser process for surface structuring by vapor pressure on Ti6Al4V. *J. Mater. Process. Technol.* **2020**, *277*, 116450. [CrossRef]

56. Spranger, F.; Hilgenberg, K. Dispersion behavior of TiB2 particles in AISI D2 tool steel surfaces during pulsed laser dispersing and their influence on material properties. *Appl. Surf. Sci.* **2019**, *467–468*, 493–504. [CrossRef]

57. Weber, R.; Graf, T.; Berger, P.; Onuseit, V.; Wiedenmann, M.; Freitag, C.; Feuer, A. Heat accumulation during pulsed laser materials processing. *Opt. Express* **2014**, *22*, 11312–11324. [CrossRef]

58. Weber, R.; Graf, T.; Freitag, C.; Feuer, A.; Kononenko, T.; Konov, V.I. Processing constraints resulting from heat accumulation during pulsed and repetitive laser materials processing. *Opt. Express* **2017**, *25*, 3966–3979. [CrossRef] [PubMed]

59. Temmler, A.; Pirch, N. Investigation on the mechanism of surface structure formation during laser remelting with modulated laser power on tool steel H11. *Appl. Surf. Sci.* **2020**, *526*, 146393. [CrossRef]

60. Temmler, A.; Pirch, N.; Luo, J.; Schleifenbaum, J.H.; Häfner, C.L. Numerical and experimental investigation on formation of surface structures in laser remelting for additive-manufactured Inconel 718. *Surf. Coat. Technol.* **2020**, *403*, 126370. [CrossRef]

61. Pfefferkorn, F.E.; Morrow, J.D. Controlling surface topography using pulsed laser micro structuring. *CIRP Ann.* **2017**, *66*, 241–244. [CrossRef]

62. Preußner, J.; Oeser, S.; Pfeiffer, W.; Temmler, A.; Willenborg, E. Microstructure and Residual Stresses of Laser Structured Surfaces. *AMR* **2014**, *996*, 568–573. [CrossRef]

63. Preußner, J.; Oeser, S.; Pfeiffer, W.; Temmler, A.; Willenborg, E. Microstructure and residual stresses of laser remelted surfaces of a hot work tool steel. *IJMR* **2014**, *105*, 328–336. [CrossRef]

64. Spranger, F.; de Oliveira Lopes, M.; Schirdewahn, S.; Degner, J.; Merklein, M.; Hilgenberg, K. Microstructural evolution and geometrical properties of TiB2 metal matrix composite protrusions on hot work tool steel surfaces manufactured by laser implantation. *Int. J. Adv. Manuf. Technol.* **2020**, *106*, 481–501. [CrossRef]

metals

Article

The Effect of Silica Sand Proportion in Laser Scabbling Process on Cement Mortar

Tam-Van Huynh [1], Youngjin Seo [1] and Dongkyoung Lee [1,2,3,*]

[1] Department of Future Convergence Engineering, Kongju National University, Cheonan 31080, Korea; tamhv.engi@gmail.com (T.-V.H.); syjvlfry1004@gmail.com (Y.S.)
[2] Department of Mechanical and Automotive Engineering, Kongju National University, Cheonan 31080, Korea
[3] Center for Advanced Powder Materials and Parts of Powder (CAMP2), Kongju National University, Cheonan 31080, Korea
*   Correspondence: ldkkinka@kongju.ac.kr; Tel.: +82-41-521-9260

Abstract: Cement mortar composite has a wide range of applications on construction sites, including masonry, plastering and concrete repair. In construction sites, scabbling process is a method to remove from a few millimeters to several centimeters of defect concrete surfaces. As a result, it is essential to investigate the scabbling characteristics for cement mortar with different silica sand proportion in laser scabbling process. In this study, 5 types of cement mortar with different silica sand proportions in mixing were fabricated and scabbled by using a high-density power laser beam. The effects of silica sand proportion in color changing and penetration depth of the samples after laser scabbling process were studied. Furthermore, the generation of micro-cracks and pores were observed by using scanning electron microscopy (SEM). In addition, chemical composition changes between processed zone and non-processed zone were also evaluated by Energy Dispersive X-ray (EDX) analysis. The results of this study are expected to provide valuable knowledge in understanding of the laser scabbling process for cement-based materials.

Keywords: cement-based material; laser scabbling; microstructural analysis; chemical analysis; thermal properties

Citation: Huynh, T.-V.; Seo, Y.; Lee, D. The Effect of Silica Sand Proportion in Laser Scabbling Process on Cement Mortar. *Metals* **2021**, *11*, 1914. https://doi.org/10.3390/met11121914

Academic Editors: Sergey N. Grigoriev, Marina A. Volosova and Anna A. Okunkova

Received: 6 November 2021
Accepted: 25 November 2021
Published: 27 November 2021

Publisher's Note: MDPI stays neutral with regard to jurisdictional claims in published maps and institutional affiliations.

1. Introduction

Scabbling method is a technique for removing thin layers of stone or concrete that can range in thickness from a few millimeters to several centimeters. Scabbling has also been used in many applications, including the decontamination of radioactive surface layers in nuclear power plants removing road markings, many other applications in traffic and construction works. There are several scabbling methods such as electro-hydraulic scabbling (EHS), robotic wall scabbler, piston scabbler, and handle scabbler. The EHS method achieves a scabbling depth of up to 1 inch and a scabbling rate of 30 ft^2/h, whereas the piston scabbler method achieves a depth of 1/8 inch and a scabbling rate of 130 ft^2/h [1,2]. However, there are several drawbacks, including the massive weight of the working system, high current, high voltage, mechanical reaction force, wear down of equipment, and negative environment impact, as well as the operator's health [3].

Laser-aided manufacturing (LAM) has been widely employed to overcome the disadvantages of previous techniques in various fields, including car manufacturing, aerospace, and semiconductor [4–6]. Due to its advantages (e.g., high precision, high speed, non-contact method, remote control capabilities), the applicability of LAM has also been investigated on concrete. Numerous studies on the interaction between laser beam and concrete were carried out. For instance, Seo et al. [7] employed a high-power fiber laser cutting for 50 mm thick cement-based material. The results demonstrated that a line energy of 1.22×10^{14} J/m^3 was required to fully cut the 50 mm thick cement paste. On the other hand, the effect of material composition on penetration depth, microstructural characteristic and chemical changes in the laser cutting of cement-based material were also studied [8,9]. Muto et al. [10] used a 7 kW fiber laser at a scanning speed of 2.5 mm/s, and a spot size of 10 mm. The results pointed out that 100 mm in thickness of concrete slabs was completely cut after 21 passes. In laser drilling concrete field, Kaori et al. [11] obtained drilling diameters of 4 mm to 6 mm and depths of around 50 mm in concrete samples with a compressive strength of 20–100 N/mm^2 by using the appropriate laser settings. Meanwhile, the use of laser for scabbling on cement-based materials has not seemed to be attractive. In addition, to examine the effect of concrete composition, curing regime and aging on the laser scabbling process, Peach et al. [12–15] exploited IPG Photonics YPS-5000 5 kW-Continuous wave laser with nominal beam diameter of 60 mm to scabble on several concrete types, the effect curing regime, aging concrete composition were reported. The authors revealed that two major mechanism of this scabbling method were thermal stress and pore stress spalling of concrete. These results were obtained due to low power density laser. However, the using of high-power density laser in scabbling concrete has not been reported.

Several studies have carried out to investigate the interaction of laser with cement-based materials including laser cutting, laser drilling, laser glazing and laser scabbling. However, the effect of silica sand proportion in laser scabbling on cement-based material has not been studied yet. As a result, it is critical to comprehend the impact of silica sand composition in laser scabbling process. The aim of this study is obtaining a fundamental understanding of the interaction between laser and cement mortar with the differences of silica sand proportions. The formation of the glassy layer and heat-affected zone (HAZ) are the major results in the laser scabbling process on cement mortar. Thus, the study of changes in each region are needed. After laser scabbling process, all samples were evaluated in morphological changes such as color changing and scabbling penetration depth. In addition, the changes in the microstructure of each region of the scabbled samples were also determined by using a scanning electron microscope (SEM). Energy dispersive X-ray (EDX) analysis was also used to analyze changes in the chemical composition and the chemical distribution in each zone. According to the findings of this study, it is believed that the using of laser scabbling on cement based materials is applicable for walls and floor at construction sites, as well as for removing the radioactive layers in nuclear facilities.

2. Materials and Mix Design

A series of cement mortars (CM) were prepared to study the effect of the proportion of silica sand in laser scabbling. The materials involved in this study were Ordinary Portland Cement (OPC), silica sand containing about 93% SiO_2 weight with a median size range of 0.15~0.53 mm. Mineral admixtures consist of cement and silica sand whose chemical compositions are shown in Table 1. In addition, Table 2 shows the proportions of the mortar samples. The label of the samples was given the same as the silica sand to cement ratio. The specimens were fabricated by mixing cement and silica sand in a laboratory mixer for 60 s before adding water and mixing for 3 min. After mixing, the mixtures were then poured into a cylindrical shape with a diameter and thickness of 53 ± 0.5 mm and 12 ± 0.35 mm, respectively. To improve hydration, the cast samples were enclosed in a plastic sheet. The specimens were removed from the mold after 24 h of storage at room temperature and immersed in a water tank for 28 days of curing. After that, the samples were then taken out and dried to eliminate the remaining water.

Table 1. Chemical compositions of cement and silica sand measured by XRF analysis.

	Chemical Component (%)									
	CaO	SiO_2	Fe_2O_3	SO_3	Al_2O_3	K_2O	MgO	P_2O5	ZnO	SrO
Cement	62.93	17.62	8.28	4.12	3.25	2.08	0.92	0.31	0.28	0.21
Silica sand	0.44	93.01	0.95	-	3.61	1.28	-	0.67	-	0.04

Table 2. Mix compositions of the mortar specimens (proportions by weight).

Samples	Materials		
	Cement	Water	Silica Sand
CM0.2	1	0.3	0.2
CM0.4	1	0.3	0.4
CM0.6	1	0.3	0.6
CM0.8	1	0.3	0.8
CM1.5	1	0.3	1.5

3. Experimental Setup

The experimental schematic for the scabbling process of cement mortar is illustrated in Figure 1a. The laser source exploited in this study was ytterbium-pulsed fiber laser (IPG-YLPM, IPG photonics, model YLP-HP IPG photonics, Southbridge, MA, USA) with a wavelength of 1065 nm, the laser pulse duration of 30 ns, and laser spot size of 40 μm. This laser has a beam quality M^2 of 1.3. Furthermore, The F-theta lens are designed to focus the laser beam transmitted through fibers from a laser source. The focal distance from the F-Theta lens to scabbling area is 180 mm. In this study, the laser power was set at 180 W and 2 mm/s for scanning speed. In Table 3, the details of each parameter are presented. In addition, depending on dimension and minimizing the cracking of tested samples, the scabbled area was chosen 10×10 mm^2 (Figure 1b). To ensure that laser irradiation could cover the working area, the scabbling path was divided into two sections: (1) a square boundary line and (2) zigzag hatching lines. The pitch of the zigzag hatching lines was chosen at 0.05 mm to advance this scabbling method, as shown in Figure 1b.

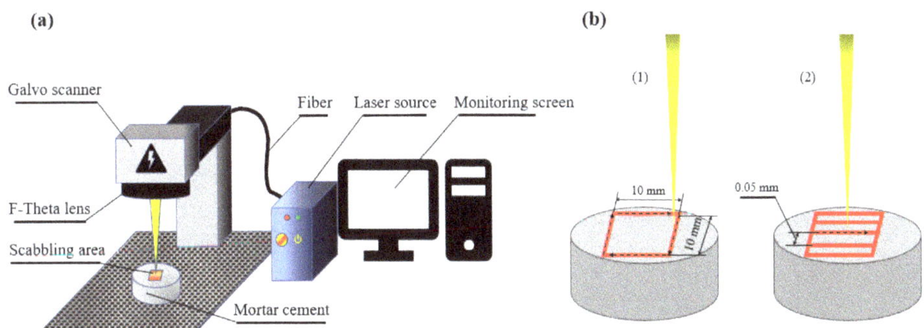

Figure 1. (**a**) Experimental setup, (**b**) scabbling path (while (1) a square boundary line and (2) zigzag hatching lines).

Table 3. Laser parameters of the scabbling experiment.

LASER Parameter	
Out-put Power [W]	180
Frequency [kHz]	403
Pulse duration [ns]	30
Laser spot diameter [μm]	40
Scanning speed [mm/s]	2
Power density [W/cm^2]	1.432×10^7
Peak power [W]	1.49×10^4
Pulse energy [mJ]	0.447

4. Results and Discussions

4.1. Morphological Analysis

Most of the OPC has a large composition of Al_2O_3, SiO_2, and Fe_2O_3, which are the fundamental elements for creating an amorphous glassy layer at high-temperature conditions [16]. The evaporation of free water molecules and the phase transition of the specimen's composition were caused by the thermal effect of laser irradiation on the surface of cement mortar, resulting in the formation of the glassy layer. Through observing the top, bottom and cross-section surfaces as shown in Figures 2 and 3, respectively. The glassy layer formed on the specimens' processed zone. Moreover, the color of the glassy layer changed from scorched (Figure 2a) to green and even turquoise as seen in Figure 2d,e, respectively. One explanation for this phenomenon could be the presence of metal transition ions in oxidation states, especially ferric ions in the Al^{3+} and Fe^{2+} oxidation states. At high temperatures, the Al^{3+} and Fe^{2+} exposed green and blue, respectively [17]. Furthermore, Wignarajah et al. [18] reported that the presence of metallic oxides in cement samples contributed to color changes on the processed zone impacted by the laser. Table 4 shows the color of the glassy layer that correspond to the metallic oxides in the material composition.

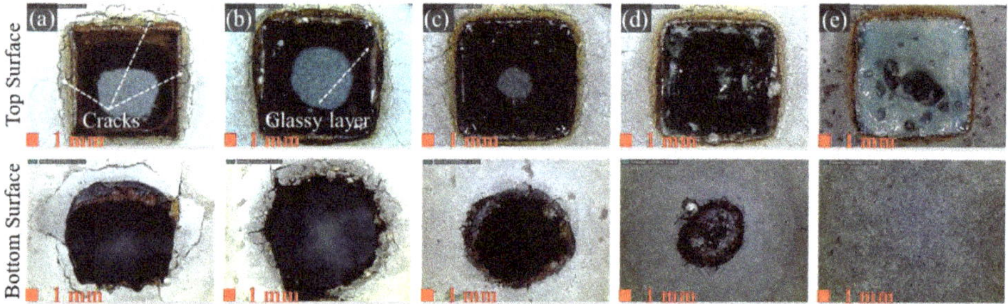

Figure 2. Photographs of the top and bottom surfaces: (**a**) CM0.2; (**b**) CM0.4; (**c**) CM0.6; (**d**) CM0.8; (**e**) CM1.5.

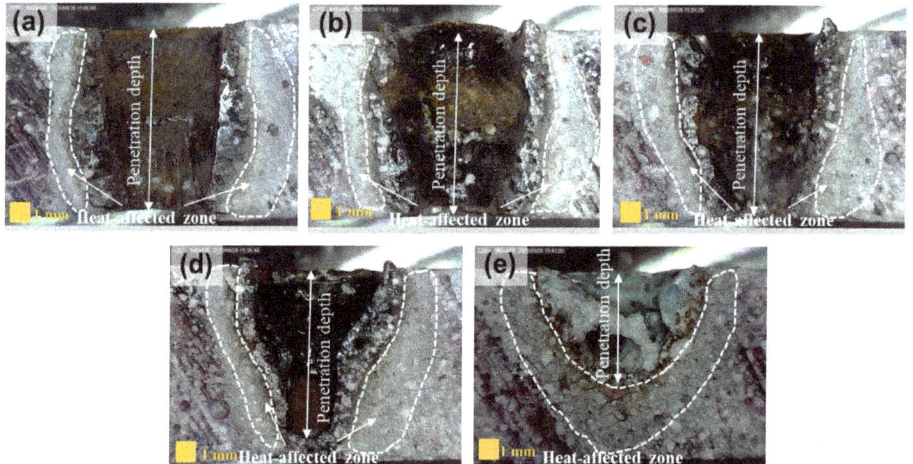

Figure 3. Photographs of cross section: (**a**) CM0.2; (**b**) CM0.4; (**c**) CM0.6; (**d**) CM0.8; (**e**) CM1.5.

Table 4. Example of colorful glassy layers produced by laser irradiation on the surface of zeolite mortar [18].

Type of Oxide	Color Produced
Cr_2O_3	Light to dark green
CoO	Light to dark blue
MnO	Brown
CuO	Brown, black, brick red
Fe_2O_3	Grey to black

During the laser process, a thermal gradient between the melting zone and the substance material was generated, resulting in the development of thermal stress. In addition, the significant temperature gradient between the laser irradiation temperature and the ambient temperature after the scabbling process caused crack formation. Furthermore, the significant reduction of cracks in the top and bottom surfaces with increasing in silica sand proportions of samples was visually observed, as can be seen in the Figure 2.

After the scabbling experiment, the samples were cut in a cross-section to observe the three main material sections: (1) non-processed zone, (2) heat affected zone (HAZ), and (3) processed zone. Furthermore, glassy layer was generated in processed zone. The procedure for obtaining the section view is shown in Figure 4.

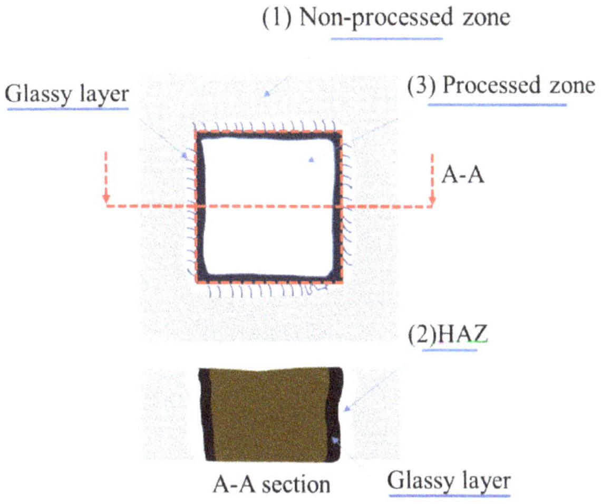

Figure 4. Illustrating image of three main material sections.

Seo et al. [8] also confirmed that adding silica sand into basic cement-based materials results in decreasing the penetration depth. Meanwhile, Figure 4 reveals a significant decrease in penetration depth while increasing silica sand in proportion from 0.2 to 1.5. In another words, thermal conductivity (κ) describes how quickly heat flows through a material from the hotter side to the colder side under steady-state conditions, and thermal diffusivity (d_t) describes how well a material can spread heat [19]. The relationship between thermal conductivity and thermal diffusivity is proportional, as shown by Equation (1). As a result, a decrease in thermal conductivity leads to a decrease in thermal diffusivity.

$$d_t = \frac{\kappa}{c_p \cdot \rho} \tag{1}$$

where, c_p [J·Kg^{-1}·k^{-1}] is specific heat, κ [W·m^{-1}·K^{-1}] is thermal conductivity, d_t [m^2·s^{-1}] is thermal diffusivity and ρ [Kg·m^{-3}] is density. According to Ganeev et al. [20] the

Equation (2) expresses the relationship between thermal penetration depth by laser beam energy and the thermophysical properties of the specimen surface as equation below.

$$L_d = \sqrt{d_t * \tau} \tag{2}$$

where, L_d [m] is thermal penetration depth, and τ [s] is pulse duration. Furthermore, higher silica proportion reduced the thermal conductivity of cement mortar [21]. As a result, increasing the silica sand proportions in cement-based materials lowered the scabbling penetration depth. Along with that, the laser beam was absorbed and limited due to the rapid formation of melting layer induced by the heat effect of laser irradiation.

Peach et al. [12] reported that two main mechanisms for laser scabbling of concrete are pore pressure spalling and thermal stress spalling. Pore pressure spalling was caused by the rapid increase in pore pressure caused by the vaporization of free water, while the formation of thermal stress as a result of severe thermal gradients induced by high heat rates and low thermal conductivity of concrete causes thermal stress spalling. However, they were able to achieve those results by using a low-power density laser of 1.768 W/cm^2. On the other hand, in present study employed a laser with a high-power density laser of 1.432×10^7 W/cm^2. the evaporation of material was the major process when using this scabbling method. In addition to the evaporation of the material in the processed zone and the formation of the glassy laser were described above.

The heat affected zone (HAZ) is a zone of the base material that was not melted but impacted by the heat generated during the laser scabbling process. According to Maruyama et al. [22], the temperature for the dehydration reaction of Ca(OH)$_2$ is 400 °C or higher. From this study it was reported that microcracks occurred in cement-based materials due to the breaking of chemical and physical bonds caused by temperature. The dehydration of calcium hydroxide is shown in Equation (3) [22]:

$$\text{Ca(OH)}_2 + \text{heat} \rightarrow \text{CaO} + \text{H}_2\text{O} \tag{3}$$

As mentioned earlier, increasing the proportion of silica sand in the specimen decreased the thermal conductivity of the samples. As a result, heat transfer from the processed zone was limited. Along with that, the rapid development of the glassy layer absorbed heat energy and reduced the transmission of the heat energy generated by the laser beam. As a result, the heat generated in the samples containing higher silica sand proportion focused on the processed zone, resulting in a larger HAZ. Moreover, the heat of laser process caused the dehydration and decomposition of cement-based materials, thus the color in the HAZ changed and appeared in a whitish grey [23,24]. In all cases, the heat-affected zone can be clearly observed in a grey-white color, as can be seen in Figure 3.

4.2. Microstructural Analysis

Scanning electronic microscopy (SEM) was used to investigate changes in microstructure in samples subjected to different silica sand proportions. As shown in Figure 5a–e, micro-cracks found by SEM on the cement paste region might explain the apparent surface cracks in samples such as CM0.2 (Figure 2a), CM0.4 (Figure 2b), and CM0.6 (Figure 2c). The dehydration of cement paste was generated by heat affected by laser irradiation. Micro-cracks were generated in the cement hydrate region as can be seen Figure 5a,b. More cracks were formed in cement paste region of samples with low silica content, such as CM0.2 and CM0.4. Furthermore, Figure 5d,e show that micro-cracks formed more densely in the silica sand particles and less densely in the cement paste region, indicating that the occurrence of the cracks on the top surface (Figure 2e) decreased as the increasing of silica sand proportion in the cement mortar. Due to the higher melting temperature of silica sand, the generation of a glassy layer with the major component being silicon in the silica sand restricted the heat transmission of laser beam.

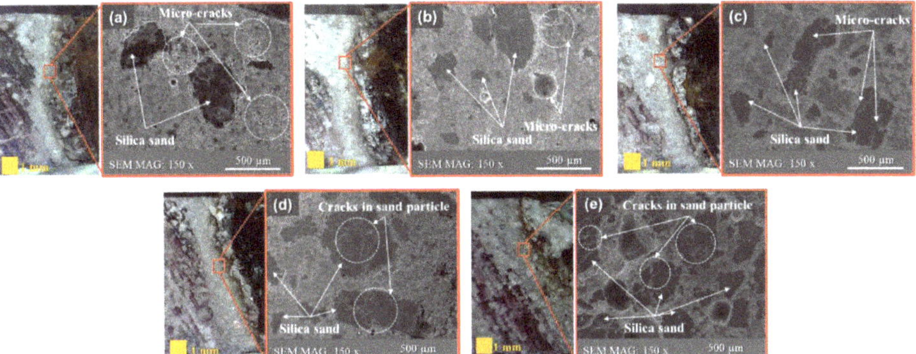

Figure 5. SEM images of heat-affected zone: (**a**) CM0.2; (**b**) CM0.4; (**c**) CM0.6; (**d**) CM0.8; (**e**) CM1.5.

Figure 6 shows the SEM images in the processed zone of cement mortar samples. The size of the pores grew with increasing of silica sand proportions in the cement mortar. Furthermore, samples with a higher silica sand content had a denser appearance of micro-cracks near the pores. Due to that, it can be assumed that the porosity of the glassy layer in samples CM0.8 and CM1.5 is higher than in other samples. It also implies that the glassy layer's strength is decreased and easily removed with an increasing of silica sand proportion.

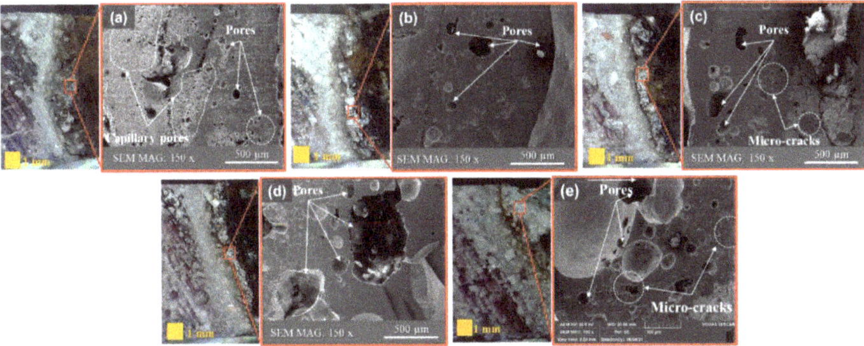

Figure 6. SEM images of processed zone: (**a**) CM0.2; (**b**) CM0.4; (**c**) CM0.6; (**d**) CM0.8 and (**e**) CM1.5.

4.3. EDX Analysis

The analytical method of energy-dispersive X-ray spectroscopy (EDX) was also used to determine the elements (elemental composition) present on any given material sample. In addition, when used in combination with SEM, allows for the analysis of near-surface elements and their amount at different positions providing a map of the sample [25]. Since the glassy layer is the zone directly affected by laser irradiation, this study investigated the changes in the chemical composition of cement-based materials by analyzing two zones, (1) the non-processed zone and (2) the processed zone (glassy layer), as shown in Figure 7. The decrease in calcium and the increase in silicon percentage are the two major changes in the specimens when comparing the chemical composition of the glassy layers and non-processed zones in the CM series, respectively. The most significant change in calcium element was observed in sample CM0.6, where the percentage of calcium element in the non-processed zone was 81.75% and decreased to 39.77% in the processed zone. In addition, at specimen CM0.8, the percentage of silicon element in the non-processed zone increased from 8.75 to 40.59% in the processed zone. Since the evaporation temperature of calcium is 1484 °C and that of silicon is 3265 °C, a part of the calcium has evaporated after solidification, which can explain the above-mentioned phenomenon. As a result of calcium evaporation, the percentage of silicon has increased while the percentage of calcium has dropped.

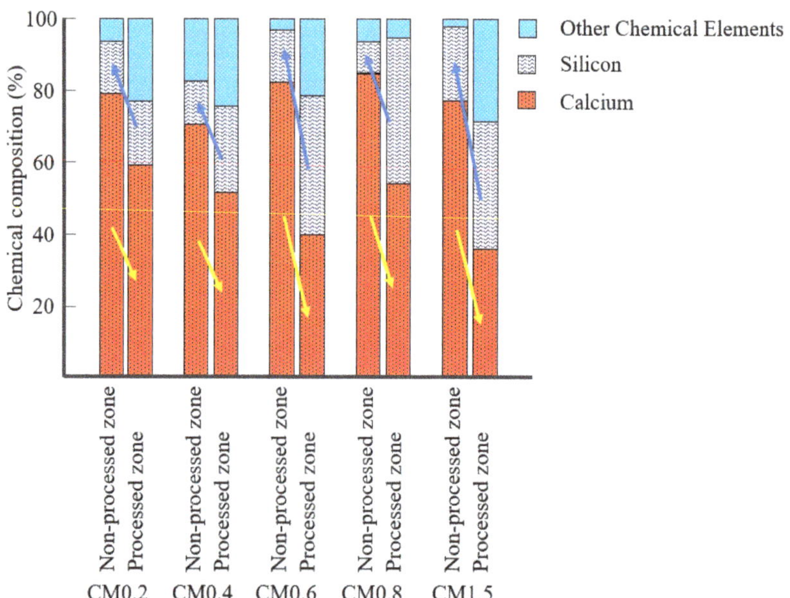

Figure 7. Energy dispersive X-ray (EDX) analysis of chemical composition change.

EDX mapping was also exploited to determine the distribution of silicon and calcium composition in non-processed and processed zone. Before the EDX mapping process, the specimens were ground after cross section cutting to achieve a high surface quality. Figure 8 shows the EDX mapping images of 2 main chemical components of silicon and calcium in each sample. The chemical mapping in non-processed zone in each sample shows the distribution of silica sand and cement paste. Meanwhile, the heat effect of laser irradiation generated a glassy layer whose major components were silica sand and cement paste, which was melted and resolidified. As a result, the heat effect of laser irradiation caused the redistribution of the element composition between the processed and non-processed zone. Furthermore, the mixing of silicon and calcium component is the obvious difference between the processed and non-processed zone. Under the direct effect of the laser beam in the processed zone, the sample components which included cement and silica sand were melted and redistributed. Thus, silica sand particles were not found in the processed zone. As can be observed in Figure 8, there is a mixing of silicon and calcium components in the processed zone. In contrast, there is no mixing or redistribution of silicon and calcium components in the non-processed zone. Silica sand particles with silicon as the major component could be detected. In addition, the chemical redistribution is significantly greater in high silica sand proportion samples such as CM0.8 and CM1.5 than in CM0.2 and CM0.4 samples. Furthermore, when comparing the distribution of silicon and calcium, the density of silicon was densely distributed in non-processed zone as the proportion of silica sand increased in high silica sand proportion samples such as CM0.8 and CM1.5.

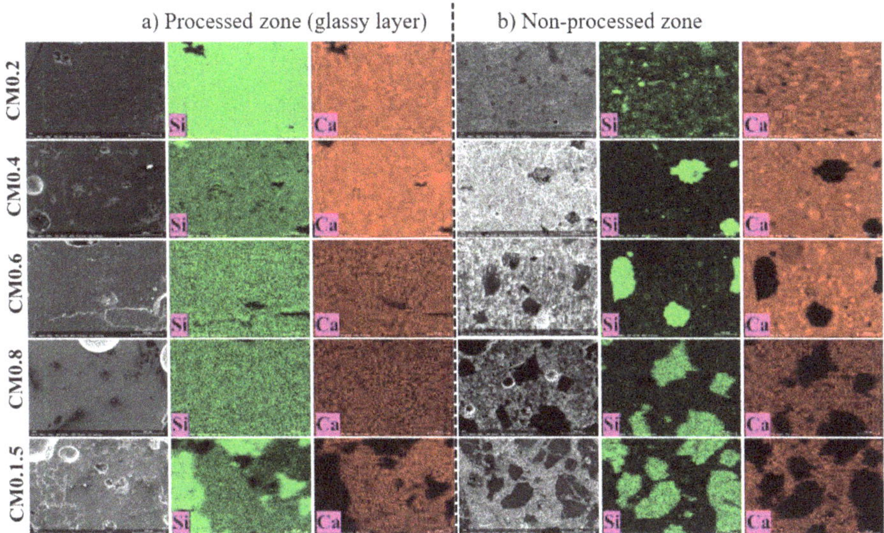

Figure 8. Elemental mapping for Ca and Si on the specimens after the laser scabbling.

5. Conclusions

Laser scabbling, which advances beyond the limitations of the previous methods, might be one of the better options for scabbling cement-based materials. Understanding the effects of silica sand on the composition of materials used in scabbling is essential to study. A pulsed laser was used in this study with a power density of 1.45×10^5 W/mm^2. Experiments on cement mortar samples CM0.2, CM0.4, CM0.6, CM0.8, and CM1.5 were conducted to investigate the effect of silica sand proportion in laser scabbling on cement mortar. Furthermore, SEM/EDX analysis explored changes in the microstructure and chemical composition of cement-based materials under the influence of laser. The main observations and results of this study are summarized as follows:

1. Increasing the silica sand proportion during laser scabbling process resulted in two primary results: (1) decreased scabbling penetration depth, and (2) fewer surface cracks on the top and bottom surfaces.
2. Samples with lower silica sand proportions such as CM0.2, CM0.4 appeared more surface-cracks than CM0.8, CM1.5 which can be explained by the more generation of microcracks in the cement paste of the HAZ.
3. Microcracks and pores were more apparent and denser in the glassy layer of the samples with higher silica sand proportions.
4. The evaporation of material was the dominant mechanism of this scabbling method due to high power density laser.
5. In laser scabbling, the chemical changes in the cement mortar were an increase in silicon percentage and a decrease in calcium percentage when compared between the processed zone and non-processed zone. The main reason for this phenomenon is that the evaporation temperature of silica sand is greater than cement.

Due to the laboratory scale, the use of silica sand and cement types was limited in this study. On the other hand, laser scabbling on cement-based material is a work that is dependent on specific silica sand and cement types in specific locations. So that, in future work, the various types of silica sand and cement in making cement mortar is needed to investigate. Furthermore, experiments are required to determine the effects of silica sand proportion on the influence of volume removal, the surface temperature of samples scabble using a high-power density laser. Furthermore, the optimization of laser parameters is considered to achieve the optimal result in depth penetration and volume removal.

Author Contributions: D.L. and Y.S. conceived and designed the experiment; D.L. and Y.S. performed the experiments; D.L., Y.S. and T.-V.H. analyzed the data; D.L., Y.S. and T.-V.H. wrote the paper. All authors have read and agreed to published version of the manuscript.

Funding: The research described herein was sponsored by the National Research Foundation of Korea (NRF) grant funded by the Korean government (MSIT; Ministry of Science and ICT) (No. 2019R1A2C1089644) and by the Korea Innovation Foundation grant funded by the Korea government (MSIT) (No. 2020-DD-SB-0159). This work was also supported by the research grant of the Kongju National University in 2021.

Institutional Review Board Statement: Not applicable.

Informed Consent Statement: Not applicable.

Data Availability Statement: Data is contained within the article.

Conflicts of Interest: The authors declare no conflict of interest.

References

1. International Atomic Energy Agency. *State of the Art Technology for Decontamination and Dismantling*; IAEA: Vienna, Austria, 1999; p. 81.
2. Information Concerning Nuclear Installations Decommissioning Projects. In *Decontamination Techniques Used in Decommissioning Activities*; Nuclear Energy Agency: Paris, France, 1999; p. 51.
3. Feltcorn, E. *Technology Reference Guide for Radiologically Contaminated Surfaces*; United States Environmental Protection Agency: Washington, DC, USA, 2006.
4. Trinh, L.N.; Lee, D. The characteristics of laser welding of a thin aluminum tab and steel battery case for lithium-ion battery. *Metals* **2020**, *10*, 842. [CrossRef]
5. Trinh, L.N.; Lee, D. The effect of using a metal tube on laser welding of the battery case and the tab for lithium-ion battery. *Materials* **2020**, *13*, 4460. [CrossRef] [PubMed]
6. Lee, D.; Mazumder, J. Effects of laser beam spatial distribution on laser-material interaction. *J. Laser Appl.* **2016**, *28*, 032003. [CrossRef]
7. Seo, Y.; Lee, D.; Pyo, S. High-Power fiber laser cutting for 50-mm-thick cement-based materials. *Materials* **2020**, *13*, 1113. [CrossRef] [PubMed]
8. Lee, D.; Seo, Y.; Pyo, S. Effect of laser speed on cutting characteristics of cement-based materials. *Materials* **2018**, *11*, 1055. [CrossRef] [PubMed]
9. Seo, Y.; Lee, D.; Pyo, S. Microstructural characteristics of cement-based materials fabricated using multi-mode fiber laser. *Materials* **2020**, *13*, 546. [CrossRef] [PubMed]
10. Muto, S.; Tei, K.; Masuda, Y.; Miyagawa, N.; Jyosui, K.; Yamaguchi, S.; Fujioka, T. Cutting Technique for Civil Engineering Using a High-Power Laser. Rev. *Laser Eng.* **2008**, *36*, 1203–1205. [CrossRef]
11. Nagai, K.; Beckemper, S.; Poprawe, R. Laser Drilling of Small Holes in Different Kinds of Concrete. *Civ. Eng. J.* **2018**, *4*, 766. [CrossRef]
12. Peach, B.; Petkovski, M.; Blackburn, J.; Engelberg, D.L. An experimental investigation of laser scabbling of concrete. *Constr. Build. Mater.* **2015**, *89*, 76–89. [CrossRef]
13. Peach, B.; Petkovski, M.; Blackburn, J.; Engelberg, D.L. Laser scabbling of mortars. *Constr. Build. Mater.* **2016**, *124*, 37–44. [CrossRef]
14. Peach, B.; Petkovski, M.; Blackburn, J.; Engelberg, D.L. The effect of concrete composition on laser scabbling. *Constr. Build. Mater.* **2016**, *111*, 461–473. [CrossRef]
15. Peach, B.; Petkovski, M.; Blackburn, J.; Engelberg, D.L. The effect of ageing and drying on laser scabbling of concrete. *Constr. Build. Mater.* **2018**, *188*, 1035–1044. [CrossRef]
16. Lawrence, J.; Li, L. A comparative study of the surface glaze characteristics of concrete treated with CO_2 and high power diode lasers Part II: Mechanical, chemical and physical properties. *Mater. Sci. Eng. A* **2000**, *287*, 25–29. [CrossRef]
17. Mungchamnankit, A.; Kittiauchawal, T.; Kaewkhao, J.; Limsuwan, P. The color change of natural green sapphires by heat treatment. *Procedia Eng.* **2012**, *32*, 950–955. [CrossRef]
18. Wignarajah, S.; Sugimoto, K.; Nagai, K. Effect of laser surface treatment on the physical characteristics and mechanical behaviour of cement base materials. *Laser Inst. Am.* **1993**, *75*, 383–392.
19. Bouguerra, A.; Aït-Mokhtar, A.; Amiri, O.; Dio, M.B. Measurement of thermal conductivity, thermal diffusivity and heat capacity of highly porous building materials using transient plane source technique. *Int. Commun. Heat Mass Transf.* **2001**, *28*, 1065–1078. [CrossRef]

20. Ganeev, R.A. *Periodic Nanoripples Formation on the Semiconductors Possessing Different Bandgaps*; Elsevier Ltd: Amsterdam, The Netherlands, 2018; Volume 1, pp. 1–38.
21. Demirboğa, R.; Gül, R. The effects of expanded perlite aggregate, silica fume and fly ash on the thermal conductivity of lightweight concrete. *Cem. Concr. Res.* **2003**, *33*, 723–727. [CrossRef]
22. Maruyama, A.; Kurosawa, R.; Ryu, J. Effect of Lithium Compound Addition on the Dehydration and Hydration of Calcium Hydroxide as a Chemical Heat Storage Material. *ACS Omega* **2020**, *5*, 9820–9829. [CrossRef] [PubMed]
23. Khoury, G.A. Compressive strength of concrete at high temperatures: A reassessment. *Mag. Concr. Res.* **1992**, *44*, 291–309. [CrossRef]
24. Hager, I. Colour Change in Heated Concrete. *Fire Technol.* **2014**, *50*, 945–958. [CrossRef]
25. Yesilkir -Baydar, S.; Oztel, O.N.; Cakir-Koc, R.; Candayan, A. *Evaluation Techniques*; Elsevier Ltd: Amsterdam, The Netherlands, 2017.

Article

Bionic Repair of Thermal Fatigue Cracks in Ductile Iron by Laser Melting with Different Laser Parameters

Siyuan Ma [1,2], **Ti Zhou** [1,3,*], **Hong Zhou** [1,2], **Geng Chang** [1,2], **Benfeng Zhi** [1,2] and **Siyang Wang** [1,2]

[1] Key Laboratory of Automobile Materials, Jilin University, Ministry of Education, Changchun 130025, China; masiyjlu@163.com (S.M.); zhoutijlu@yeah.net (H.Z.); changgengjlu@163.com (G.C.); zhibf@163.com (B.Z.); siyangwangjlu@163.com (S.W.)

[2] College of Material Science and Engineering, Jilin University, Changchun 130025, China

[3] School of Mechanical Science and Aerospace Engineering, Jilin University, Changchun 130025, China

* Correspondence: masy17@mails.jlu.edu.cn; Tel.: +86-155-2685-4687

Received: 12 December 2019; Accepted: 3 January 2020; Published: 9 January 2020

Abstract: Nodular iron brake discs typically fail due to serious thermal fatigue cracking, and the presence of graphite complicates the repair of crack defects in ductile iron. This study presents a novel method for remanufacturing ductile iron brake discs based on coupled bionics to repair thermal fatigue cracks discontinuously using bio-inspired crack blocking units fabricated by laser remelting at various laser energy inputs. Then, the ultimate tensile force and thermal fatigue crack resistance of the obtained units were tested. The microhardness, microstructure, and phases of the units were characterized using a digital microhardness meter, optical microscopy, scanning electron microscopy, and X-ray diffraction. It was found that the units without defects positively impacted both the thermal fatigue resistance and tensile strength. The unit fabricated at a laser energy of 165.6^{+19}_{-15} J/mm^2 had sufficient depth to fully close the crack, and exhibited superior anti-cracking and tensile properties. When the unit distance is 3 mm, the sample has excellent thermal fatigue resistance. In addition, the anti-crack mechanism of the units was analysed.

Keywords: laser remelting; ductile iron; bionic crack blocked unit; repair discontinuously; thermal fatigue crack

---

1. Introduction

Ductile iron with good wear resistance, heat conductivity, and castability has become a popular cast metal material that is widely used to manufacture train brake discs [1]. Brake discs are important components of trains, and they mainly fail due to thermal fatigue cracks [2–4]. Analyzing the condition of railway train brake discs suggests that their surfaces experienced significant pressures and friction, and they often operate in hot/cold alternating environments due to frequent braking which generates large thermal stresses [5]. When the stress exceeds the limit of the material strength, thermal fatigue cracks appear and expand on the disc surface, leading to the failure and discarding of brake discs and huge wastes of resources [6]. Therefore, it is of great significance to take effective technical measures to repair the discarded brake discs.

Many efforts have been made to identify the causes of thermal fatigue damage to nodular iron brake discs and repair thermal fatigue cracks on brake disc surfaces. Goo et al. [7] found that the graphite in ductile iron was the source of thermal fatigue cracks and improved the thermal fatigue resistance by regulating its composition and metallurgical structures. Traditionally, during brake disc repair, cracks are first melted with metals using welding equipment and then refilled with liquid metals through spontaneous flow, which completely welds the cracks together. For example, according to the research of Yang et al. [8], after a brake disc was preheated to 300 °C, thermal cracks could be repaired by surfacing welding. Li et al. [9] repaired thermal fatigue cracks with CO$_2$ gas shielded welding, and immediately after welding, the workpiece was heated to about 630 °C and held for 4–6 h to eliminate welding stresses.

Such traditional repair processes are tedious, require complex procedures, long repair cycles, and thermal stresses are generated after repair, which decreased the weld strength. In addition, the use of laser cladding technology was shown to repair thermal fatigue cracks by adding powder with the same chemical composition and content as the substrate in a layer-by-layer manner [10]. Gao et al. [11] exploited the electromagnetic heating effect to repair hot fatigue cracks inside a hot extrusion die of 3Cr2W8V steel using pulse discharge. However, using this method to repair thermal fatigue cracks on ductile iron brake discs is expensive, which limits its application.

Bionics is a new interdisciplinary field that uses biological mechanisms and laws to solve human needs that has rapidly developed since the 20th century [12,13]. Bionics has had a profound impact on materials science over the past two decades, as the unique structure, composition, and corresponding superior properties of biology provide inspiration for researchers to improve the properties of materials or the reliability of structural components [14,15]. For example, the leaves of plants can resist fatigue cracking caused by storms, which is closely related to the phenomenon of biological coupling. The veins of leaves are composed of vascular bundles and mechanical tissues with strong textures, while the mesophyll is composed of parenchyma cells with soft textures [16]. The force required to sustain the propagation of cracks on the mesophyll is not sufficient to tear the veins, and when a crack encounters a hard vein, the cracking behavior is terminated or the propagation is deflected at an angle (Figure 1a). Coincidentally, the same phenomenon is observed dragonfly wings, whose membranes are covered with crisscross veins that are composed of multi-layer compound mechanisms and are the main force bearing units (Figure 1b) [17]. It has been found that cracks do not propagate along a straight line, but frequently deflect. It shows that the crack growth on the dragonfly wings is greatly hindered.

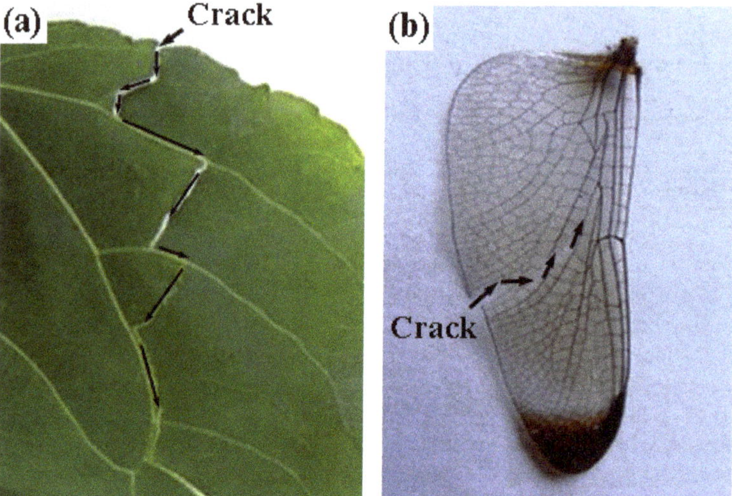

Figure 1. Deflection of crack in the leaf (**a**) and butterfly wing (**b**).

The design of coupled bionic functional materials based on leaves and wings of dragonfly can be used to improve the crack growth resistance of nodular cast iron, which provides a novel idea to repair thermal fatigue cracks. Material surfaces can be laser-treated to produce a region with a different microstructure than the substrate at a predetermined position to form a non-smooth surface in which the softer and harder points are alternately distributed. The surface also has special functions similar to biological surfaces [18]. Laser bionics has been recommended for processing points, strips, and nets on material surfaces to prepare non-smooth surfaces with improved thermal fatigue resistance and tensile properties [19]. According to Zhou et al. [20,21], lasers with different parameters have different effects on improving the thermal fatigue properties of materials.

All the aforementioned studies were aimed at improving the thermal fatigue resistance of materials, however, how to repair the thermal fatigue cracks of nodular iron brake discs is not mentioned. In this study, inspired by bionics, a new method was proposed to repair thermal fatigue cracks in brake discs using discontinuous laser melting treatment on cracks, which the service life of brake disc is prolonged and the efficiency of brake disc repair is improved.

2. Method and Materials

2.1. Experiment Materials

Material was taken from a train brake disc cracked due to thermal fatigue, which was made of nodular iron. Additionally, the chemical composition is shown in Table 1, and the microstructure is shown in Figure 2. The microstructure was composed of spheroidal graphite (G), pearlite (P), and ferrite (F). Figure 3 shows that the surface crack of the damaged brake disc was about 0.33 mm wide and about 0.54 mm deep.

Table 1. Chemical compositions of nodular cast iron.

Element	C	Si	Mn	P	S	Mg	Fe
Composition (wt%)	3.65	2.42	0.60	0.05	0.02	0.05	Bal.

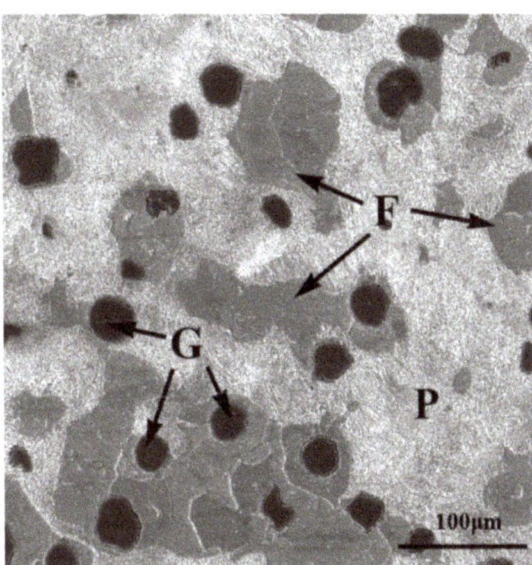

Figure 2. Microstructure of ductile iron.

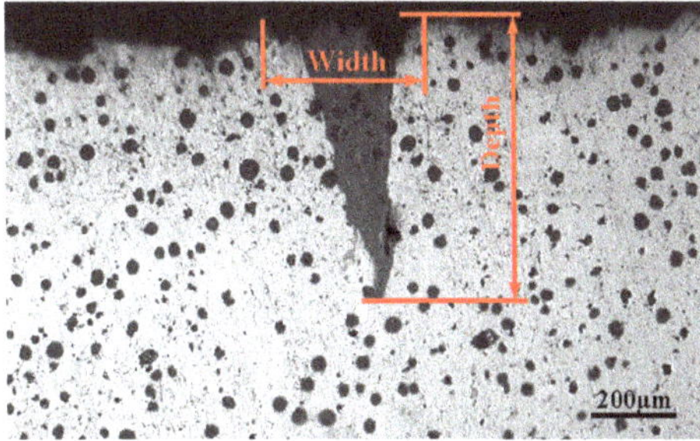

Figure 3. Crack of unrepaired thermal fatigue sample.

2.2. Sample Preparations

Figure 4 is a schematic diagram of the experimental sample, where Figure 4a is the tensile sample with a gauge length of 30 mm, a width of 20 mm, and a thickness of 5 mm. Considering the operability of the experiment, a notch with a width of about 0.3 mm and a depth of about 0.5 mm was preset in the middle of the tensile sample specimens to replace the crack equivalent. Figure 4b shows a sample for thermal fatigue test, with a size of 30 mm × 20 mm × 5 mm. A circular hole with a radius of 1.5 mm was formed 1 mm away from the upper edge to suspend the samples during thermal fatigue tests. In order to study the effect of unit spacing on the thermal fatigue performance, samples with unit spacings of 3, 5, 7 mm were prepared and labeled as T_1, T_2, and T_3. The aforementioned samples were cut using an electric spark machine (DK7732, Huadong Group, Hangzhou, China). To prevent thermal fatigue and tensile property reduction, due to the surface and side-face roughness, the specimens were mechanically polished progressively by various grits of silicon-carbide-impregnated emery papers prior to laser biomimetic treatment. Oil stains were cleaned with an ultrasonic cleaner containing an acetone solution to form smooth and clean surfaces.

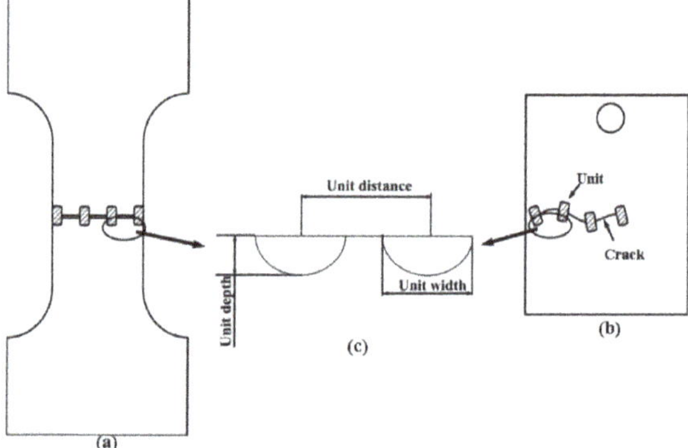

Figure 4. Sketch of crack repair. (**a**) Tensile test sample, (**b**) thermal fatigue test sample, and (**c**) the cross section of the sample.

All test samples imitating the anti-crack structure of leaves were fabricated using a solid-state Nd-YAG pulsed laser (XL-500WF, Rofin, Munich, Germany) with a wavelength of 1064 nm and a maximum rated output power of 500 W (Figure 5) [22]. Then, long cracks were segmented into several smaller cracks. In contrast, the structure of the laser remelting zone was obviously different from that of the substrate; thus, the laser remelting zone was defined as a bionic crack blocked unit (unit) for a more convenient discussion. The units were fabricated by a pulse laser with different laser parameters, which are shown in Table 2. Through the work-bench movement, a unit with a length of 3 mm processed by a laser beam was perpendicular to the direction of crack propagation. The unit distance was 2.5 mm. In addition, the untreated specimens were also compared to the laser-treated specimens.

Table 2. Details of the biomimetic specimens and laser processing parameters.

Sample	Electric Current (A)	Pulse Duration (ms)	Frequency (Hz)	Laser Spot Diameter (mm)	Laser Energy Density (J/mm^2)	Laser Power (W)
NO. 1	95	8	10	1	80.4	212.8
NO. 2	110	8	10	1	96.3	246.4
NO. 3	125	8	10	1	116.4	280.0
NO. 4	140	8	10	1	144.8	313.6
NO. 5	155	8	10	1	165.5	347.2

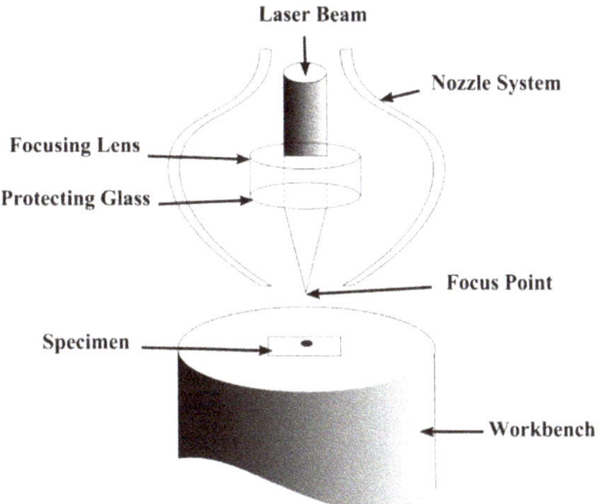

Figure 5. The schematic diagram of laser processing.

2.3. Experimental Methods

After laser processing, the cross-sections of each unit were cut parallel to the laser direction, and their width and depth were observed under an optical microscope (OLYMPUS, PMG3, Tokyo, Japan). The microstructure was characterized on a scanning electron microscope (SEM, Zeiss, Evo18, Oberkochen, Germany), when the unit cross-sections were smoothed with sandpaper, polished with woolen polished cloth and then corroded for 40 s with a 4% nitric acid alcohol solution. The microhardness of the units was measured using a digital micro hardness meter (Hua Yin, HVS-1000A, Beijing, China) with a 200 g applied load. In addition, phase structures were determined by X-ray diffraction (D/Max, 2500PC, Tokyo, Japan) equipped with Cu Ka radiation at an operating voltage of 40 kV, and a current of 40 mA and a scanning speed of 4°/min.

2.4. Tensile Tests

Tensile tests were conducted using a mechanical testing machine (MTS 810, MTS Systems Corporation, Minnesota, MN, USA), controlled by a servo-controlled hydraulic testing system at a strain rate of 2 mm min^{-1} at room temperature.

2.5. Thermal Fatigue Test

Thermal fatigue tests were carried out by using a self-made thermal fatigue testing machine. Samples were heated in a high-temperature induction resistance furnace and cooled with tap water at room temperature. The heat cycle testing machine scheme is shown in Figure 6. Figure 7 illustrates the average surface-temperature variety of the test sample during one thermal cycle. Test samples were heated from room temperature to 700 °C in 160 s and then cooled to 24 °C in 5 s in one cycle. During thermal test, no samples had any externally applied loads. After tests, samples were taken out, and changes in the crack widths were measured, and the number of cold and hot cycles experienced by each sample was recorded. In this part, the thermal fatigue test was performed five times, and the average values are reported.

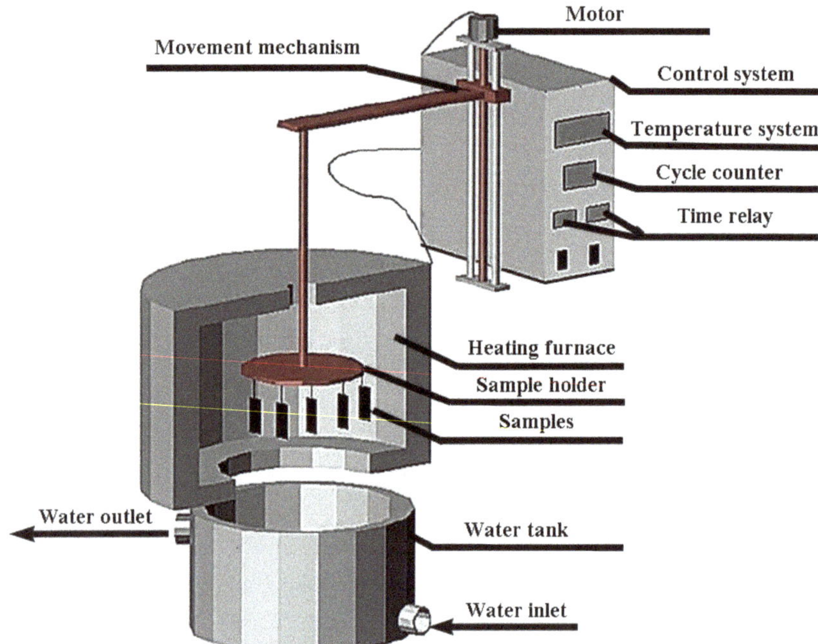

Figure 6. Sketch of the thermal-cycle testing rig.

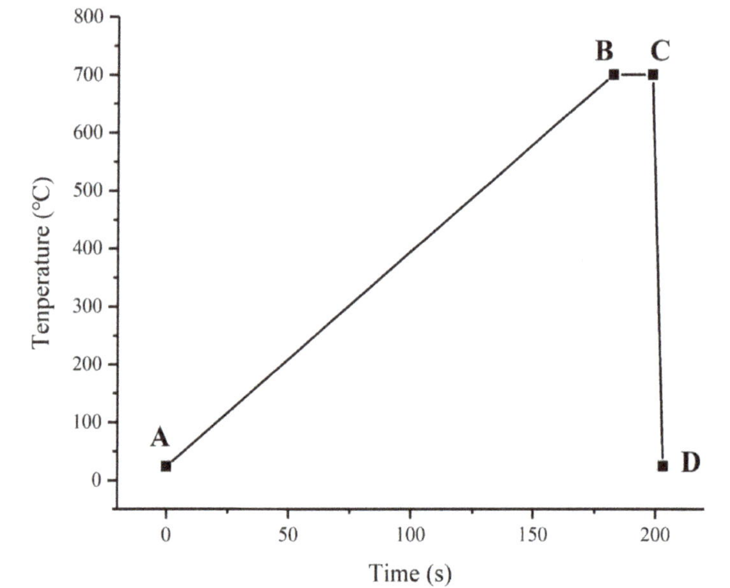

Figure 7. The actual temperature profile of one thermal cycle (Point A and D—25 °C; B and C—700 °C).

3. Results and Discussion

3.1. Microstructure and Sizes of Units

Figure 8 shows optical microscopy images of the cross-sectional morphology of units after cracks were repaired using different laser parameters. It can be clearly seen that there were obvious differences between the unit and the matrix after laser melting. The unit was parabolic, and there was no spheroidal graphite in the matrix. Due to the different laser parameters, the units had different shapes and sizes. The laser energy density used during laser treatment and the size of the prepared units are shown in Table 3.

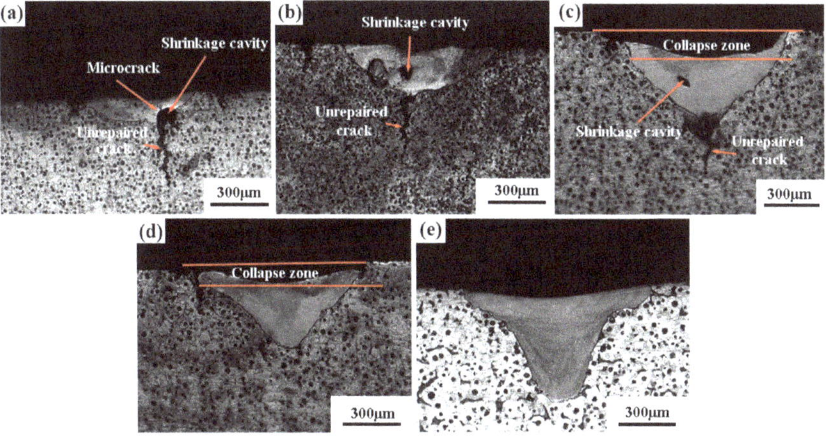

Figure 8. Optical microscopy images of the cross-sectional morphology of units: (**a**) No. 1, (**b**) No. 2, (**c**) No. 3, (**d**) No. 4, and (**e**) No. 5.

Table 3. The dimensions and microhardness of the units.

Sample	No. 1	No. 2	No. 3	No. 4	No. 5	Untreated Sample
Depth (mm)	0.11	0.25	0.37	0.44	0.59	-
Width (mm)	0.69	0.75	0.89	0.95	1.03	-
Microhardness ($HV_{0.2}$)	501	545	581	637	680	298

The sizes of units were measured from each sample using an optical microscope. Consequently, along with an increase in the laser energy input, the width of the units were in the order of No. 1 < No. 2 < No. 3 < No. 4 < No. 5, while the unit depths were in the order No. 1 < No. 2 < No. 3 < No. 4 < No. 5. The laser beam energy obeys a Gaussian function, and the energy is mainly concentrated in the central part of the laser spot [23]. The higher the laser energy density, the deeper the heat transferred to the sample, and the depth gradually increased. The width of units was mainly related to the spot diameter, and since it was the same, no significant changes in width were observed. The units gradually changed from a flat crescent shape to a U-shape. Due to the low laser energy densities, samples No. 1–4 had remelting areas that were too small to completely bridge the cracks. Since less metal was melted by heat, the flowing liquid metal could not completely fill the cracks. After solidification, cracks and holes appeared in the remelting zone, which may form new cracks and reduce the thermal fatigue resistance of the material. There were no microcracks or holes in the remelting zone of sample No. 5, and the thermal fatigue cracks were completely repaired.

Figure 9 shows the microstructure of the units under SEM. Due to the rapid increase in the sample surface temperature during laser treatment and then a rapid decline after laser treatment, the microstructure of the melting zone was a dense dendritic structure. Thus, the nucleation of the molten metal during rapid cooling occurred much faster than grain growth, which resulted in a small micron-sized microstructure. Obviously, the microstructure of samples No. 1–5 showed similar dendritic crystals but with different dendrite densities. As the laser energy density increased, the dendrite densification of No. 1–5 cells gradually increased, and the grains were gradually refined. The smaller interdendritic spacing resulted in the higher micro-hardness and the better mechanical properties theoretically [24].

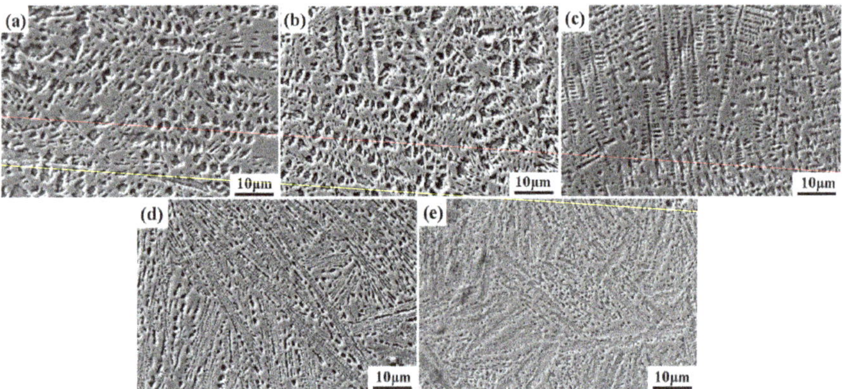

Figure 9. SEM images of the microstructure of units: (**a**) No. 1, (**b**) No. 2, (**c**) No. 3, (**d**) No. 4, and (**e**) No. 5.

X-ray diffraction was used to analyze the phase of the matrix and units on the surface of untreated and laser melted samples. The results are shown in Figure 10. Due to the large temperature gradient between the molten pool and the surrounding substrate, liquid metal was supercooled and thus recrystallization occurred at a higher cooling speed to form martensite (M). Graphite and iron formed cementite (Fe_3C), and tough residual austenite (γ-Fe) was also found in the unit. In the heat-affected zone (HAZ) between the substrate and the melted zone, a slightly lower temperature below the melting point of the cast iron led to a faster austenitization and partial dissolution of graphite. Due to rapid cooling, carbon could not evenly spread, and the carbon concentration of austenite near graphite increased, causing high-carbon needle martensite (M) and residual austenite to form after rapid cooling. Additionally, a good metallurgical bond was formed between the HAZ and the melting zone, as shown in Figure 11.

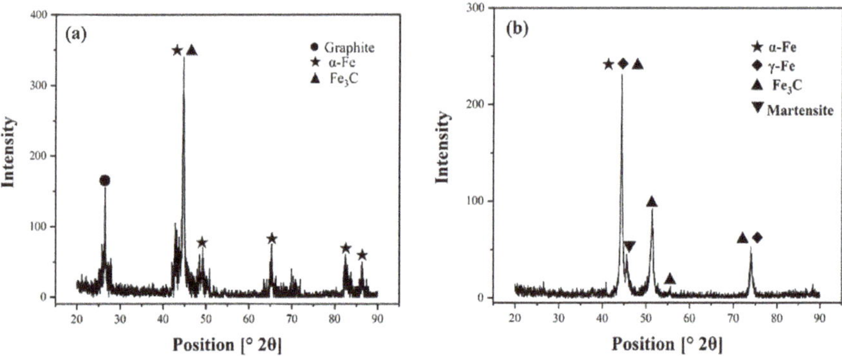

Figure 10. X-ray diffraction pattern taken from the surface of (**a**) untreated sample and (**b**) bionic blocked unit.

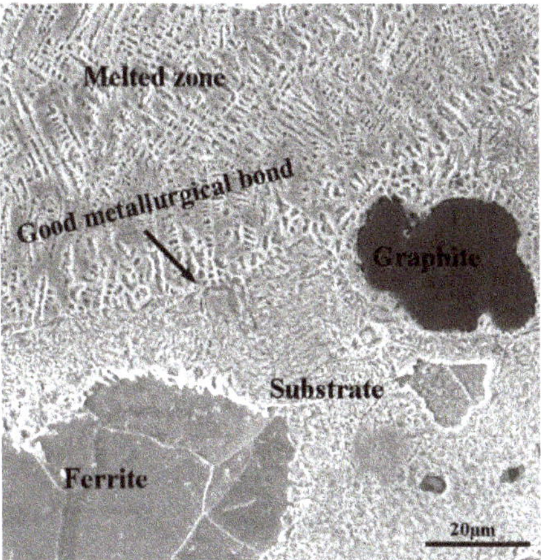

Figure 11. The interface between the melting zone and the substrate.

The reported microhardness values in Table 3 are the average of ten measurements taken at different locations in the cross-section of each unit. As the laser energy input increased, the micro-hardness of units followed the order No. 5 > No. 4 > No. 3 > No. 2 > No. 1, and the microhardness of all units were significantly higher than that of the untreated sample. Sample No. 5 displayed the maximum microhardness of 680 $HV_{0.2}$, which was 128.2% higher than the untreated sample. Fine-grain strengthening was the main reason for this order. Smaller grain sizes resulted in a larger grain boundary area, which increased the resistance of grain plastic deformation, while smaller plastic deformation leads to a higher microhardness [25]. Phase transition strengthening was also involved, as evidenced by the increased dislocation density due to the formation of martensite after laser treatment. High-density dislocations are prone to winding and plugging, which results in deformation hardening. As the crystal defects and microstructural fragmentation increased, the distribution of high pressure on the surface increased the microhardness.

3.2. Tensile Tests with Various Laser Parameters

The stress-strain curves of the tested samples are shown in Figure 12. Samples No. 1–5 showed notably better tensile properties than the unrepaired sample. As the laser energy increased, the ultimate tensile force (UTF) of specimens No. 1 (31.90 kN) < No. 2 (33.24 kN) < No. 3 (35.02 kN) < No. 4 (38.15 kN) < No. 5 (40.68 kN), while the ultimate tension of the unrepaired sample is 29.67 kN. Specifically, when the laser energy density was 165.6^{+19}_{-15} J/mm^2, the specimen reached the highest UTF of 40.68 kN, which was 37.11% higher than the untreated sample, demonstrating the best tensile property.

3.3. Tensile Mechanism

In the laser melted zone, the metal melted and solidified at an extremely rapid rate and formed a smaller and denser microstructure. According to the research of Aqida et al. [26], microstructure improvements enhanced the mechanical properties. As can be seen from Figure 13, during the initial stage of tension application, the units resisted external force, and the stress acting on the crack was reduced. After fracture, the propagation path of the crack inside the unit presented a folded line, which indicated that during tensile tests, units with good toughness concentrated most of the tensile stresses, which hindered crack propagation; therefore, units with greater strengths led to specimens with greater tensile strengths. The relationship between grain size and yield strength can be expressed by the Hall–Petch formula [27]:

$$\delta_y = \delta_i + k_y d^{-1/2} \tag{1}$$

where δ_y is the yield strength of the material, δ_i represents the resistance to dislocation movement; k_y is a constant related to the grain size, and d is the average diameter of grains. The formula indicates that the finer the grain size, the higher the strength. In addition, the larger effective cross-section size of the units led to the more significant enhancement effect on the tensile strength of specimens. For samples No. 1 to No. 5, as the laser energy input increased, the unit size was increasingly enlarged, but the grain size gradually decreased, so the strength exhibited a trend of increasing.

As shown in Figure 5, the units of samples No. 1–4 did not completely lock the cracks, and microcracks and shrinkage cavities were observed in the sample. On the one hand, these defects reduced the effective cross-sectional area of the unit; on the other hand, when samples were subjected to loads, these residual defects will serve to create new cracks, thus reducing the tensile strength of the sample. However, No. 5 displayed the highest tensile strength because there were no defects in the unit, and the cracks were completely locked.

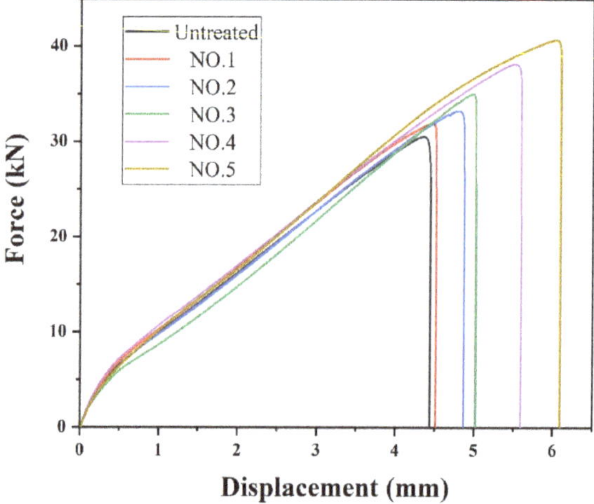

Figure 12. The stress-strain curves of the tested samples.

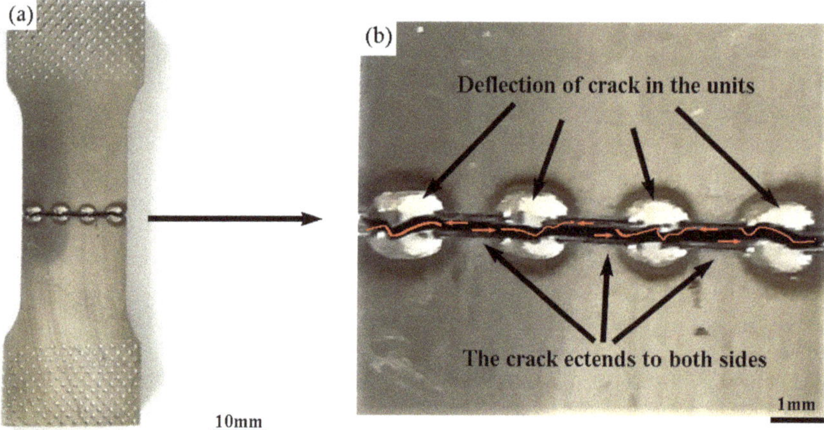

Figure 13. The fracture morphology of units. (**a**) Tensile specimen after fracture; (**b**) partial magnification of the tensile fracture.

3.4. Thermal Fatigue Tests

Since samples were heated in a heating furnace, the surface temperature of the samples slowly increased, so there was almost no temperature gradient between the sample surface and interior during heating; however, the samples expanded during heating. When samples were subsequently rapidly cooled, the surface immediately shrank, while the internal temperature was still high. Therefore, the shrinkage of the surface was limited by the internal material, and the sample surface generated tensile thermal stresses, which caused the cracks to expand.

Figure 14 shows the W-N curve which is drawn with the number of cycles (N) taken as the horizontal coordinate and the width of the thermal crack on each sample surface (W) taken as the vertical coordinate.

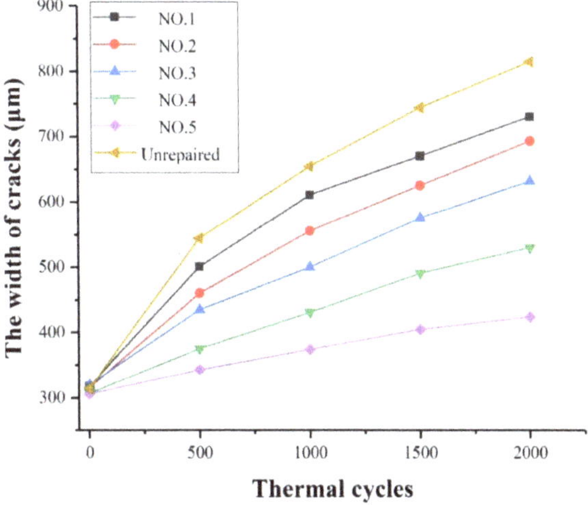

Figure 14. The W-N curve (W is the width of the thermal crack; N is the number of thermal cycles).

It is obvious that the crack width in the untreated specimen after 2000 thermal cycles was 499.21 μm, which was larger than the main crack of the repaired specimens. This shows that the locking unit can effectively prevent crack propagation. An increase in the rate of crack expansion in repaired samples was indicated by the different slopes of their curves in Figure 14, which shows that the crack width increase rates followed the order: No. 5 < No. 4 < No. 3 < No. 2 < No. 1. Figure 14 also shows that the increase in the crack width of sample No. 5 was the smallest, with a value of 118.31 μm. The specimen No. 1 with the largest crack width increment of 412.34 μm is 248.53% larger than sample No. 5.

Specimens T_1, T_2, and T_3 were subjected to go through 0 thermal cycles and 2000 thermal cycles respectively, as shown in Figure 15, which shows that the increase in the crack width increment was obviously different after 2000 thermal cycles and followed the order T_1 (57.68 μm) < T_2 (118.31 μm) < T_3 (150.62 μm) as can be seen from Figure 16. Therefore, it can be concluded that as the distance between units increased, the crack width increased gradually, and the locking effect of units on cracks gradually decreased.

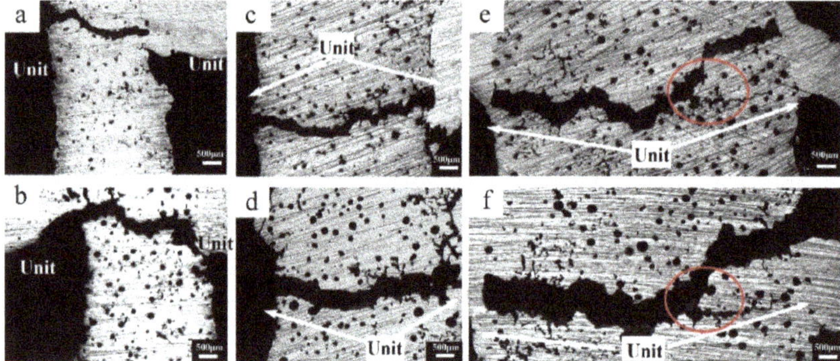

Figure 15. The change of cracks width of sample T_1, T_2, and T_3. (**a**,**b**) Sample T_1; (**c**,**d**) sample T_2; (**e**,**f**) sample T_3. (**a**,**c**,**e**) 0 thermal cycles; (**b**,**d**,**f**) 2000 thermal cycles.

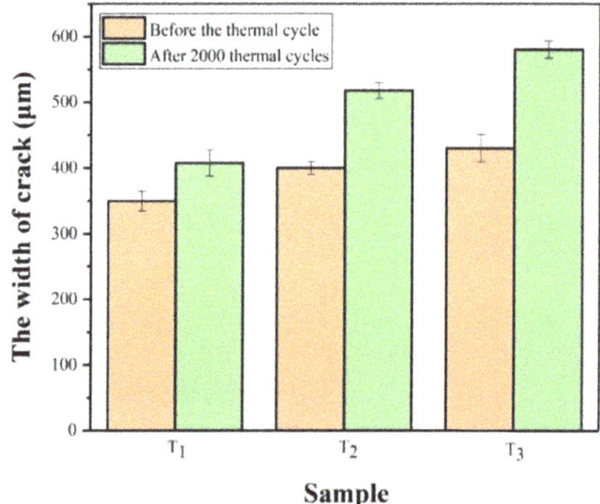

Figure 16. The crack width of sample T_1, T_2, and T_3 under different thermal cycles.

3.5. Blocking Mechanisms of Units

3.5.1. Effects of Unit Microstructure

Fatigue cracks generally occur at heterogeneous nucleation sites, such as inclusions, pores, or soft spots in microstructures [28]. It can be seen from Figure 15 above that there was more spheroidal graphite in the matrix of nodular cast iron. Similar to a hole, the graphite among the metallic matrix has no intensity. During thermal cycling, cracks are always initiated at the graphite phase, and the main crack often propagates along the graphite and matrix between the nearest graphite. Therefore, the presence of graphite reduces the thermal fatigue resistance of the material matrix to a certain extent. However, no graphite phase was found in the units, hence no bridge connection was produced between the main crack and the microcracks, whose growth depends on graphite, i.e., the crack propagation route was cut off by the unit. Therefore, the units had better thermal fatigue resistance and prevented cracks from propagating. In addition, laser remelting can refine the grains of the structure and improve the unit strength, which prevents crack propagation.

3.5.2. Effects of the Effective Size and Distance of Units

Schematic diagrams of different-sized units which prevented crack propagation are shown in Figure 17 (Figure 17A$_1$–D$_1$ is the sample without thermal cycling, and Figure 17A$_2$–D$_2$ is the sample after 2000 thermal cycles). Since the units had similar surface areas, the depth determined the size of its cross-sectional area. A greater depth of a unit indicates a greater bridging degree of a crack and also a larger cross-sectional area of the strengthened unit; however, the existence of microcracks or holes in the unit reduces the effective depth and cross-sectional area. In addition, the existence of such defects greatly reduces the strength of the unit. Under the action of thermal stress, thermal fatigue cracks will continue to expand into the unit, preventing the unit from effectively locking cracks. Therefore, when the laser energy density was increased, the unit with a larger effective size completely bridged the cracks, and since there were no cracks or holes in the interior, it had a larger tensile strength and could withstand larger thermal tension caused by thermal cycling. This allowed it to greatly reduce the thermal tension on the cracks, thus effectively preventing cracks from propagating.

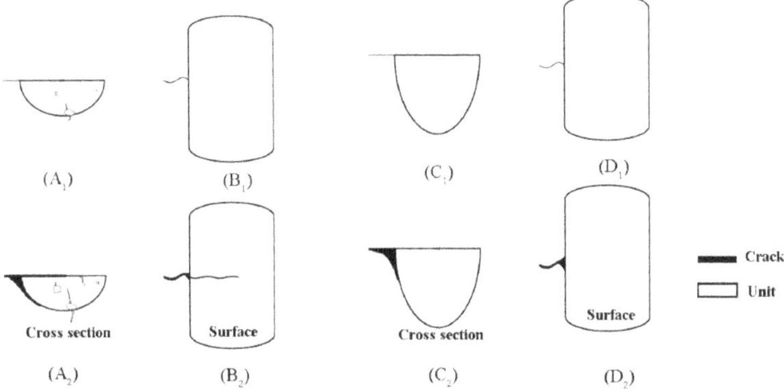

Figure 17. Schematic description for cracks blocked by units with different sizes. (**A$_1$**–**D$_1$**) before thermal cycles; (**A$_2$**–**D$_2$**) after thermal cycles.

Since thermal stress was uniformly distributed on the crack, the total force of the thermal stress on the crack depends mainly on the crack length. The smaller the distance between adjacent units, the shorter the length of the crack that is blocked by the unit, and therefore the smaller the resultant thermal stress on the crack. In addition, a smaller unit spacing increases the number of units that can share thermal stresses. When subjected to the same magnitude of thermal stress, the crack in specimens with a smaller unit spacing was subject to less thermal stress, and the cracking speed was slower. In summary, a smaller unit spacing resulted in a better crack blocking effect.

4. Conclusions

This study provides an effective and convenient method to repair brake disc cracks and to extend the service life of brake discs and save resources. Thermal fatigue cracks of ductile iron were repaired discontinuously using biomimetic locking units. The effects of size, microstructure, and spacing of the units on thermal fatigue crack repair were studied. The following conclusions were obtained:

(1) The units without defects improved both the thermal fatigue resistance and tensile strength. As the laser energy increased, the depth and microhardness of the units gradually increased, while the grain size gradually decreased. The microhardness and depths of the units reached a maximum at a laser input energy of 165.6^{+19}_{-15} J/mm^2. In addition, the sample treated at this laser energy exhibited the highest tensile force of 40.68 kN, which was 37.11% higher than the unrepaired specimen.

(2)   Compared with the substrate, the units without graphite effectively prevented crack propagation. In this experiment, after 2000 thermal fatigue cycles, the crack width of the unrepaired specimen increased by 499.21 μm, while the crack width of the repaired specimen increased between 118.31 and 412.34 μm. The crack width of the sample with a laser energy density of 165.6^{+19}_{-15} J/mm^2 increased by 118.31 μm, which was 23.70% of the unrepaired sample, showing the best crack blocking effect.

(3)   A larger effective depth resulted in a better blocking effect of thermal fatigue cracks. The presence of cracks and holes reduced the strength of the unit and weakened its crack arresting effect, which was also affected by the distance between adjacent units. In this experiment, after 2000 thermal fatigue cycles, the increase in the crack width of the unit sample with a spacing of 7 mm was 150.62 μm, while the crack width in the unit sample with a spacing of 3 mm increased by 57.68 μm, which was 61.70% smaller than that of the unrepaired sample. This demonstrates the beneficial effects of reducing the spacing of units on inhibiting thermal fatigue crack propagation.

Author Contributions: Conceptualization, T.Z. and B.Z.; data curation, S.M. and G.C.; funding acquisition, H.Z.; methodology, S.M.; project administration, H.Z.; validation, S.W.; writing—original draft, S.M.; writing—review and editing, S.M. All authors have read and agreed to the published version of the manuscript.

Funding: This work was supported by Project 985-High Performance Materials of Jilin University, Project 985-Bionic Engineering Science and Technology Innovation, National Natural Science Foundation of China (U1601203), and double first-class project by Jilin Province and Jilin University (SXGJXX2017-14).

Acknowledgments: Acknowledgments are made to the Key Laboratory of Automobile Materials of Jilin University for the use of scanning electron microscopy and X-ray diffraction.

Conflicts of Interest: The authors declare no conflicts of interest regarding the publication of this paper. I would like to declare on behalf of my co-authors that the work described was original research that has not been published previously, and not under consideration for publication elsewhere, in whole or in part. All the authors listed have approved the manuscript that is enclosed.

References

1.   Fragassa, C.; Radovic, N.; Pavlovic, A.; Minak, G. Comparison of mechanical properties in compacted and spheroidal graphite irons. *Tribol. Ind.* **2016**, *38*, 49–59.

2.   Šamec, B.; Potrč, I.; Šraml, M. Low cycle fatigue of nodular cast iron used for railway brake discs. *Eng. Fail. Anal.* **2011**, *18*, 1424–1434. [CrossRef]

3.   Chen, H. Dynamic failure mechanism and reverse design and preparation of new high-speed brake disc composite materials for high-speed trains above 350 km/h. *Acad. Trends* **2011**, *4*, 20–24.

4.   Bagnoli, F.; Dolce, F.; Bernabei, M. Thermal fatigue cracks of fire fighting vehicles gray iron brake discs. *Eng. Fail. Anal.* **2009**, *16*, 152–163. [CrossRef]

5.   Li, Z.; Han, J.; Yang, Z.; Pan, L. The effect of braking energy on the fatigue crack propagation in railway brake discs. *Eng. Fail. Anal.* **2014**, *44*, 272–284. [CrossRef]

6.   Mazánová, V.; Polák, J. Initiation and growth of short fatigue cracks in austenitic Sanicro 25 steel. *Fatigue Fract. Eng. Mater. Struct.* **2018**, *41*, 321–330. [CrossRef]

7.   Goo, B.-C.; Lim, C.-H. Thermal fatigue of cast iron brake disk materials. *J. Mech. Sci. Technol.* **2012**, *26*, 1719–1724. [CrossRef]

8.   Yang, F.T. Repair of brake disc of new series winch. *Coal Mine Mach.* **1989**, *8*, 14–17.

9.   Li, Y.H.; Li, X.F.; Niu, M.L. Repair of front wheel brake plate of electric wheel dumper. *Eng. Mach. Maint.* **2012**, *2*, 146.

10.   Ge, M.Z.; Xiang, J.; Zhen, F. Effect of laser cladding repair on fatigue crack propagation rate of TC4 titanium alloy. *Mater. Bull.* **2108**, *32*, 98–103.

11.   Gao, D.K.; Fu, Y.M.; Wang, P. Crack prevention and repair of 3Cr2W8V thermal fatigue crack. *China Surf. Eng.* **2001**, *14*, 42–44.

12.   Liu, Y.; Zhou, H.; Yang, C.Y.; Chen, J.Y. Thermal Fatigue Resistance of Bionic Compacted Graphite Cast Iron Treated with the Twice Laser Process in Water. *Strength Mater.* **2015**, *47*, 170–176. [CrossRef]

13. Xin, T.; Hong, Z.; Li, C.; Zhang, Z.-H.; Ren, L. Effects of c content on the thermal fatigue resistance of cast iron with biomimetic non-smooth surface. *Int. J. Fatigue* **2008**, *30*, 1125–1133. [CrossRef]

14. Liang, Y.; Huang, H.; Li, X.; Ren, L. Fabrication and Analysis of the Multi-Coupling Bionic Wear-Resistant Material. *J. Bionic Eng.* **2010**, *7*, S24–S29. [CrossRef]

15. Ren, L.Q.; Liang, Y.H. Biological couplings: Classification and characteristic rules. *Sci. China Ser. E Technol. Sci.* **2009**, *52*, 2791–2800. [CrossRef]

16. Kamat, S.; Su, X.; Ballarini, R.; Heuer, A.H. Structural Basis for the Fracture Toughness of the Shell of the Conch Strombus Gigas. *Nature* **2000**, *405*, 1036–1040. [CrossRef]

17. Ren, L.Q.; Li, X. Functional characteristics of dragonfly wings and progress in bionic research. *Sci. China Sci. Technol.* **2013**, *56*, 11–25. [CrossRef]

18. Chen, Z.; Zhu, Q.; Wang, J.; Yun, X.; He, B.; Luo, J. Behaviors of 40Cr steel treated by laser quenching on impact abrasive wear. *Opt. Laser Technol.* **2018**, *103*, 118–125. [CrossRef]

19. Faisal, T.R.; Abad, E.M.K.; Hristozov, N.; Pasini, D. The Impact of Tissue Morphology, Cross-Section and Turgor Pressure on the Mechanical Properties of the Leaf Petiole in Plants. *J. Bionic Eng.* **2010**, *7*, S11–S23. [CrossRef]

20. Zhou, H.; Tong, X.; Zhang, Z.; Li, X.; Ren, L. The thermal fatigue resistance of cast iron with biomimetic non-smooth surface processed by laser with different parameters. *Mater. Sci. Eng. A* **2006**, *428*, 141–147. [CrossRef]

21. Chang, G.; Zhou, T.; Zhou, H.; Zhang, P.; Ma, S.; Zhi, B.; Wang, S. Effect of Composition on the Mechanical Properties and Wear Resistance of Low and Medium Carbon Steels with a Biomimetic Non-Smooth Surface Processed by Laser Remelting. *Metals* **2020**, *10*, 37. [CrossRef]

22. Wang, Y.; Huang, G.; Su, Y.; Zhang, M.; Tong, Z.; Cui, C. Numerical Analysis of the Effects of Pulsed Laser Spot Heating Parameters on Brazing of Diamond Tools. *Metals* **2019**, *9*, 612. [CrossRef]

23. Lu, H.; Liu, M.; Yu, D.; Zhou, T.; Zhou, H.; Zhang, P.; Bo, H.; Su, W.; Zhang, Z.; Bao, H. Effects of Different Graphite Types on the Thermal Fatigue Behavior of Bionic Laser-Processed Gray Cast Iron. *Metall. Mater. Trans. A* **2018**, *49*, 5848–5857. [CrossRef]

24. Babenko, A.A.; Zhuchkov, V.I.; Selmenskih, N.I. Effect of Boron on the Microstructure and Mechanical Properties of Low-Carbon Tube Steel. *Mater. Sci. Forum* **2019**, *946*, 374–379. [CrossRef]

25. Yuan, Y.; Zhang, P.; Zhao, G.; Gao, Y.; Tao, L.; Chen, H.; Zhang, J.; Zhou, H. Effects of Laser Energies on Wear and Tensile Properties of Biomimetic 7075 Aluminum Alloy. *J. Mater. Eng. Perform.* **2018**, *27*, 1361–1368. [CrossRef]

26. Aqida, S.N.; Brabazon, D.; Naher, S. An investigation of phase transformation and crystallinity in laser surface modified H13 steel. *Appl. Phys. A Mater. Sci. Process.* **2013**, *110*, 673–678. [CrossRef]

27. Busisiwe, J.M.; Ntombizodwa, R.M.; Tshabalala, L.C.; Popoola, P.A.I. The Effect of Stress Relief on the Mechanical and Fatigue Properties of Additively Manufactured AlSi10Mg Parts. *Metals* **2019**, *9*, 1216.

28. Pan, S.; Gang, Y.; Li, S. Experimental and numerical study of crack damage under variable amplitude thermal fatigue for compacted graphite iron EN-GJV-450. *Int. J. Fatigue* **2018**, *113*, 184–192. [CrossRef]

Article

Effect of Composition on the Mechanical Properties and Wear Resistance of Low and Medium Carbon Steels with a Biomimetic Non-Smooth Surface Processed by Laser Remelting

Geng Chang [1,2], Ti Zhou [1,3,*], Hong Zhou [1,2], Peng Zhang [1,2], Siyuan Ma [1,2], Benfeng Zhi [1,2] and Siyang Wang [1,2]

[1] Key Laboratory of Automobile Materials, Jilin University, Ministry of Education, Changchun 130025, China; changgengjlu@163.com (G.C.); qingqingzijinyo@sina.com (H.Z.); pengzhangjlu@sina.com (P.Z.); masiyjlu@126.com (S.M.); zhibf17@163.com (B.Z.); siyangwangjlu@163.com (S.W.)

[2] School of Material Science and Technology, Jilin University, Changchun 130025, China

[3] School of Mechanical Science and Aerospace Engineering, Jilin University, Changchun 130025, China

* Correspondence: changgeng17@mails.jlu.edu.cn

Received: 19 November 2019; Accepted: 18 December 2019; Published: 24 December 2019

Abstract: To study the effect of laser biomimetic treatment on different material compositions, five kinds of steels with different carbon element contents were studied by laser remelting. The characteristics (depth, width), microstructure, hardness, tensile properties, and wear resistance of the samples were compared. The results show that when the laser processing parameters are fixed, the characteristics of the unit increase with an increase of carbon element content. Moreover, the hardness of the unit also increases. Compared with the untreated samples, when the carbon content is 0.15–0.45%, the tensile strength of the laser biomimetic samples is higher than that of the untreated samples. For the biomimetic samples with different carbon content, with an increase of carbon content, the tensile strength increases first and then decreases, while the plasticity of the biomimetic samples decreases continuously. The bionic samples have better wear resistance than that of the untreated samples. For bionic specimens with different carbon elements, wear resistance increases with an increase of carbon element content.

Keywords: composition; laser bionic unit; tensile properties; wear resistance

---

1. Introduction

Natural organisms live in harsh and complex environments for long periods, so their body surfaces must possess a strong ability to resist external damage. A non-smooth morphology, non-smooth structure, and different material compositions on the surface of the organism are typical biological coupling characteristics [1–5]. Various non-smooth features provide an organism with an appropriate combination of strength and toughness and play a unique role in resisting external alternating stress or direct damage. For example, dragonfly wings [6,7] are often in a vibrational state without failing, which is the result of the collaborative effect of the wing surface's grid shape and the wing vein's sandwich structure composed of multiphase materials. The desert scorpion [8] has been subjected to the erosion by sandstone permanently without damage, which is the result of the collaborative effect of its non-smooth surface and multiple layers of soft connective tissue under the skin; shellfish [9–11] have been eroded by seawater frequently without crack initiation, which is the result of their non-smooth surface composed of an alternating soft and hard structure and the coupling effect of the composite materials. Thus, inspired by the coupling characteristics of the surfaces of organisms, constructing a bionic non-smooth structure on the surface of mechanical parts to obtain specific properties has become a new surface modification method.

Laser remelting treatment [12–15], as a new technology, can be combined with coupled bionics to offer a new approach for manufacturing bionic non-smooth structures. In this process, a high energy laser beam irradiates the surface of the workpiece, making the laser radiation region of the workpiece rapidly melt. Then, the region is quickly solidified via heat conduction from the substrate, so the obtained microstructure is different from the matrix, and when the selected laser processing parameters are appropriate, the grain will be refined. The material's hardness is also much improved based on the pattern [16–18]. In this way, the non-smooth bionic structure with soft and hard phases is processed on the surface. Among these structures, the hard structure processed by laser fusion is called the bionic unit [19].

Previous studies [20–23] have shown that the improvement of material properties is closely related to the biomimetic unit. Materials with non-smooth structural biomimetic units have good wear resistance and fatigue resistance. The properties of the biomimetic unit are closely related to their composition, the microstructure of their matrix material, and their laser processing parameters. For example, Meng [24] studied the unit microstructure and thermal fatigue resistance of different kinds of die steel after laser biomimetic treatment. Zang [25] compared the microstructure and mechanical properties of H13 steel with different microstructures after laser biomimetic treatment. At present, the effect of the material composition on laser biomimetic treatment is rarely discussed in the literature. Therefore, in this paper, the low and medium carbon steels commonly used in industry were chosen as the research object, and we studied the influence of carbon content on the characteristics, microstructure, and mechanical properties of the bionic unit by machining the strip bionic unit on its surface. Finally, the wear experiments of samples with different carbon content were demonstrated. The purpose of this study is to provide a reliable theoretical reference for laser remelting strengthening of different carbon steel parts in actual industrial production and to serve industrial production practices.

2. Experimentation

2.1. Materials

Samples of 15 steel, 25 steel, 35 steel, 45 steel, and 55 steel with carbon content of 0.15%, 0.25%, 0.37%, 0.45%, and 0.58%, respectively, were selected. The matrix samples are represented by A1, A2, A3, A4, and A5, respectively, and their chemical components are shown in Table 1. Laser biomimetic samples are represented by LR-A1, LR-A2, LR-A3, LR-A4, and LR-A5.

Table 1. The compositions of the samples (wt.%).

Groups	C	Si	Mn	S	P	Fe
A1	0.15	0.11	0.28	<0.01	<0.01	Bal
A2	0.25	0.21	0.53	<0.01	<0.01	Bal
A3	0.37	0.18	0.61	<0.01	<0.01	Bal
A4	0.45	0.20	0.55	<0.01	<0.01	Bal
A5	0.58	0.25	0.75	<0.01	<0.01	Bal

2.2. Experimental Methods

The wear samples were cut into the samples to be processed (with a volume size of $20 \times 30 \times 6\ \text{mm}^3$) using an electric spark machine (DK7732, Huadong Group, Hangzhou, China), and the samples were processed successively with waterproof abrasive paper of different granularities to remove the traces of wire cutting. Then, an ultrasonic cleaner was used to clean the samples, thereby removing the residual oil and debris on the surface of the sample to achieve a smooth and clean surface.

A bionic strip specimen was fabricated using a solid-state pulsed Nd: YAG laser with a wavelength of 1.064 μm and maximum output power of 500 W. The processing schematic diagram is shown in Figure 1. The laser processing parameters are shown in Table 2.

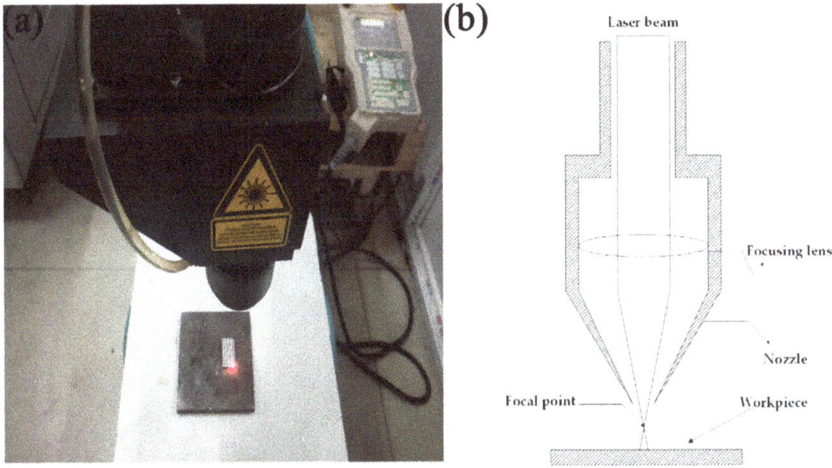

Figure 1. The schematic diagram of laser bionic processing.

Table 2. Laser processing parameters.

Energy (J)	Pulse Duration (ms)	Frequency (Hz)	Scanning Speed (mm/s)	Beam Diameter (mm)
16.88	8	5	1	1.59

The sample was cut in the direction perpendicular to the incident direction of the laser using an electric spark machine. The cross-section was sequentially milled and polished. Pickling was carried out using a 4% nitric acid solution. Characterizations of the cross-section of the unit were carried out using an optical microscope (Zeiss, Axio Image A2m, Oberkochen, Germany). The microstructure of the unit and the substrate were observed using an electron scanning microscope (Zeiss, EV018, Oberkochen, Germany). The X-ray diffractometer (Rigaku D/Max 2500PC, Tokyo, Japan) was used for phase analysis.

Microhardness testing was performed on a microhardness tester (HVS-1000A, Beijing, China). The load was 1.962 N, and the holding time was 10 s. A schematic diagram of the hardness is shown in Figure 2.

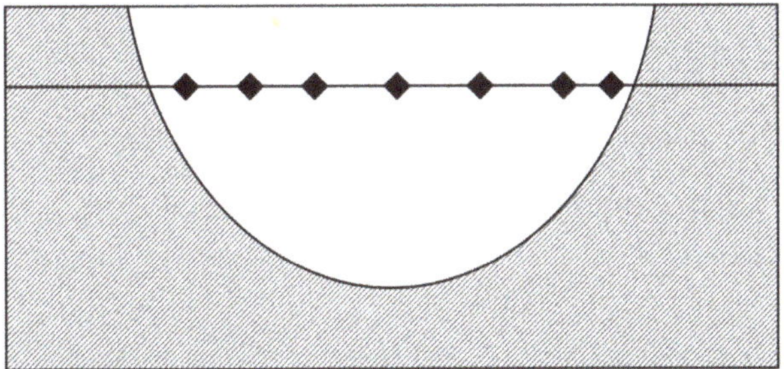

Figure 2. The position of the hardness measurement points.

The tensile test was carried out on a hydraulic servo test machine (MTS 810, MTS Systems Corporation, Eden Prairie, MN, USA) at room temperature and the crosshead speed is 1 mm/s. Figure 3a shows the dimensions of the tensile specimen after laser bionic treatment. Three test specimens for each set were tested and averaged as the final result. After the test, the tensile fracture morphology for all samples were analyzed.

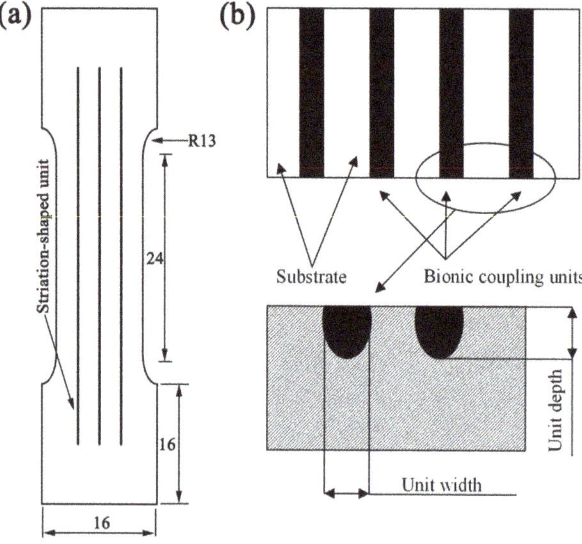

Figure 3. The bionic specimens used for (**a**) tensile and (**b**) wear testing.

Figure 3b shows a schematic of the wear sample after laser bionic treatment. The schematic of the reciprocating sliding wear testing machine developed by the laboratory is shown in Figure 4. During the experiment, the friction pair was fixed, and the sample located in the card slot was linearly reciprocated. The friction pair material is made of quenched gray iron (800 HV), the eccentric speed is 690 r/min, the wear time is 600 min, and the load is 100 N. All experiments were carried out at room temperature under dry conditions.

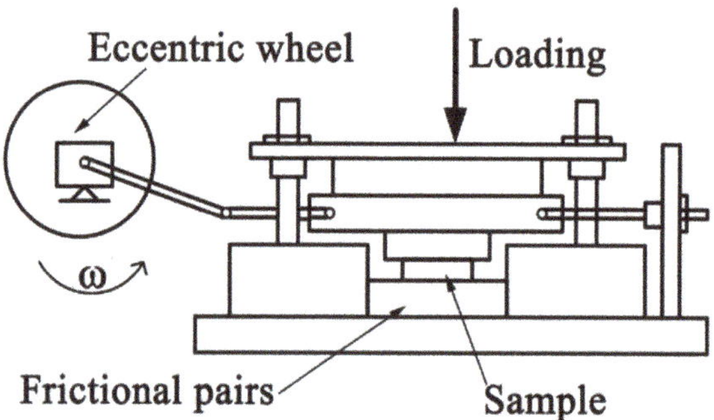

Figure 4. Schematic of the siding wear tester.

Before and after the wear test, the samples were cleaned with an ultrasonic cleaner, dried for half an hour, and then weighed before and after wear by an electronic balance (FA2400, Shanghai jinmin instrument equipment co. LTD, Changshu city, China); then, the weight loss was calculated. Each experiment was repeated three times, and the final results were averaged. Finally, the wear surface morphology was observed with a surface profiler (NT9100, Brock, Germany).

3. Results and Discussion

3.1. Cross-Section Morphology and Size of the Biomimetic Units with Different Carbon Content

Figure 5 shows the cross-sectional morphology of biomimetic unit with different carbon content. It can be seen that the cross-sectional morphology is parabolic, and the unit is composed of two parts: One part is a fusion zone with a bright white color, and the other part is a heat-affected zone (HAZ) in contact with the matrix. Table 3 lists the cross-sectional dimensions of the bionic unit with different carbon contents. It can be concluded from Figure 5 and Table 3 that when the laser processing parameters are the same, as the carbon content increases, the unit characteristics are also increased. The unit characteristics are mainly related to the laser processing parameters and the thermophysical properties of the material [26]. When the laser processing parameters are the same, the thermal conductivity is the determining factor of the unit's characteristics. For carbon steel materials, thermal conductivity increases as the carbon content increases. A material with high thermal conductivity propagates heat at a faster rate, and its fusion zone is wider and deeper. The unit characteristics with different carbon content show this rule.

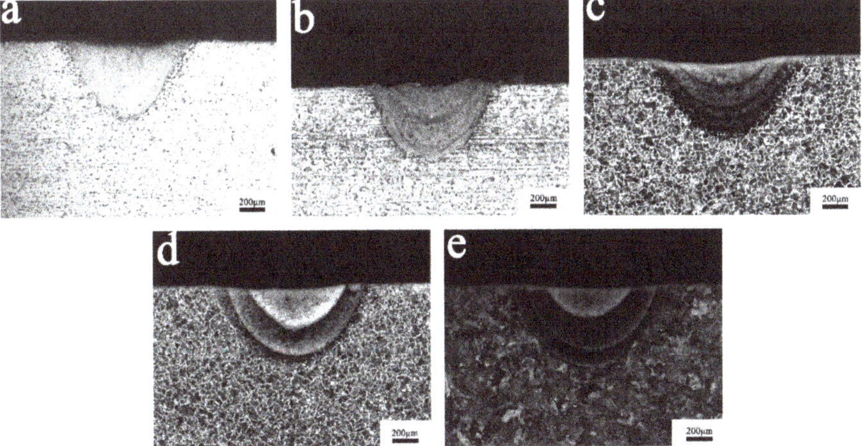

Figure 5. The cross-section of the biomimetic units: (**a**) LR-A1, (**b**) LR-A2, (**c**) LR-A3, (**d**) LR-A4, and (**e**) LR-A5.

Table 3. Unit characteristics of different materials.

Groups	Unit Width (μm)	Unit Depth (μm)
LR-A1	947.30	543.32
LR-A2	1116.43.84	579.24
LR-A3	1241.64	607.29
LR-A4	1265.4	663.37
LR-A5	1346.02	726.47

3.2. Microstructure and Phase Composition of Different Carbon Content Matrixes and Bionic Samples

3.2.1. The Microstructure of the Matrix

Figure 6a–e shows the microstructure of the matrix materials. It can be seen that the matrix structure of the five materials is the same, consisting of pearlite and ferrite; with an increase of carbon content, the ferrite content decreases, and the pearlite content increases.

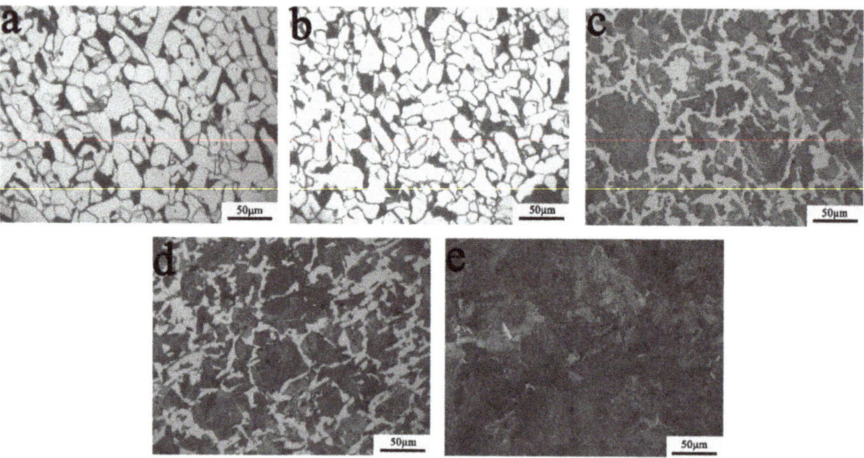

Figure 6. The microstructure of base material samples: (**a**) A1, (**b**) A2, (**c**) A3, (**d**) A4, and (**e**) A5.

3.2.2. The microstructure of the unit

Figure 7a–e shows the microstructure of the unit remelting zone of the laser biomimetic sample. Although the carbon content of the base material is different, the microstructure of the laser fusion zone is the same (which is composed of lath martensite and plate martensite). With an increase of carbon content, the content of the plate martensite gradually increases, and the content of lath martensite decreases. This is due to the different carbon content of the material.

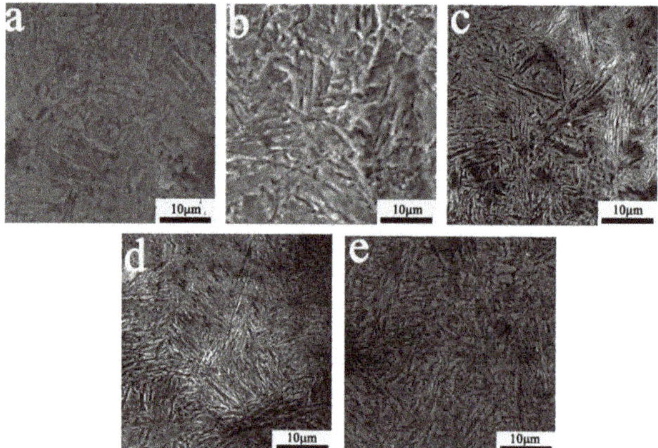

Figure 7. The microstructure of remelting zone of biomimetic unit: (**a**) LR-A1, (**b**) LR-A2, (**c**) LR-A3, (**d**) LR-A4, and (**e**) LR-A5.

3.2.3. Phase Analysis

Figure 8a–e shows a comparison of the XRD between the matrix (A1–A5) and the biomimetic unit (LR-A1–LR-A5), and Figure 8f shows an XRD comparison diagram of the biomimetic unit (LR-A1–LR-A5). The results show that the main phase of the matrix is ferrite, and the phase of the laser biomimetic unit is martensite.

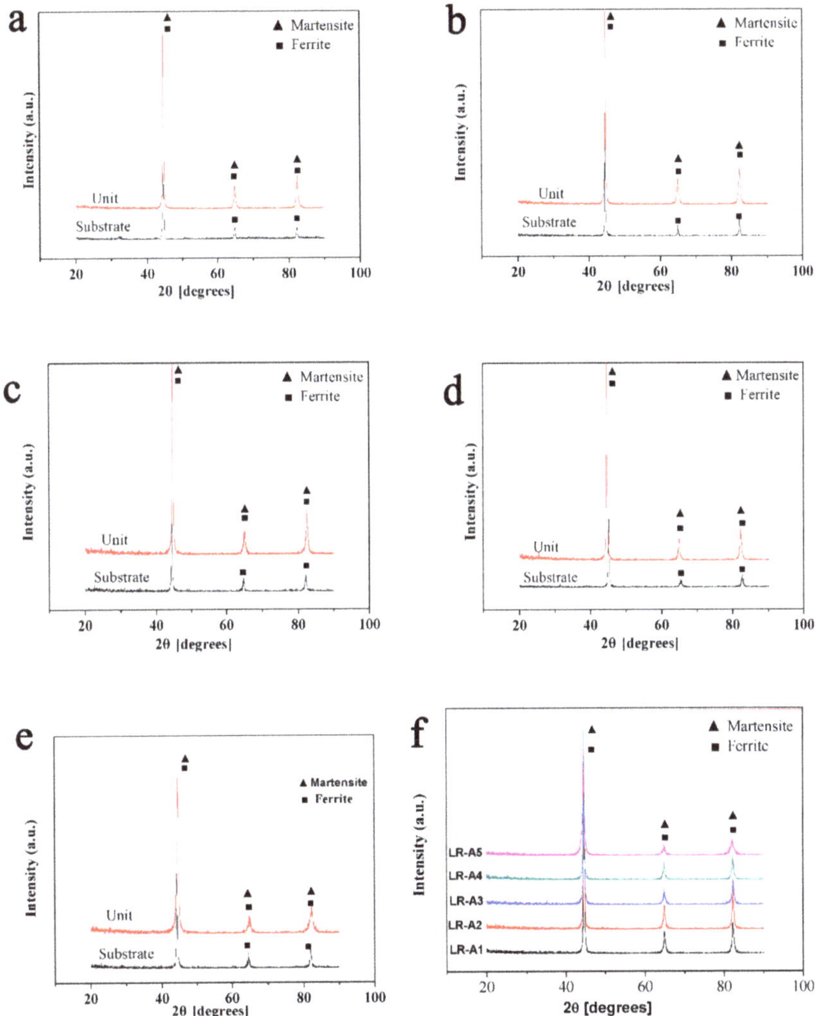

Figure 8. X-ray diffraction profiles of the biomimetic units and matrixes: (**a**) A1, (**b**) A2, (**c**) A3, (**d**) -A4, (**e**) A5; (**f**) comparative diffraction curves of LR-A1-LR-A5.

At the same time, Jade was used to analyze the half-height and width of the diffractive peak (FWHM) of the matrix and unit with different carbon content. The results show that the FWHM of the matrix materials are A1: 0.140°, A2: 0.219°, A3: 0.239°, A4: 0.249°, and A5: 0.300°; the laser biomimetic samples' FWHM is 0.239°, 0.243°, 0.282°, 0.286°, and 0.519°. The FWHM of the bionic unit of the same material is larger than that of the matrix material. Here we use Scherrer's formula [27]:

$$D = \frac{K\lambda}{\beta\cos\theta},$$

(1)

where D is the grain size, K (K value is 0.89) is the Scherrer constant, λ is the wavelength of the X-ray, β is the full width at half maximum, and θ is the Bragg diffraction angle. It can be seen that under the same testing conditions, the grain size of the unit has been refined to different degrees, and the order of the grain size of the unit is LR-A5(67 nm) < LR-A4(121 nm) < LR-A3(123 nm) < LR-A2(143 nm) < LR-A1(145 nm).

3.3. Mechanical Properties of Varying the Carbon Content Matrix and Bionic Samples

3.3.1. Microhardness of Matrix and Unit

The microhardness of the base materials A1, A2, A3, A4, and A5 are 126 HV, 176 HV, 250 HV, 270 HV, and 320 HV, respectively. Figure 9a is the microhardness curve of the bionic unit along the X-axis direction, and Figure 9b is the microhardness curve of the bionic unit along the Y-axis direction. Figure 9a,b show that the hardness of the remelting zone and the HAZ of the biomimetic unit are higher than those of the matrix, and the hardness of the bionic unit is gradually increased as the carbon content increases. This also echoes the microstructure changes in Figure 7. After calculation, the average hardness of the bionic unit after laser fusion treatment is 321, 343, 466, 493, and 659 HV. The degree of hardness improvement is 155%, 89%, 86%, 83%, and 106%, relative to the respective matrix.

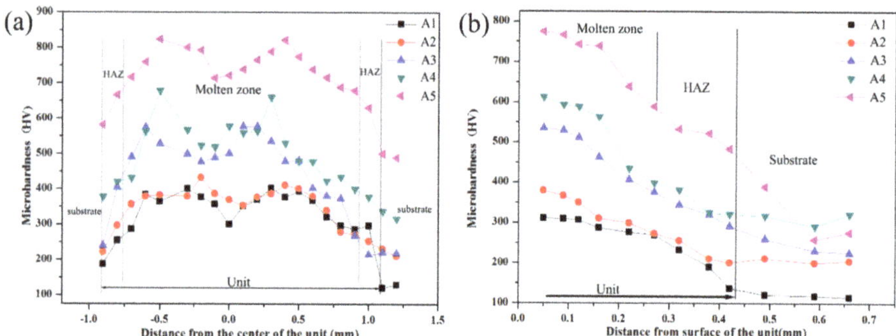

Figure 9. The hardness of the biomimetic unit: (**a**) section profile along the X-axis and (**b**) section profile along the Y-axis.

3.3.2. Tensile Properties of Specimens

To compare the tensile properties of the samples with different carbon contents after laser bionic treatment, the bionic units with the same volume proportions were processed on the surface of the base material. The effect of carbon content on the tensile properties of the biomimetic samples was investigated. Figure 10a–e shows the stress—strain curves of the different carbon-containing untreated samples and laser bionic samples; the corresponding yield strength (YS), tensile strength (TS), and elongation (EL) are shown Table 4. To compare the tensile properties of each biomimetic sample, the tensile property improvement ratio [28] is used to analyze the strengthening effect of different biomimetic samples. The tensile property change ratio refers to the percentage of yield strength, tensile strength, and elongation change of the biomimetic sample compared to the untreated sample, which are represented by *YSC*, *TSC*, and *ELC*, respectively.

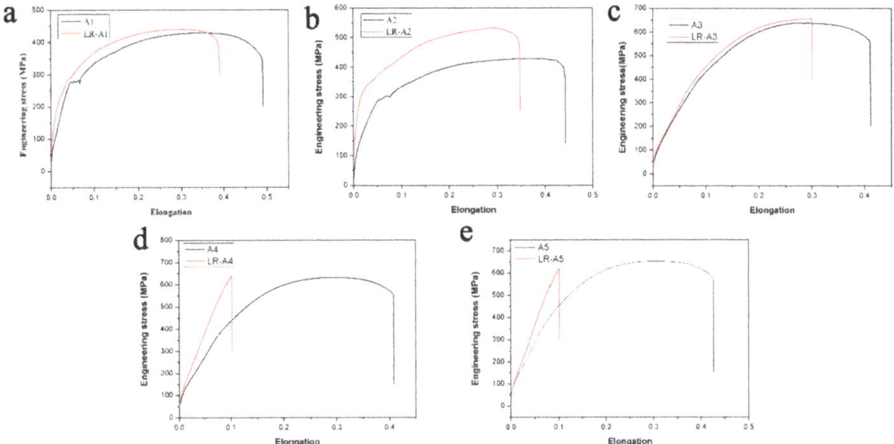

Figure 10. Tensile curves of bionic samples and untreated samples with different carbon contents: (**a**) A1, (**b**) A2, (**c**) A3, (**d**) A4, and (**e**) A5.

Table 4. Tensile test results of the bionic treatment samples and untreated samples.

Groups	YS (MPa)		YSC (%)	TS (MPa)		TSC (%)	EL (%)		ELC (%)
	Untreated	Treated		Untreated	Treated		Untreated	Treated	
A1	275	342	24.36	428	440	2.80	39	19	−34.48
A2	285	387	35.79	440	532	20.91	24	14	−41.67
A3	295	390	32.20	636	656	3.14	21	10	−52.38
A4	268			631	639	1.26	18	3	−83.33
A5	414			652	618	−5.21	15	2	−86.67

From Figure 10 and Table 4, it can be concluded that when the carbon content is in the range of 0.15–0.37%, the bionic sample shows a ductile fracture, and when the carbon content is in the range of 0.45–0.58%, the bionic sample shows a brittle fracture. When the carbon content is in the range of 0.15–0.37%, the tensile strength and yield strength of the samples after laser biomimetic treatment are better improved compared to the untreated sample. However, the tensile strength of the bionic samples with carbon content in the range of 0.45–0.58% decreased, especially the tensile strength of the bionic sample with a carbon content of 0.58%, which is even lower than that of the untreated sample.

Figure 11 shows a comparison of the tensile curve of bionic samples with different carbon contents. It can be seen that when the carbon content is between 0.15% and 0.37%, the strength of the biomimetic sample increases, and when the carbon content increases to 0.45%, the tensile strength begins to decline. The elongation of the biomimetic samples decreases continuously with increasing carbon content.

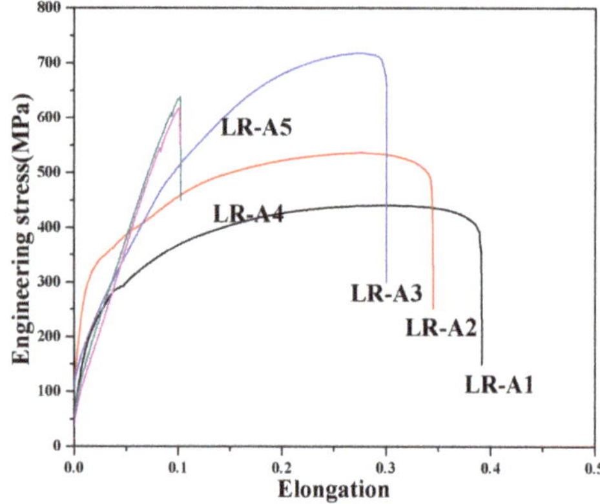

Figure 11. Comparative tensile curves of different bionic samples.

3.3.3. Tensile Fracture Morphology of Samples

Figure 12a–e shows the fracture morphology of untreated samples with different carbon contents. After plastic deformation, dimples and tearing ridges appear in the fracture of the sample. Furthermore, as the carbon content increases, the number of dimples increases, indicating that the strength of the samples also increase.

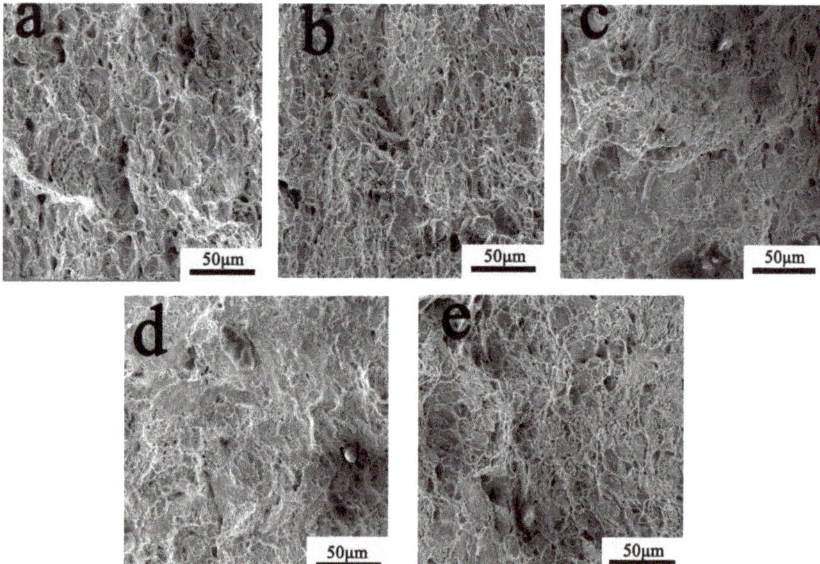

Figure 12. Fracture surfaces of the untreated samples: (**a**) A1, (**b**) A2, (**c**) A3, (**d**) A4, and (**e**) A5.

Figure 13a–e shows the fracture morphologies of the units. It can be seen that the fracture morphologies are different from those of the untreated samples. When the carbon content is between 0.15% and 0.37%, more dimples appear in the fracture of the bionic unit, and the dimples become smaller and deeper. Furthermore, as the carbon content increases, the dimples gradually decrease. For the tensile properties of the specimen, the smaller and deeper the dimple, the more energy required to break the sample. When the carbon content is 0.37%, tearing ridges and cleavage planes appear at the fracture of the biomimetic unit, indicating that the sample has a brittle fracture tendency. When the carbon content increases to 0.45–0.55%, the sample experiences a brittle fracture, and the fracture morphology has obvious tearing ridges, cleavage step, and a river pattern, further verifying that the tensile properties are inferior.

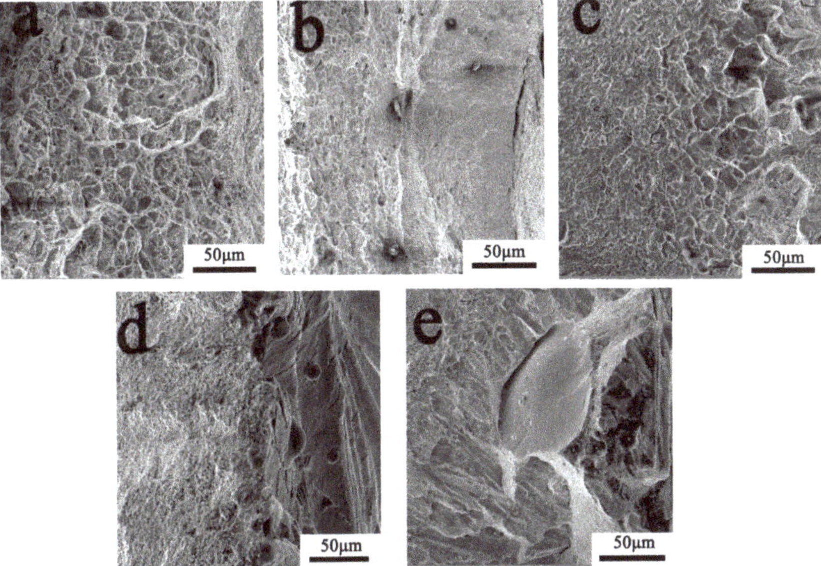

Figure 13. Fracture surfaces of the units: (**a**) LR-A1, (**b**) LR-A2, (**c**) LR-A3, (**d**) LR-A4, and (**e**) LR-A5.

3.4. Wear Properties of Samples

3.4.1. Analysis of Wear Loss

To compare the wear performance of the samples with different carbon content after laser biomimetic treatment, the biomimetic unit with the same volume proportions was processed on the surface of the base metal. The effect of carbon content on the wear properties of the bionic samples was studied. Figure 14 shows the weight loss of the bionic samples and untreated samples with different carbon content. It can be seen that the wear loss weight of the laser biomimetic samples is significantly reduced compared to that of the untreated samples. To compare the antiwear properties of the biomimetic samples, the weight loss reduction percentage [29] (*WLRP*) is defined, as shown in Equation (2):

$$\mathrm{WLRP} = \frac{WL_{Untreated} - WL_{treated}}{WL_{Untreated}} \times 100\% \tag{2}$$

where $WL_{Untreated}$ is the wear-loss weight of the untreated sample, and $WL_{Treated}$ is the wear-loss weight of the treated sample.

The WLRP order of the laser biomimetic samples with different carbon content is: LR-A1 (31.01%) < LR-A2 (31.72%) < LR-A3 (36.08%) < LR-A4 (48%) < LR-A5 (67.76%). This order is due to the different hardness between the materials leading to differences in wear resistance. When the hardness of the sample is higher, the wear resistance is better, and the wear loss is smaller.

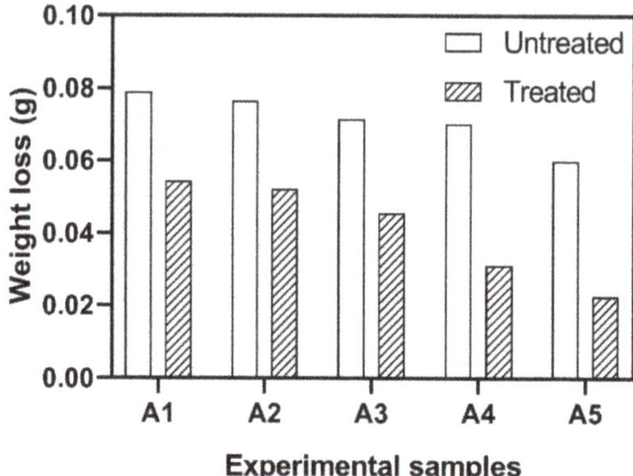

Figure 14. Mass loss of samples.

3.4.2. Analysis of Wear Morphology

Figure 15a–e shows the wear morphology of the untreated samples with different carbon content. It can be seen that a large amount of spalling and deep furrows occurs during the wear process, especially when the carbon content is low. Therefore, the main wear forms of the untreated sample are adhesive wear and abrasive wear.

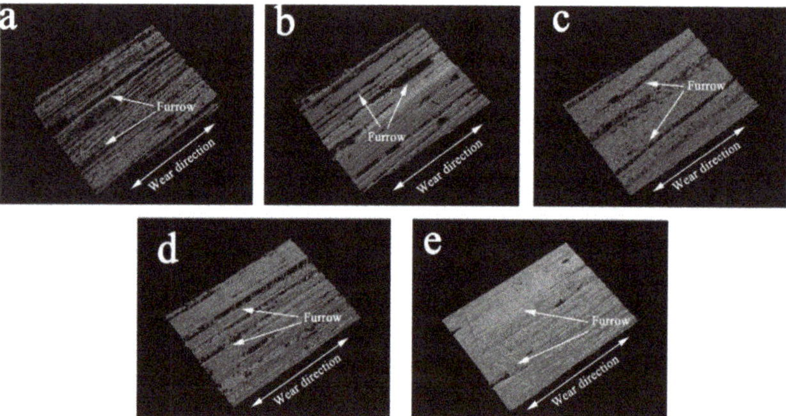

Figure 15. Wear morphology of the untreated samples: (**a**) A1, (**b**) A2, (**c**) A3, (**d**) A4, and (**e**) A5.

Figure 16a–e shows the wear morphology of the laser biomimetic samples with different carbon content. Compared with the wear morphology of the untreated sample, it can be seen in the difference between the unit and the matrix that the unit does not have large peeling or deep furrow phenomena, indicating that the wear resistance of the unit is better than that of the matrix. By comparing the wear morphology of LR-A1-LR-A5, it can be seen that the density and depth of the furrows on the units also gradually reduced, indicating that with an increase of carbon content, the wear resistance of the laser biomimetic samples also gradually improves. This process corresponds to the weight loss and the hardness of the unit.

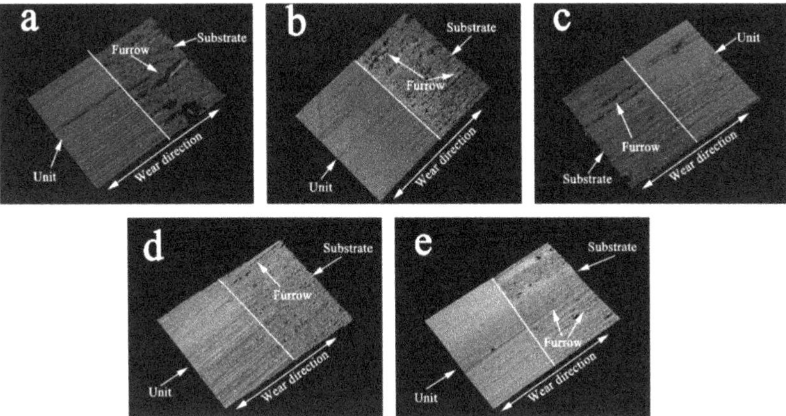

Figure 16. Wear morphology of the bionic samples: (**a**) LR-A1, (**b**) LR-A2, (**c**) LR-A3, (**d**) LR-A4, and (**e**) LR-A5.

4. Conclusions

In this paper, the effects of carbon content on the characteristics, microstructure, hardness, and properties of the bionic unit were studied, and the response of the base material, with different compositions, to laser bionic treatment was obtained. The conclusions are as follows:

1. The unit characteristics (depth and width) are affected by the laser parameters and the parent material's composition. When the laser processing parameters are constant, the unit characteristics increase as the carbon content increases. The order of the unit characteristics is LR-A1 < LR-A2 < LR-A3 < LR-A4 < LR-A5.

2. Under the same laser processing parameters, the microstructure of the remelting zone with different carbon content is the same—composed of martensite. With an increase of carbon content, the content of lath martensite decreases, and the content of plate martensite increases, which increases the hardness of the biomimetic unit.

3. Compared with untreated samples, the tensile strength of the biomimetic sample with the same volume fraction of the unit increases when the carbon content is 0.15–0.45%; when the carbon content is higher than 0.45, the tensile strength is reduced. For laser biomimetic samples, as the carbon content increases, the tensile strength of the biomimetic sample first increases and then decreases, while the plasticity decreases. the degree of improvement in tensile strength is LR-A5 < LRA-4 < LR-A1 < LR-A3 < LR-A2; the degree of plasticity reduction is LR-A1 < LR-A2 < LR-A3 < LR-A4 < LR-A5.

4. Compared to the untreated samples, the wear resistance of the bionic specimens with the same volume fraction of the unit shows better improvement. For the laser biomimetic samples, as the carbon content increases, the wear resistance increases.

5. This study indicates the application feasibility of laser remelting strengthening for different carbon steel parts in manufacturing industry.

Author Contributions: G.C. Study design, Data collection, Data analysis, Writing; T.Z. Literature search. H.Z. Study design; P.Z. Study design, Literature Search; S.M. Data analysis; B.Z. Literature search; S.W.: Data analysis. All authors have read and agreed to the published version of the manuscript.

Funding: This work was supported by Project 985-High Performance Materials of Jilin University, Project 985-Bionic Engineering Science and Technology Innovation, and the double first-class project by Jilin Province and Jilin University (SXGJXX2017-14), Natural Science Foundation of Liaoning Province (No. 20170540421) and National Natural Science Foundation of China (U1601203).

Conflicts of Interest: It is submitted to be considered for publication as an "Original research work" in your journal. This paper is new. Neither the entire paper nor any part of its content has been published or has been accepted elsewhere. It is not being submitted to any other journal. And there is no conflict of interest in this article.

References

1. Ruys, A. 10—Alumina in bionic feedthroughs: The bionic eye and the future. In *Alumina Ceramics*; Ruys, A., Ed.; Woodhead Publishing: Cambridge, UK, 2019; pp. 283–319.
2. Safyannikov, N.M.; Bureneva, O.I.; Aleksanyan, Z.A. Non-invasive Method of Intelligent Sensory Control of Hands' Motor Functions for Bionic Systems. *Procedia Comput. Sci.* **2019**, *150*, 333–339. [CrossRef]
3. Wu, J.; Ma, G. Imitation of nature: Bionic design in the study of particle adjuvants. *J. Controll. Release* **2019**, *303*, 101–108. [CrossRef] [PubMed]
4. Zhang, T.; Wang, A.; Wang, Q.; Guan, F. Bending characteristics analysis and lightweight design of a bionic beam inspired by bamboo structures. *Thin Walled Struct.* **2019**, *142*, 476–498. [CrossRef]
5. Wang, J.; Cheng, C.; Zeng, X.; Zheng, J.; Zhou, Z. Bionic-tribology design of tooth surface of grinding head based on the bovine molar. *Tribol. Int.* **2020**, *143*, 106066. [CrossRef]
6. Rajabi, H.; Moghadami, M.; Darvizeh, A. Investigation of microstructure, natural frequencies and vibration modes of dragonfly wing. *J. Bionic Eng.* **2011**, *8*, 165–173. [CrossRef]
7. Ren, L.; Li, X. Functional characteristics of dragonfly wings and its bionic investigation progress. *Sci. China Technol. Sci.* **2013**, *56*, 884–897. [CrossRef]
8. Huang, H.; Zhang, Y.; Ren, L. Particle Erosion Resistance of Bionic Samples Inspired from Skin Structure of Desert Lizard, Laudakin stoliczkana. *J. Bionic Eng.* **2012**, *9*, 465–469. [CrossRef]
9. Jagota, A.; Hui, C.Y. Adhesion, friction, and compliance of bio-mimetic and bio-inspired structured interfaces. *Mater. Sci. Eng. R Rep.* **2011**, *72*, 253–292. [CrossRef]
10. Kamat, S.; Su, X.; Ballarini, R.; Heuer, A.H. Structural basis for the fracture toughness of the shell of the conch Strombus gigas. *Nature* **2000**, *405*, 1036–1040. [CrossRef]
11. Meldrum, F.C.; Cölfen, H. Controlling mineral morphologies and structures in biological and synthetic systems. *Chem. Rev.* **2008**, *108*, 4332–4432. [CrossRef]
12. Etsion, I. State of the art in laser surface texturing. *J. Tribol.* **2005**, *127*, 248–253. [CrossRef]
13. Kannatey-Asibu, E. *Principles of Laser Materials Processing*; John Wiley & Sons: Hoboken, NJ, USA, 2008; pp. 1–819.
14. Mazumder, J.; Steen, W.M. Heat transfer model for cw laser material processing. *J. Appl. Phys.* **1980**, *51*, 941–947. [CrossRef]
15. Meijer, J. Laser beam machining (LBM), state of the art and new opportunities. *J. Mater. Proc. Technol.* **2004**, *149*, 2–17. [CrossRef]
16. Shi, W.; Wang, P.; Liu, Y.; Han, G. Experiment of Process Strategy of Selective Laser Melting Forming Metal Nonhorizontal Overhanging Structure. *Metals* **2019**, *9*, 385. [CrossRef]
17. Yu, Y.; Zhang, M.; Guan, Y.; Wu, P.; Chong, X.; Li, Y.; Tan, Z. The Effects of Laser Remelting on the Microstructure and Performance of Bainitic Steel. *Metals* **2019**, *9*, 912. [CrossRef]
18. Zhang, Y.; Guo, Y.; Chen, Y.; Kang, L.; Cao, Y.; Qi, H.; Yang, S. Ultrasonic-Assisted Laser Metal Deposition of the Al 4047 Alloy. *Metals* **2019**, *9*, 1111. [CrossRef]
19. Kurella, A.; Dahotre, N.B. Review paper: Surface modification for bioimplants: The role of laser surface engineering. *J. Biomater. Appl.* **2005**, *20*, 5–50. [CrossRef]
20. Li, C.; Yang, L.J.; Yan, C.C.; Chen, W.; Cheng, G.H. Biomimetic anti-adhesive surface micro-structures of electrosurgical knife fabricated by fibre laser. *J. Laser Micro Nanoeng.* **2018**, *13*, 309–313.
21. Yu, D.; Zhou, T.; Zhou, H.; Bo, H.; Lu, H. Non-single bionic coupling model for thermal fatigue and wear resistance of gray cast iron drum brake. *Opt. Laser Technol.* **2019**, *111*, 781–788. [CrossRef]
22. Yuan, Y.; Zhao, G.; Zhang, P.; Zhou, H. Effects of shapes of biomimetic coupling units on wear resistance of 7075 aluminum alloy. *Opt. Laser Technol.* **2020**, *121*, 105786. [CrossRef]
23. Ji, M.; Xu, J.; Chen, M.; El Mansori, M. Enhanced hydrophilicity and tribological behavior of dental zirconia ceramics based on picosecond laser surface texturing. *Ceram. Int.* **2019**, *30*, 106–116. [CrossRef]

24. Meng, C.; Zhou, H.; Tong, X.; Cong, D.L.; Wang, C.W.; Ren, L.Q. Comparison of thermal fatigue behaviour and microstructure of different hot work tool steels processed by biomimetic couple laser remelting process. *Mater. Sci. Technol.* **2013**, *29*, 496–503. [CrossRef]

25. Zang, C.; Zhou, T.; Zhou, H.; Yuan, Y.; Zhang, P.; Meng, C.; Zhang, Z. Effects of substrate microstructure on biomimetic unit properties and wear resistance of H13 steel processed by laser remelting. *Opt. Laser Technol.* **2018**, *106*, 299–310. [CrossRef]

26. Wang, W.; Kodur, V. Chapter 2—Thermal properties of steel at elevated temperature. In *Material Properties of Steel in Fire Conditions*; Wang, W., Kodur, V., Eds.; Academic Press: Cambridge, MA, USA, 2020; pp. 29–41.

27. Epp, J. 4—X-ray diffraction (XRD) techniques for materials characterization. In *Materials Characterization Using Nondestructive Evaluation (NDE) Methods*; Hübschen, G., Altpeter, I., Tschuncky, R., Herrmann, H.-G., Eds.; Woodhead Publishing: Cambridge, UK, 2016; pp. 81–124.

28. Wang, C.; Zhou, H.; Liang, N.; Wang, C.; Cong, D.; Meng, C.; Ren, L. Mechanical properties of several laser remelting processed steels with different unit spacings. *Appl. Surf. Sci.* **2014**, *313*, 333–340. [CrossRef]

29. Su, W.; Zhou, T.; Zhang, P.; Zhou, H.; Li, H. Effect of distribution of striated laser hardening tracks on dry sliding wear resistance of biomimetic surface. *Opt. Laser Technol.* **2018**, *98*, 281–290. [CrossRef]

Article

Effects of Laser Melting Distribution on Wear Resistance and Fatigue Resistance of Gray Cast Iron

Haiyang Yang [1], Ti Zhou [2,*], Qingnian Wang [1] and Hong Zhou [3]

[1] College of Automotive Engineering, Jilin University, Changchun 130025, China; yhyang@163.com (H.Y.); zhoutijlu@yeah.net (Q.W.)
[2] School of Mechanical Science and Aerospace Engineering, Jilin University, Changchun 130025, China
[3] College of Material Science and Engineering, Jilin University, Changchun 130025, China; zhouti@jlu.edu.cn
* Correspondence: masy17@mails.jlu.edu.cn

Received: 6 July 2020; Accepted: 14 September 2020; Published: 17 September 2020

Abstract: The coupling bionic surface is generally prepared by laser melting on the surface of a gray iron brake hub, which can allow the brake hub to achieve excellent wear resistance and fatigue resistance. The designs of most previous experiments have been based on independent units that were uniform in their distribution patterns. Although some progress has been made in the optimization of cell features, there is still room for further improvement with respect to bionics and experimental optimization methods. Here, experiments on units with non-uniform distributions of different distances were used to rearrange and combine the bionic elements. This paper is that the original uniform distribution laser melting strengthening model was designed as a non-uniform distribution model, and the heat preservation and tempering strengthening effect of continuous multiple melting strengthening on the microstructure of the melting zone is discussed. The mechanism of crack initiation and the mode of crack propagation were analyzed. The relationship between the internal stress in the melting zone and the crack initiation resistance was also discussed. In this paper, the mechanism of different spacing distribution on the surface of gray cast iron by laser remelting is put forward innovatively and verified by experiments, which provides a solid theoretical basis for the follow-up industrial application.

Keywords: laser melting; biomimetic model; brake pads; surface wear

---

1. Introduction

Results have shown that the base metal on the brake drum surface of gray cast iron may be melted rapidly by laser technology. This can change the structure of the brake drum base metal, refine the grain, and greatly improve hardness, strength, and toughness [1–3]. By imitating the biological characteristics of organisms that exist in nature that show excellent wear and fatigue resistance [4,5], a bionic functional surface similar to the surface of such organisms can be fabricated by applying a laser to the inner wall of a gray cast iron brake drum [6,7]. The structure of the base metal of the brake drum can be irradiated by a laser [8,9], causing it to melt rapidly and then solidify again instantaneously [10]. The new structure obtained has a strength and toughness far beyond that of the gray cast iron base metal [11,12]. The locus of the melted structure is distributed on the inner surface of the brake drum with a certain shape [13,14]. Thus, the hard unit and the base metal are combined to form a soft/hard interphase bionic surface composed of different structures and shapes. Previous test results showed that the hard element embedded in the matrix has a "dike" and "nail pile" effect [15,16]. This hard element can effectively prevent the growth of cracks in the brake drum, reduce the growth rate of surface cracks in the gray cast iron, and improve the service life of the brake drum.

The experiments conducted in this study attempted to address human needs by learning and applying the mechanisms and laws of the biological world that have been discovered by the application of the bionics principle [17,18]. The phenomenon of biological coupling is an inherent property of living things that has increased the vitality of organisms throughout their evolutionary history. The researchers in this study found that the ability of natural organisms to adapt to their environment did not simply involve changes in a single factor; instead, this adaptability resulted from the synergy of two or more different parts or the coupling of different factors. Some examples of this adaptability include the self-cleaning function of the leaves of plants such as the lotus leaf and reed and the anti-sticking property exhibited by the wings of insects such as the night moth [19,20]. These functionalities are all realized by the coupling of various factors such as the non-smooth shape of a surface or the micro/nanocomposite structure of low-energy materials. For example, the non-smooth hard scales on the backs of lizards, rock lizards, and scorpions are coupled with multiple layers of flexible connective tissue that lie just under the skin, which allows these species to have excellent resistance to erosion in desert environments. The excellent wear resistance of conch and other seashells depends on the coupling of non-smooth composite morphology, multilayer structures, and special materials [21,22].

Similarity science points out that the principles of certain biological structures and functions can be used to construct technical systems and make the characteristics of these technical systems like those of biological systems. The systems formed have functions that are like those of the original system. The surface structures of organisms with excellent wear resistance and fatigue resistance share many similar characteristics [23]. First, they all have alternating structures made up of hard and soft elements. The distribution of hard elements can take various forms. The relatively high degree of hardness of the hard elements comes from the difference in structure or material between the hard and soft elements. The coupling of morphology, structure, and material gives such surfaces excellent wear resistance and fatigue resistance. The properties of a surface comprising soft phases and hard phases can be brought fully into play in biology [24]. The hard phase structure can play a supporting role by, for example, preventing crack initiation and propagation; in addition, this structure can improve the wear resistance of materials. According to the principle of biological coupling, the bionic coupling wear-resistant and thermal fatigue-resistant model is a type of hard element with certain shapes distributed on the soft base metal. This hard element derives different microstructural or constituent materials from the soft base metal, and the two constitute a structure with soft and hard intersections [25]. Coupling bionic wear-resistance and anti-fatigue properties has been proposed as a way to theoretically solve the problem of part failure caused by fatigue and wear on the surface of materials and engineering application problems such as surface adhesion and drag reduction.

Much research and achievements in applications of bionic coupling wear-resistance and fatigue resistance have been made using laser treatment [26]. Laser technology also has a lot of applications in the surface strengthening of alloy parts. Qiuyue Su et al. studied the influence of nano layer depth etched by femtosecond and nanosecond laser on the precision of resistance modulation [27]. This method is applied to wear-resistant parts under various working conditions. The hard phases are processed on their parent bodies by laser melting or cladding. These hard phases and the parent bodies form different types of soft and hard interphase structures on the surface of the parts, replacing the original surface and greatly improving their thermal fatigue resistance, wear-resistant performance, and the service life of the parts. According to the principle of biological coupling, different functions can be obtained by different coupling and factor combinations, and different functional requirements can be obtained by changing the parameters of each coupling element to form corresponding models [28]. All these studies are based on the experimental optimization design of various bionic coupling models, which are composed of coupling elements that include various shapes, structures, and materials. After the effects of different coupling bionic treatments on the thermal fatigue resistance and wear resistance of materials are determined, they can be applied to each wear-resistant component to improve the service life of components [29].

The application of coupled biomimetic theory has achieved many breakthroughs in terms of wear resistance and fatigue resistance. However, in the process of biomimetic model design, research has primarily been based on a single type of spacing [30]. Although the performance achieved by the biomimetic model depends on the characteristic parameters of the coupling element, the spacing distribution of the unit always maintains an average distribution [31]. In nature, the non-uniform distribution of various bio-coupling elements on the surface of organisms can further improve the performance of organisms by enhancing their ability to cope with more complex biological environments. In the process of braking, the surface of the brake drum is constantly subjected to tension and compression stress in the process of brake pad wear [32]. At the same time, because of the large amount of heat generated in the process of friction, the interior of the brake drum is constantly subjected to the internal stress transformations caused by the alternation of cold and hot temperatures. Therefore, the fatigue failure and crack growth mode of the brake drum represent locally concentrated non-uniform sudden changes in growth. This study designed a coupling bionic model of the surface of a non-uniform gray cast iron brake drum to further enhance the reliability of the wear-resistance and anti-fatigue characteristics of the bionic brake drum. In this paper, based on the above research, the uniform distribution of the bionic surface was designed as a non-uniform distribution. The microstructure and internal stress of laser melting changed from discontinuous processing to continuous processing of the local area. Therefore, the wear resistance and fatigue resistance of the non-uniform distribution bionic surface also change correspondingly. This paper hopes to further optimize the wear resistance and fatigue resistance of the gray iron coupling bionic surface and provide a more practical model. The model that we designed used a different combination of multiple units rather than a single unit to enhance the wear-resistance and anti-fatigue properties [33].

2. Experimental

2.1. Experiment Materials

The cast iron used in the experiment was cut from the brake drum of a heavy truck provided by the First Automotive Work shop Group in China. The cast iron brand of the brake drum production was HT250, and the structure of cast iron material is shown in Figure 1. The graphite type contained a small amount of type B flake graphite (see Figure 1a); the matrix structure was pearlite (more than 90%) + a small amount of ferrite (see Figure 1b,c); and the main chemical composition is shown in Table 1.

Figure 1. Microstructure of cast iron material. (**a**) Graphite structure with cast iron. (**b**) Matrix structure of matched cast iron. (**c**) Pearlite structure amplification with cast iron.

Table 1. Chemical compositions of nodular cast iron.

Element	C	Si	Mn	P	S	Cu	Cr	Fe
Content	3.41	1.61	0.96	0.02	0.01	0.315	0.180	Bal.

The wear pair selected for this test was a pair of brake pads commonly used in the brake drum of heavy trucks—semi-metallic brake pads—as shown in Figure 2

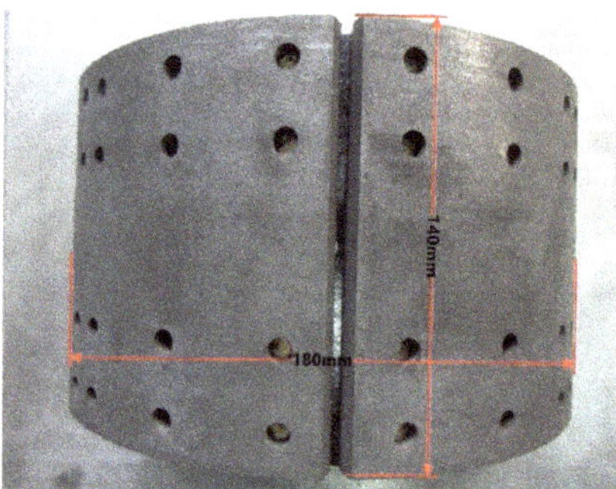

Figure 2. Semi-metallic brake pads.

2.2. Experimental Method

In this experiment, a brake hub manufactured by the FAW casting company was studied where the brake hub was destroyed, and the sample was prepared. Striped biomimetic samples were used in the experiments as previous research has shown that striped biomimetic samples have better performance/price ratios. The coupling bionic specimen with a non-uniform distribution described in this paper is shown in Figure 3. This was designed as a model with either a two, three, or four parallel offset distribution of a biomimetic unit. Each model in this paper was defined as P2, P3, and P4, and the average distribution of a single cell was defined as P1. To ensure the robustness of the tests, the total number of units on the same sample and the distribution/area ratio between the unit and the parent remained unchanged. Each unit–body combination was arranged in strips with a spacing of 2.4 mm, 3.6 mm, and 4.8 mm. After the design was complete, the laser melting method was used to complete the processing. Thermal fatigue tests were conducted on different unit combinations, and the wear tests were conducted under different thermal fatigue times. The thermal fatigue wear characteristics of different coupled bionic models were studied. Finally, the corresponding model was adjusted based on the wear and fatigue of the brake hub occurring during actual use, and the adjusted and optimized model was retested to verify that the adjustment improved the model.

2.3. Sample Preparation

Using a DK77 electric spark cutting machine produced by Donghua CNC company in China, the fragments of the grey cast iron brake hub were processed into $100 \times 20 \times 10$ mm^3 and $40 \times 20 \times 10$ mm^3 rectangular samples, and a 3 mm round hole was drilled on the upper part of 40 rectangular samples, as shown in Figure 3. The surfaces of the samples were polished with 80#, 360#, 600#, and 1000# sandpaper to remove machining traces.

The bionic sample processing was carried out by the system of controlling the laser source's moving track with a six degree of freedom manipulator (Figure 4). The main part of the system was the Nd: YAG pulse laser generator (XL-500WF, Rofin, Munich, Germany). A pulse beam with a wavelength of 1064 nm and a maximum output power of 300 W was emitted through the YAG crystal. The laser energy transmitted by the optical fiber was dispersed with a Gaussian distribution on the sample surface. The entire experimental laser setup was comprised of a cooling system, a two-dimensional rotating table, a six degree of freedom manipulator, and a servo control system. By adjusting the position coordinates of the manipulator in the z-axis direction, precise control of the defocusing amount of the laser spot can be realized, and the spot with the required size can meet the needs of the unit machining. The first mock exam was carried out to adjust the output laser parameters through the laser control cabinet, and ensure that the laser output energy was in accordance with the test requirements and that the size, shape, and organizational structure of the bionic cells in each model laser processing were the same. The laser parameters are shown in Table 2.

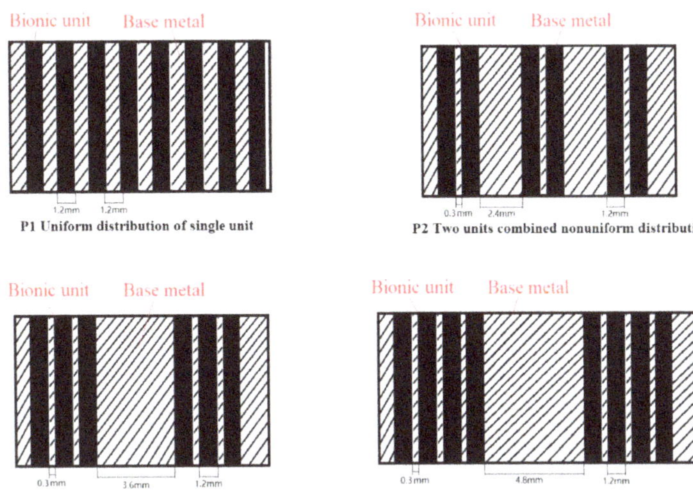

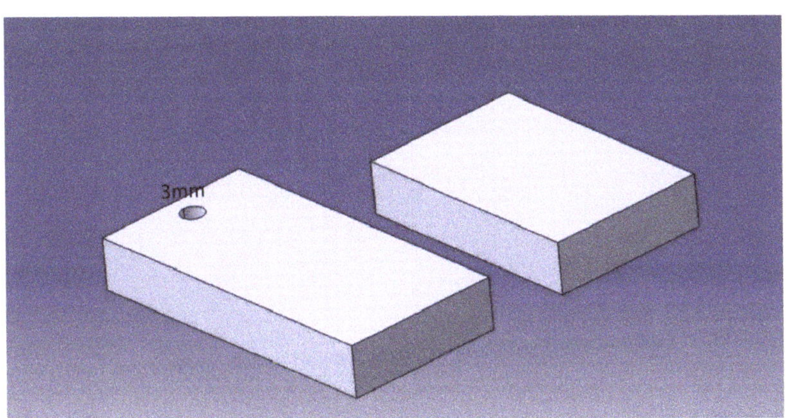

Figure 3. Bionic specimens with different distance distributions.

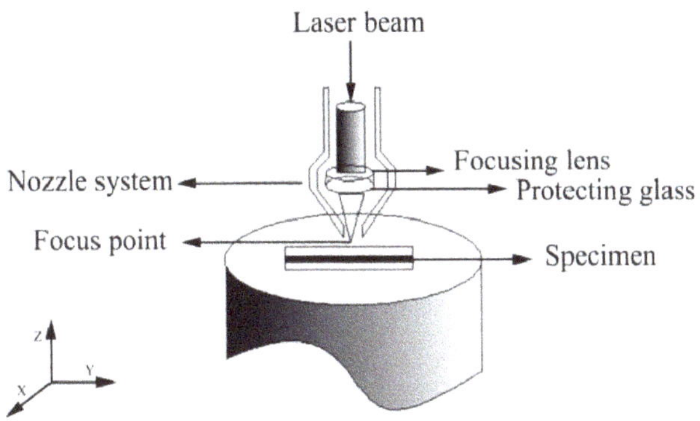

Figure 4. Schematic diagram of laser processing process.

Table 2. Laser parameters.

Sample	Electric Current (A)	Pulse Duration (ms)	Frequency (Hz)	Defocus Amount (mm)	Laser Energy Density (J/mm^2)
No.1	120	7	15	155	144.7

2.4. Thermal Fatigue Test

The equipment used in the thermal fatigue test was a self-made self-restrained thermal fatigue test machine, as shown in Figure 5. The heating temperature was 700 °C, and the cooling medium was running tap water. The heating time was 160 s, and the cooling time was 20 s. The melted sample and the untreated sample were placed in the thermal fatigue test machine for the thermal fatigue test. The samples were removed after the test, the number of cold and hot cycles was recorded, and finally the length and number of cracks were measured.

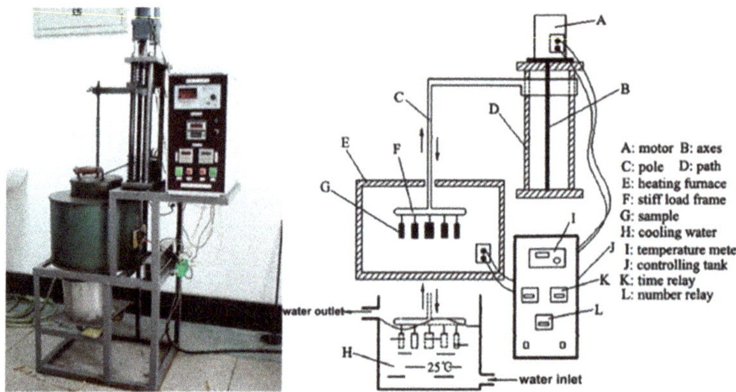

Figure 5. Shape and test process of the thermal fatigue testing machine.

2.5. Wear Tests

Due to the obvious directionality of the friction between the brake drum and the brake pad, in order to better simulate the friction conditions during the actual operation of the drum brake, the matching wear test was carried out on the self-made linear reciprocating wear test machine, and the structural diagram of the test machine is shown in Figure 6. The wear testing machine is composed of a controller, a counter, a servo motor, a linkage driving mechanism, and a wear testing area. The rotation of the servo motor drives the connecting rod to create reciprocating motion. The rotation speed of the servo motor controls the relative motion speed between the friction pairs. The slideway is coated with lubricating oil to ensure smooth reciprocating motion and maintain the level. The slider is equipped with a metal bracket to adjust the load by adding different weights to the bracket. The size specifications of the friction pairs used in this test were as follows: bionic sample was $120 \times 30 \times 10$ mm^3 rectangular, the ground cast iron sample was $100 \times 20 \times 10$ mm^3 rectangular, and the edge was chamfered by 1 mm. The brake pad sample was placed under the wear test area and kept stationary, and the cast iron sample was placed in the fixture on the sliding block. The bionic friction surface processed on the sample was opposite to the brake pad. After loading, it makes contact with the brake pad and follows the guide rod for reciprocating motion. In the experiment, by adjusting the weight on the metal plate and adjusting the speed of the motor with the controller, the sliding friction process under different loads and different friction speeds was created. All the experiments were sliding dry friction experiments conducted at room temperature. The test load was 100 N, and the motor speed was 70 r/min. The stroke of the connecting rod was 0.07 m, and the wear time was set to 40 h for each rotation of the motor. After the wear experiment, the matching cast iron samples were cleaned with an ultrasonic cleaning instrument, and the wear debris on the surface of the brake pad samples was cleaned and removed. All samples were weighed by an electronic balance with an accuracy of 0.001 g. Each experiment was repeated three times, and the average weight loss of the three experiments was taken as the final experimental result.

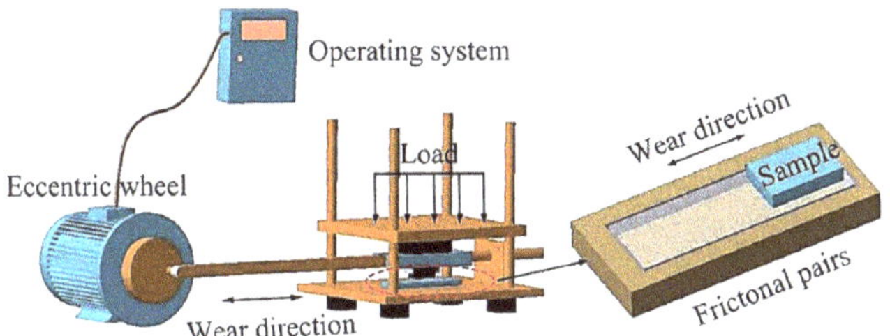

Figure 6. Linear reciprocating sliding friction and wear tester.

2.6. Microstructure Observation and Wear Morphology Observation

A DK77 Electrical discharge machining was used to cut the laser-melted sample along its cross-section. Processing traces and oil stains on the cross-section were polished with different sandpaper grades. The samples were etched with a 4% nitric acid/alcohol solution to prepare them for metallographic observation. The morphology of biomimetic cells was observed under an optical microscope, and a cell without cracks and pores was selected for microstructure observation. A JEOL JSM-5600lv (SEM, Zeiss, Evo18, Oberkochen, Germany) scanning electron microscope (SEM) was used to observe the cross-section structure of the bionic unit bodies. In addition, after the unit body sample and brake pad samples were worn, their surfaces were also observed under SEM and their wear morphology was recorded. In this process, the samples did not require polishing and corrosion.

To characterize the wear performance, in addition to measuring the mass change before and after wear, the wear morphology of different biomimetic unit models and different brake pads were compared and analyzed using three-dimensional laser confocal microscopy (LEXT-OLS 3000 Olympus).

3. Results and Discussion

3.1. Structural Analysis of Non-Uniform Models of Bionic Unit of Grey Cast Iron

According to the discussion on the formation process of the unit [29,30] in the literature, the ambient temperature around the unit formation is relatively low. After remelting, a grain boundary structure with higher dislocation density is formed. This process of remelting and rapid cooling can be compared to a self-quenching process. As the entire process is continuous, the processing of the unit will increase the temperature of the surrounding material environment. For a unit that has been processed, it is equivalent to the process of heat preservation and tempering after the self-quenching process. These two processes occur one after another, which means that the unit created by laser melting and solidification has increased dislocation density and hardness and enhanced ductility and toughness. However, the distribution mode of the cell changes the interaction degree of the adjacent cell forming process. Figure 7 depicts cross-sectional morphology photos of four different biomimetic cell combinations, and Figure 8 shows thee SEM photos of the cell microstructure of different cell combinations.

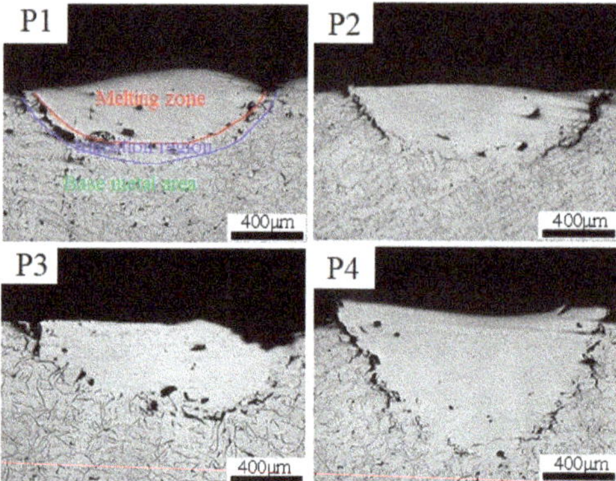

Figure 7. Photographs of cross section shapes of four different biomimetic unit combinations. The section of the bionic unit body is composed of the melting zone, transition zone, and base metal.

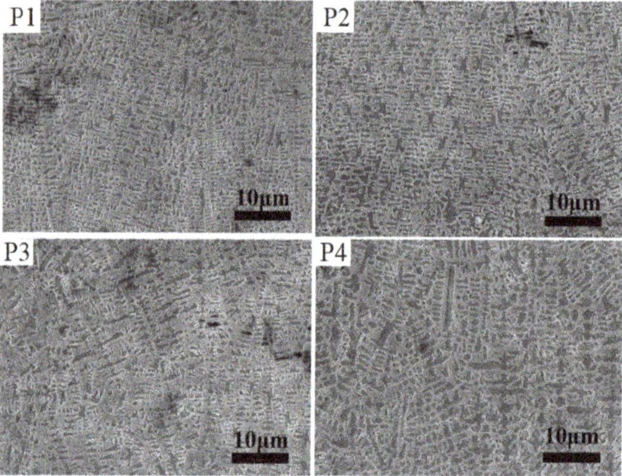

Figure 8. Electron micrograph of microstructure in melting and transition zone of biomimetic unit.

It can be seen from Figure 7 that there were some differences in cell size among the four different combinations. The maximum depth of P4 was 1.089 mm, and the minimum depth of P1 was 0.878 mm. The matrix structure of the base metal was pearlite with a small amount of ferrite. During laser treatment, the heat is transferred from the surface to the interior of the base metal. The lower the ambient temperature of the base metal, the more quickly the heat input from the laser is transferred to the surrounding area, and the more heat is transferred out. The input heat can quickly diffuse to the surrounding area, and the degree of heat accumulation in the molten pool becomes weak. Subsequently, the superheated temperature of the liquid metal in the molten pool decreases, so that the laser processing area on the sample surface reaches the melting point temperature of the gray cast iron, and the melting area becomes smaller. The heat transfer in each direction of the heating area of the four distribution modes is different. The heat transfer on the surface occurs through radiation and natural convection. As more units are processed continuously, the unit melts more slowly until it eventually solidifies. The distribution mode of the unit changes the surrounding environment of the base metal, which has an impact on the shape, size, and other characteristics of the unit body. The increase in the number of unit body combinations increases the depth of the unit body embedded in the base metal, and the thickness of the unit body increases.

The change in the unit distribution had no effect on the structure of the unit, which was composed of deformed ledeburite and a small amount of residual austenite. The deformed ledeburite had the structure of martensite distributed on the cementite substrate. Normally, the pearlite is distributed on the cementite substrate, but the unit body was formed under the condition of rapid cooling. At high temperatures, the carbon dissolve in the ferrite too late to precipitate, and the supersaturated solid solution becomes martensite in ferrite. According to the X-ray (D/Max, 2500PC, Tokyo, Japan) diffraction analysis and SEM photos of the units with different distribution modes shown in Figure 9, it can be seen that the higher the ambient temperature of the base metal during the laser processing, the lower the initial austenite volume, and the higher the amount of eutectic carbide in the unit body; meanwhile, the dendrite spacing of the carbide becomes denser and finer. As shown in Figure 7, the dendrite spacing of P4 was relatively large, while that of P1 was the densest and finest.

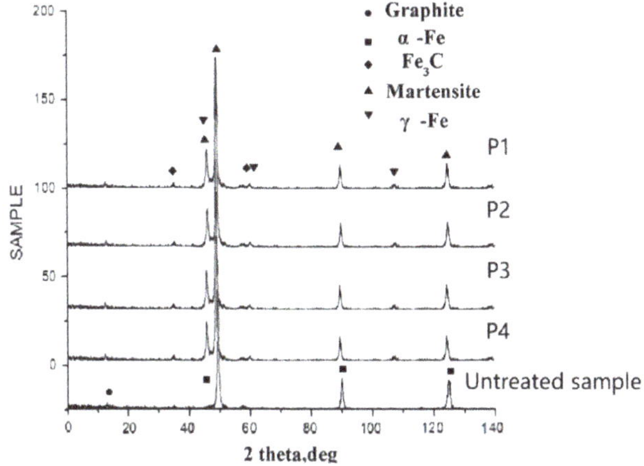

Figure 9. X-ray diffraction analysis of elements with different distribution.

The transverse and longitudinal microhardness distribution of the unit is shown in Figure 10b. The points are taken from the surface to the depth along the horizontal line at 25 um under the matrix and symmetrical center line of the unit, respectively. The measurement results are shown in Figure 10a,c. The microhardness of the surface of the two kinds of units was greater than that of the matrix. Due to the uniform structure on the same horizontal line, the hardness of a single unit showed little change, but there was an obvious hardness difference between the units arranged in different ways. The microhardness range of P1 was 765–820 HV, P2 was 675–725 HV, P3 was 590–640 HV, and P4 was 570–620 HV. The main reason for this hardness difference was the uneven heat transfer in the different arrangements, which resulted in different cooling rates across the entire molten pool. The top of the molten pool can radiate heat, and its cooling speed is faster. More carbide, martensite, and a small amount of austenite were formed during cooling. The temperature at the bottom of the molten pool was close to the melting point, and there was a large temperature gradient between the molten pool and the substrate. The cooling speed was slow. The austenite dendrite formed first, and then the austenite dendrite transformed into martensite. Therefore, in the depth direction, there were some differences in the microhardness of the unit, and the microhardness decreased gradually along the depth direction.

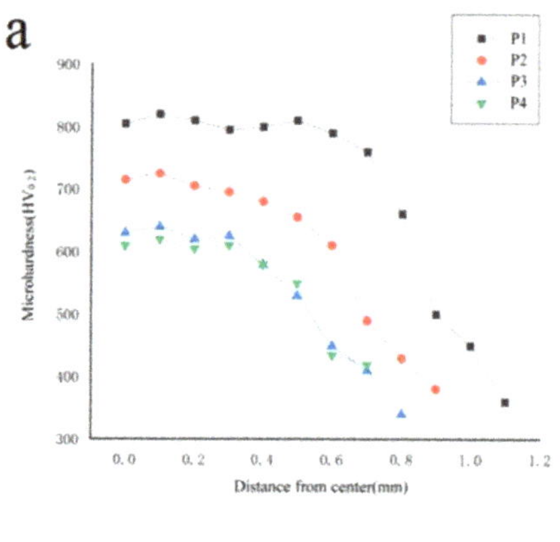

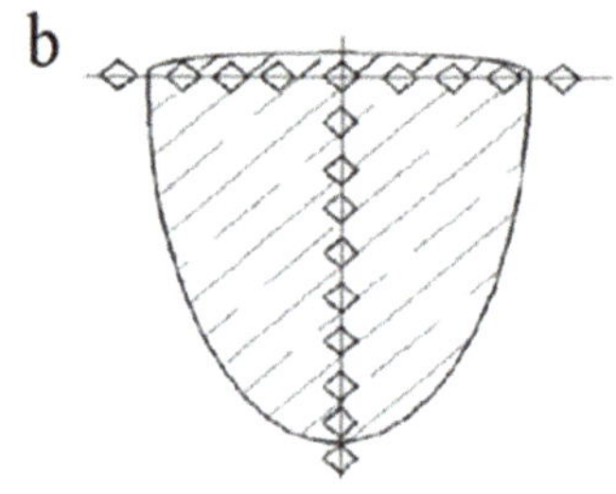

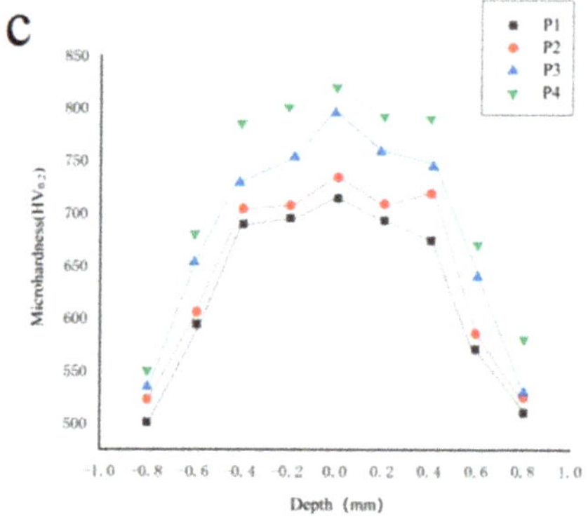

Figure 10. Distribution of microhardness in transverse and longitudinal direction of unit. (**a**) Hardness of unit in transverse direction (**b**) Hardness distribution of unit (**c**) Depth hardness distribution of unit.

3.2. Thermal Fatigue Resistance Analysis of Non-Uniform Model of Biomimetic Unit

Under the action of cyclic thermal stress and thermal strain, the surface defects tend to produce stress concentration, thus becoming the location of crack nucleation. Due to the large amount of flake graphite in gray cast iron, each flake graphite can approximately form a micro notch in the matrix. In the process of the thermal cycle, because of thermal expansion and contraction, the internal stress is concentrated at the tip of the graphite, leading to crack initiation. As shown in Figure 11, after 20 thermal cycles, microcracks initiated at the tip of the two graphite samples and propagated along the interface, with poor bonding properties and grain boundaries. The contact area between the crack and the air was large, allowing for easy oxidation. The formation of loose oxide further promoted the crack growth. Energy Dispersive Spectrometer analysis showed that the oxide from the iron distributed into the cracks. The results showed that the stress concentration effect and oxidation corrosion promoted crack initiation and propagation.

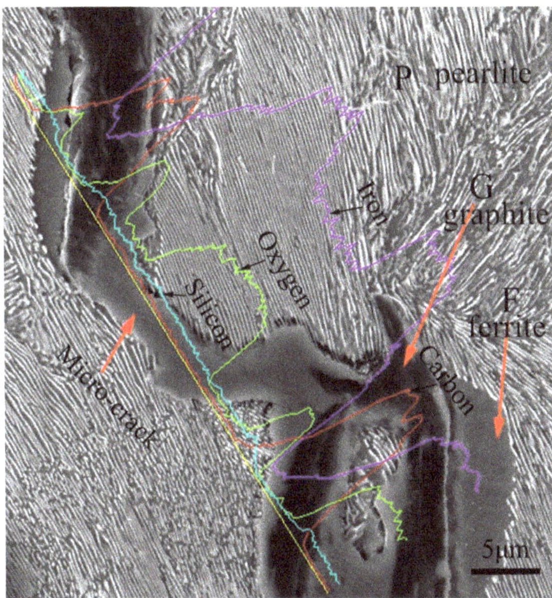

Figure 11. Crack initiation source of gray cast iron and EDS analysis of the elemental composition.

The biomimetic unit embedded in the base metal acted as a crack-arresting unit that blocked crack growth. The local structural transformation of the base metal and the change in the temperature difference between the laser irradiation zone and the base metal generate certain types of stress that are conducive to improving surface wear resistance and fatigue resistance. At the same time, the existence of residual stress can reduce the sensitivity of the crack tip to stress as well as offset part of the driving force of the crack. The residual stress values of different models were different when they were processed; consequently, different models were used to measure residual stress. As the penetration depth of x-ray to the metal was about 20 μm, the stress value measured by the test should be the plane stress value of 20 μm deep on the surface of the unit. During the measurement, three points were selected along the laser scanning direction along the laser scanning direction in the middle of the melting unit, which were marked as A1, A2 and A3, and three points were selected as B1, B2, and B3 in the phase transition region of the unit. The D φ-sin2 φ curve of each measuring point is shown in Figure 12. Residual stress and its average values were calculated by the stress measuring instrument's own software according to the slope of the regression line of D φ to sin2 φ. Residual tensile stress in the melting area of the biomimetic coupling unit in the laser scanning direction was approximately 202.8 MPa, and the residual compressive stress in the phase transformation area of the biomimetic coupling unit in the laser scanning direction was approximately 103.4 MPa. The molten metal in the molten pool shrank because of solidification through movement of the light beam. The melting layer and its surrounding transformation zone were combined metallurgically, which resulted in the contraction of the melting zone and the surrounding transformation zone, increasing the formation of residual tensile stress in the melting zone. Figure 13 shows the surface stress values of the unit in the four different distance distribution models. Surface stress values of the unit were lower as the number of biomimetic units in the model increased, which reflects the function of heat preservation and tempering in the processing of the unit.

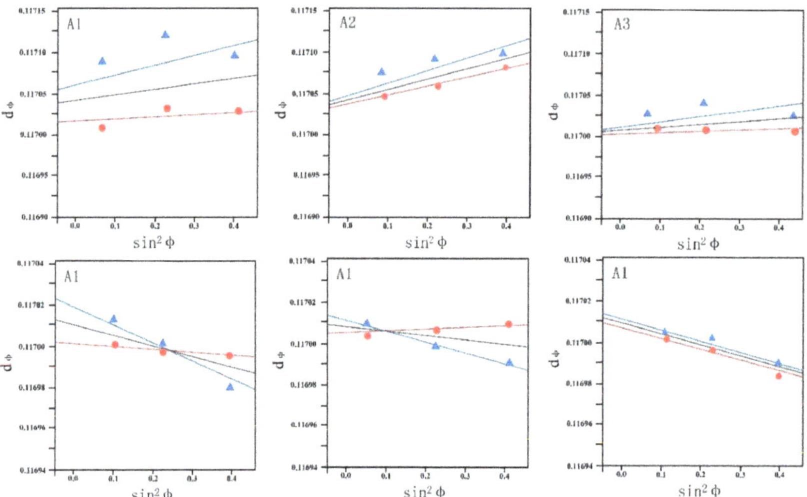

Figure 12. The D φ—sin2 φ curve of each measuring point.

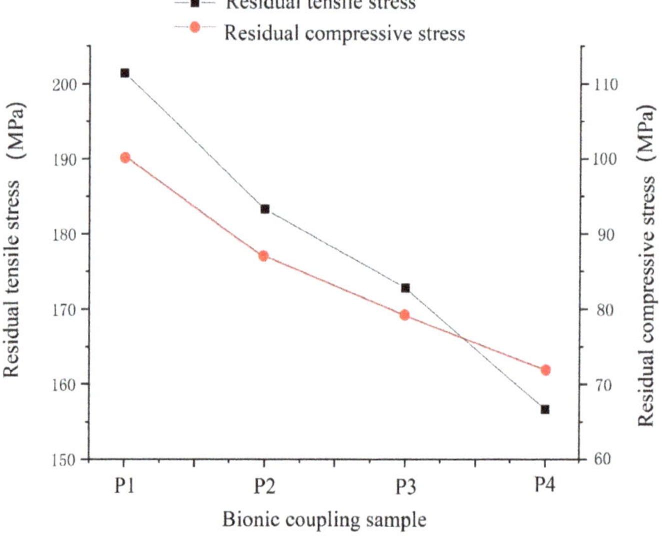

Figure 13. Surface stress values of elements in four models of different distance distribution.

When the microcrack growth is no longer dependent on the surface conditions of the material, the crack initiation phase ends, and the crack growth resistance depends on the overall properties of the material. We found that during the thermal cycle, due to the interaction between the surface layer and the thermal medium, the oxygen in the air reacts with the matrix to oxidize the surface layer. Additionally, a large amount of pearlite decomposes into ferrite and graphite, which reduces the hardness and strength of the material. As shown in Figure 14, due to the decomposition of cementite, the pearlite content was reduced, allowing the crack to expand into the matrix more easily. In the gray cast iron, the presence of graphite in the matrix was an excellent bridge for crack growth. Figure 15 shows the crack growth morphology in the macro state under the micro state. It can be seen from the figure that rapid crack growth was achieved by bridging the graphite in the matrix. Therefore, different types of graphite lead to different growth levels of macro thermal fatigue cracking.

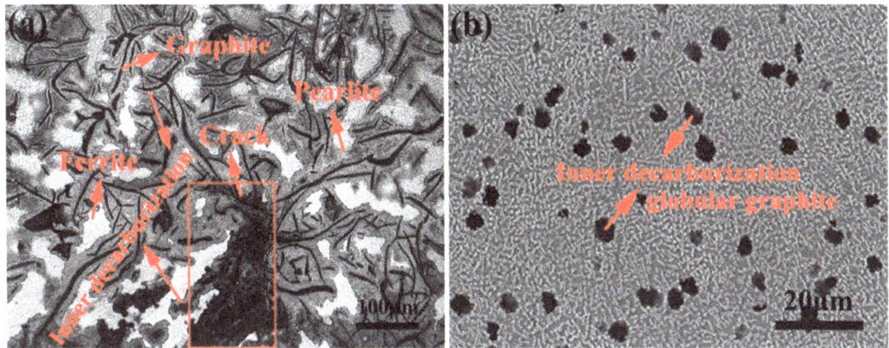

Figure 14. Microstructure of the cross sections of specimen after thermal 600 cycles. (**a**) the substrate; (**b**) the bionic unit.

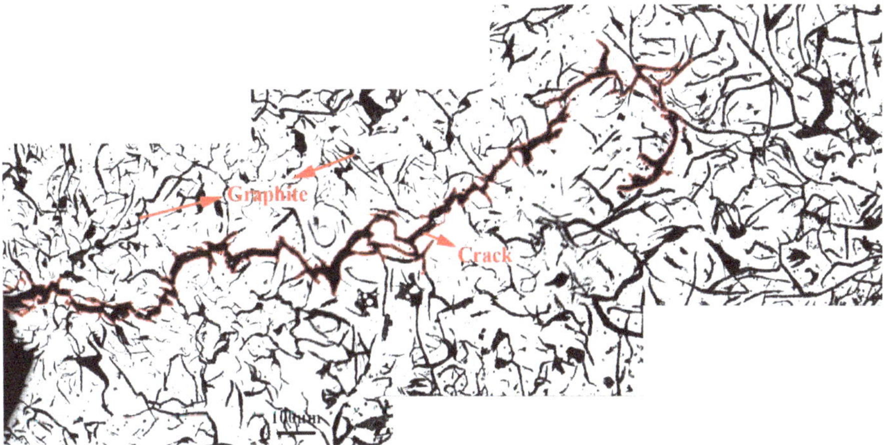

Figure 15. Micromorphology of macro thermal fatigue crack propagation.

Figure 16 shows the statistical curve of the thermal fatigue crack length and cycles of bionic units arranged in different ways from 0 to 600 times. It can be found that the fatigue resistance of the bionic model was better with the increase in the amount of bionic cell arrangement, which occurs because the bionic cell embedded in the substrate surface hinders the crack growth. With the increase in the number of thermal cycles and the continuous decarburization of the unit, the carbide and martensite phases in the unit were greatly reduced, the hardness of the unit was reduced, and the blocking effect was reduced. Therefore, in terms of the number of cracks, when the number of cycles reached 600, the number of cracks in P4 was 80, while that of P1 was 96. However, when P4 was 600 times, the maximum crack length was 2.8 mm, while that of P1 was 6.9 mm. This is due to the accumulation of energy at the crack tip because of the continuous thermal cycling, which allows the crack to break through the potential barrier of the unit. When breaking through the cell barrier, the crack deflects along the cell edge, which means that the crack propagation is interrupted. Therefore, with the increase in the number of elements in the arrangement mode, the maximum length of the crack was effectively shortened in the thermal cycle test of the sample, and the surface crack mode changed from a large, long crack to a fine, small crack (although the difference in the number of cracks was not particularly significant).

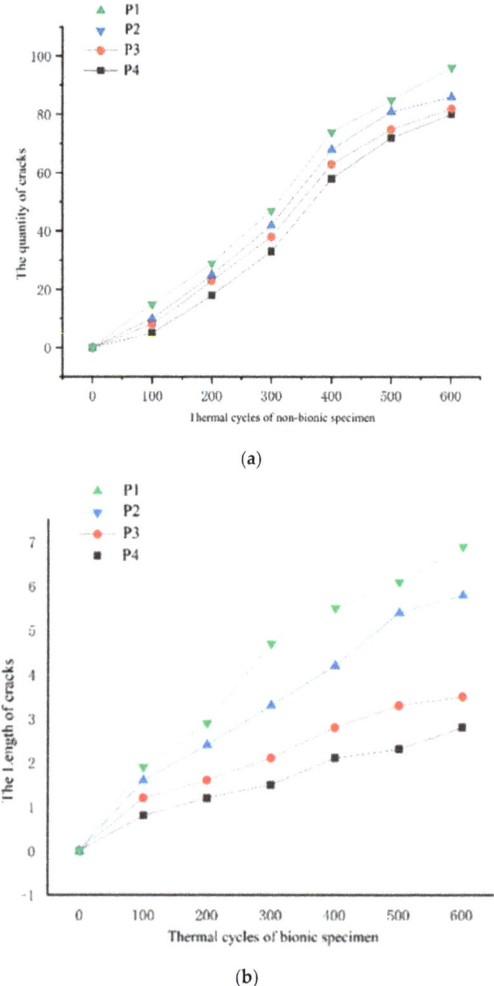

Figure 16. Statistical curves of the length and number of thermal fatigue cracks of biomimetic units arranged in different ways from 0 to 600 times. (**a**) The quantity of cracks, (**b**)The length of cracks.

3.3. Thermal Fatigue Resistance Analysis of Non-Uniform Models of Biomimetic Units

Figure 14 depicts the bionic unit structure after 600 fatigue tests. From Figure 14b, many dispersed point graphite points can be seen. This shows that in the process of thermal fatigue, the cyclic thermal process causes the martensite phase to decompose and the carbon element to precipitate. In the biomimetic unit, tiny point like and strip like cementite could also be seen, and were distributed in the basic phase composed of ferrite. Figure 17 shows the weight loss results of the bionic samples with different distribution combinations and the semi-metallic brake pads after 150, 300, 450, and 600 thermal cycles. It can be seen from the results that with the increase in fatigue time, the wear amount of the bionic samples also increased. Compared with the blank sample, the wear loss of the bionic sample was much less important. Thus, in the early stages of fatigue, the bionic unit in the bionic sample effectively resists wear. With the increase in the fatigue time, the structure of the unit and the matrix change to varying degrees. The number of non-equilibrium phases with higher hardness in the unit decreases, which makes the hardness of the unit decrease.

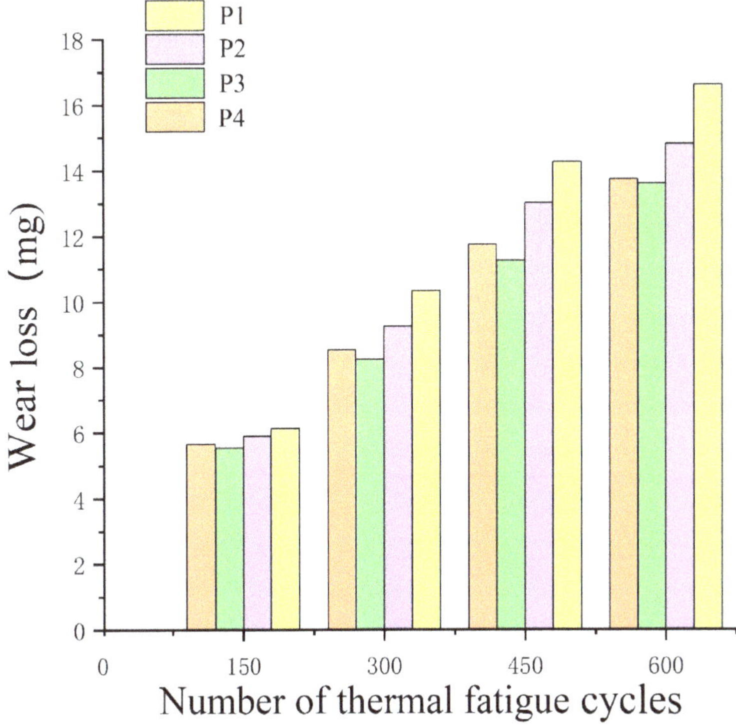

Figure 17. Weight loss results of thee bionic samples in the wear test with semi-metallic brake pads after thermal cycling.

There are different numbers of microcracks on the surface of the fatigue specimen. During the wear process, the edge of these microcracks is easier to peel off, which increases the wear amount of the specimen. Therefore, the number of microcracks in the sample surface has a great influence on the wear resistance of the sample. According to the results of wear loss, the samples with different unit distributions have different wear surfaces. P1 had the largest wear loss weight, and it also had the largest wear loss compared to the worn brake pads. This is because during the process of wear, with the increase in fatigue times, the unit is gradually impacted by the parent crack, which causes it to crack, and a large number of microcracks are produced on the surface. Moreover, with the increase in wear on the surrounding parent metal, the unit embedded in the parent metal will gradually be exposed to the friction surface. In this case, the microcracks on the surface of the unit will cause the unit to wear off and increase the amount of wear. The wear of P4 was most similar to that of P1. This is because the principle of the same unit area was adopted in the design of the bionic surface, so the surface of the base metal with a relatively large area on the P4 surface was exposed. In the process of wear, the area loses the protection of the unit body, so it also experiences relatively significant wear loss. P3 had the lowest amount of wear of all of the samples due to many factors. When the unit had the P3 distribution combination, the heat preservation and tempering process of the unit produced during the processing process increased the toughness of the unit. At the same time, the combined distribution of the units can protect each other during the wear process, which effectively blocks the development of cracks, reduces the micro cracks on the surface of the unit, and prevents excessive wear of the brake pads on the unit.

Figure 18 shows the surface wear morphology of different unit combinations after 600 thermal fatigue cycles. With the increase in fatigue times, different degrees of wear appeared on the surface of the unit, caused by the micro cracks peeling off the surface of the unit. With the increase in the number of units in each combination of units, the degree of wear was lighter, because the combined units protect each other, resulting in the relatively uniform wear that occurred between the units. Due to the wear and loss of support and protection of the matrix structure around P1, the brake pads caused more damage to P1 during the wear process, and the surface peeling was more serious.

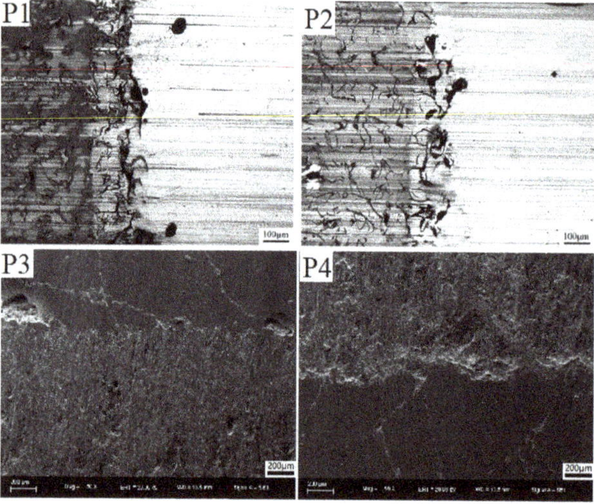

Figure 18. Surface wear morphology of different unit combinations after 600 thermal fatigue cycles.

4. Conclusions

In order to improve the wear resistance and thermal fatigue resistance of gray cast iron surface. As a result of previous studies, this paper makes some innovations and improvements. In order to improve the performance of the whole surface, the original average distribution hardening model was designed as a non-uniform distribution hardening model. The results showed that the structure, hardness, and grain size of the melting zone can be strengthened by the model of non-uniform strengthening, and the wear resistance and fatigue resistance of the whole gray cast iron surface can be improved. The specific innovative conclusions of this paper are as follows:

1. In the process of continuous laser remelting strengthening on the surface of gray cast iron, the adjacent melting zones will affect each other, resulting in heat preservation and a tempering effect, so that the area of the melting zone increases, but the grain size also increases accordingly. The hardness decreased from 765–820 HV (Hardness of Vickers) to 570–620 HV.

2. The mechanism of crack initiation in the early stage of thermal fatigue of gray iron was analyzed by EDS, and the phase transformation law of the microstructure in the melting zone was determined at the later stage of thermal fatigue.

3. The maximum residual tensile stress was 204 mpa in the melting zone and 103.4 mpa in the phase transformation zone

4. The thermal fatigue test showed that the greater the number of distribution strips in the melting zone, the longer the crack initiation time near the remelting zone, and the slower the propagation speed.

5. According to the wear test results of different blocks, it can be determined that the P3 model was the best, and the wear resistance of P1 was increased by 21.3%. The wear resistance test results were P3 > P2 > P4 > P1.

Author Contributions: Conceptualization, H.Y. and T.Z.; data curation, H.Z. and Q.W.; funding acquisition, H.Z.; methodology, H.Y.; project administration, H.Z.; validation, H.Y.; writing—original draft, T.Z.; writing—review and editing, H.Y. All authors have read and agreed to the published version of the manuscript.

Funding: This work was supported by Project 985-High Performance Materials of Jilin University, Project 985-Bionic Engineering Science and Technology Innovation, National Natural Science Foundation of China (U1601203), and double first-class project by Jilin Province and Jilin University (SXGJXX2017-14).

Acknowledgments: Acknowledgments are made to the Key Laboratory of Automobile Materials of Jilin University for the use of scanning electron microscopy and X-ray diffraction.

Conflicts of Interest: The authors declare no conflict of interest regarding the publication of this paper. I would like to declare on behalf of my co-authors that the work described was original research that has not been published previously, and not under consideration for publication elsewhere, in whole or in part. All the authors listed have approved the manuscript that is enclosed.

References

1. Hussain, A.; Ahmad, I.; Hamdani, A.H.; Nussair, A.; Shahdin, S. Laser surface alloying of Ni-plated steel with CO_2 laser. *Appl. Surf. Sci.* **2007**, *253*, 4947–4950. [CrossRef]

2. Siyang, W.; Hong, Z. Effect of different hardness units on fatigue wear resistance of low hardenability steel. *Opt. Laser Technol.* **2020**, *130*, 180–184.

3. Ren, L.Q.; Han, Z.W.; Li, J.J.; Tong, J. Effects of non-smooth characteristics on bionic bulldozer blades in resistance reduction against soil. *J. Terrramech.* **2003**, *39*, 22–30. [CrossRef]

4. Tong, J.; Almobarak, M.A.M.; Ma, Y.H.; Ye, W.; Zheng, S. Biomimetic anti-abrasion surface of a cone form component against soil. *J. Bionic Eng.* **2010**, *7*, S36–S42. [CrossRef]

5. Qingchun, G.; Hong, Z. Effect of medium on friction and wear properties of compacted graphite cast iron processed by biomimetic coupling laser remelting process. *Appl. Surf. Sci.* **2009**, *255*, 6266–6273.

6. Zang, C.; Zhou, T.; Zhou, H.; Yuan, Y.; Zhang, P.; Meng, C.; Zhang, Z. Effects of substrate microstructure on biomimetic unit properties and wear resistance of H13 steel processed by laser remelting. *Opt. Laser Technol.* **2018**, *106*, 299–310. [CrossRef]

7. Meng, C.; Zhou, H.; Zhang, H.; Tong, X.; Cong, D.; Wang, C.; Ren, L. The comparative study of thermal fatigue behavior of H13 die steel with biomimetic non-smooth surface processed by laser surface melting and laser cladding. *Mater. Des.* **2013**, *51*, 886–893. [CrossRef]

8. Cong, D.; Zhou, H.; Yang, M.; Zhang, Z.; Zhang, P.; Meng, C.; Wang, C. The mechanical properties of H13 die steel repaired by a biomimetic laser technique. *Opt. Laser Technol.* **2013**, *53*, 1–8. [CrossRef]

9. Zhou, H.; Zhang, P.; Sun, N.; Wang, C.T.; Lin, P.Y.; Ren, L.Q. Wear properties of compact graphite cast iron with bionic units processed by deep laser cladding WC. *Appl. Surf. Sci.* **2010**, *256*, 6413–6419. [CrossRef]

10. Tong, X.; Zhou, H.; Chen, L.; Zhang, Z.-H.; Ren, L.-Q. Effects of c content on the thermal fatigue resistance of cast iron with biomimetic non-smooth surface. *Int. J. Fatigue* **2008**, *30*, 1125–1133.

11. Almeida, A.; Carvalho, F.; Carvalho, P.A.; Vilar, R. Laser developed Al-Mo surface alloys: Microstructure, mechanical and wear behaviour. *Surf. Coat. Technol.* **2006**, *200*, 4782–4790. [CrossRef]

12. Ren, L.Q.; Liang, Y.H. Biological couplings: Classification and characteristic rules. *Sci. China Ser. E Technol. Sci.* **2009**, *52*, 2791–2800. [CrossRef]

13. Kamat, S.; Su, X.; Ballarini, R.; Heuer, A.H. Structural Basis for the Fracture Toughness of the Shell of the Conch Strombus Gigas. *Nature* **2000**, *405*, 1036–1040. [CrossRef] [PubMed]

14. Ren, L.Q.; Liang, Y.H. Functional characteristics of dragonfly wings and progress in bionic research. *Sci. China Sci. Technol.* **2013**, *56*, 11–25. [CrossRef]

15. Yang, H.; Wang, Q.; Zhou, T.; Zhou, H. The Relationship between the Model of the Laser Biomimetic Strengthening of Gray Cast Iron and Matching between Different Brake Pads. *Metals* **2020**, *10*, 184. [CrossRef]

16. Chen, Z.; Zhu, Q.; Wang, J.; Yun, X.; He, B.; Luo, J. Behaviors of 40Cr steel treated by laser quenching on impact abrasive wear. *Opt. Laser Technol.* **2018**, *103*, 118–125. [CrossRef]

17. Lu, H.; Liu, M.; Yu, D.; Zhou, T.; Zhou, H.; Zhang, P.; Bo, H.; Su, W.; Zhang, Z.; Bao, H. Effects of Different Graphite Types on the Thermal Fatigue Behavior of Bionic Laser-Processed Gray Cast Iron. *Metall. Mater. Trans. A* **2018**, *49*, 5848–5857. [CrossRef]

18. Zhou, H.; Tong, X.; Zhang, Z.; Li, X.; Ren, L. The thermal fatigue resistance of cast iron with biomimetic non-smooth surface processed by laser with different parameters. *Mater. Sci. Eng. A* **2006**, *428*, 141–147. [CrossRef]

19. Gan, L.; Liang, G.Y. Sliding abrasion resistance of gray cast iron surfaces with bio-inspired groove lines and groove stripes. *Mater. Res. Express* **2018**, *5*, 096525. [CrossRef]

20. Yu, D.; Zhou, T.; Zhou, H. Non-single bionic coupling model for thermal fatigue and wear resistance of gray cast iron drum brake. *Opt. Laser Technol.* **2019**, *111*, 781–788. [CrossRef]

21. Chang, G.; Zhou, T.; Zhou, H.; Zhang, P.; Ma, S.; Zhi, B.; Wang, S. Effect of Composition on the Mechanical Properties and Wear Resistance of Low and Medium Carbon Steels with a Biomimetic Non-Smooth Surface Processed by Laser Remelting. *Metals* **2020**, *10*, 37. [CrossRef]

22. Chen, Z.K.; Zhou, T.; Zhang, H.F.; Yang, W.S.; Zhou, H. Influence of Orientations of Bionic Unit Fabricated by Laser Remelting on Fatigue Wear Resistance of Gray Cast Iron. *J. Mater. Eng. Perform.* **2015**, *24*, 2511–2520. [CrossRef]

23. Su, W.; Zhou, T.; Sui, Q.; Zhang, P.; Zhou, H.; Li, H.; Zhang, Z.H. Study on the relationship between intervals among laser stripes and the abrasion resistance of biomimetic laser textured surfaces. *Opt. Laser Technol.* **2018**, *107*, 380–388. [CrossRef]

24. Su, W.; Zhou, T.; Zhang, P.; Zhou, H.; Li, H. Effect of distribution of striated laser hardening tracks on dry sliding wear resistance of biomimetic surface. *Opt. Laser Technol.* **2018**, *98*, 281–290. [CrossRef]

25. Zhang, H.; Zhang, P.; Sui, Q.; Zhao, K.; Zhou, H.; Ren, L. Influence of Multiple Bionic Unit Coupling on Sliding Wear of Laser-Processed Gray Cast Iron. *J. Mater. Eng. Perform.* **2017**, *26*, 1614–1625. [CrossRef]

26. Chen, Z.K.; Zhou, T.; Zhao, R.Y.; Zhang, H.F.; Lu, S.C.; Yang, W.S.; Zhou, H. Improved fatigue wear resistance of gray cast iron by localized laser carburizing. *Mater. Sci. Eng. A* **2015**, *644*, 1–9. [CrossRef]

27. Su, Q.; Bai, S.; Han, J.; Ma, Y.; Yu, Y.; Deng, Y.; Wu, M.; Zheng, C.; Hu, A. Precise laser trimming of alloy strip resistor: A comparative study with femtosecond laser and nanosecond laser. *J. Laser Appl.* **2020**, *32*, 022013. [CrossRef]

28. Sui, Q.; Zhang, P.; Zhou, H.; Liu, Y.; Ren, L. Influence of Cycle Temperature on the Wear Resistance of Vermicular Iron Derivatized with Bionic Surfaces. *Metall. Mater. Trans. A* **2016**, *47*, 5534–5547. [CrossRef]

29. Wang, C.-T.; Zhou, H.; Zhang, Z.; Zhao, Y.; Cong, D.-L.; Meng, C.; Zhang, P.; Ren, L.-Q. Mechanical property of a low carbon steel with biomimetic units in different shapes. *Opt. Laser Technol.* **2013**, *47*, 114–120. [CrossRef]

30. Jing, Z.; Zhou, H.; Zhang, P.; Wang, C.; Meng, C.; Cong, D. Effect of thermal fatigue on the wear resistance of graphite cast iron with bionic units processed by laser cladding WC. *Appl. Surf. Sci.* **2013**, *271*, 329–336. [CrossRef]

31. Xu, L.W. Application and analysis of surface engineering technology in mold manufacturing. *Telecommun. Power Technol.* **2017**, *34*, 114–115.

32. Ma, S.; Zhou, T.; Zhou, H.; Chang, G.; Zhi, B.; Wang, S. Bionic Repair of Thermal Fatigue Cracks in Ductile Iron by Laser Melting with Different Laser Parameters. *Metals* **2020**, *10*, 101. [CrossRef]

33. Zhi, B.; Zhou, T.; Zhou, H.; Zhang, P.; Ma, S.; Chang, G. Improved localized fatigue wear resistance of large forging tools using a combination of multiple coupled bionic models. *SN Appl. Sci.* **2019**, *122*, 63–68. [CrossRef]

Article

Effect of Cavitation Erosion Wear, Vibration Tumbling, and Heat Treatment on Additively Manufactured Surface Quality and Properties

Sergey N. Grigoriev [1], Alexander S. Metel [1], Tatiana V. Tarasova [1], Anastasia A. Filatova [1,*], Sergey K. Sundukov [2], Marina A. Volosova [1], Anna A. Okunkova [1,*], Yury A. Melnik [1] and Pavel A. Podrabinnik [1]

[1]   Department of High-Efficiency Processing Technologies, Moscow State University of Technology «STANKIN», Vadkovsky per. 1, 127055 Moscow, Russia; s.grigoriev@stankin.ru (S.N.G.); a.metel@stankin.ru (A.S.M.); t.tarasova@stankin.ru (T.V.T.); m.volosova@stankin.ru (M.A.V.); yu.melnik@stankin.ru (Y.A.M.); p.podrabinnik@stankin.ru (P.A.P.)

[2]   Department "Technology of Construction Materials", Moscow Automobile and Road Construction State Technical University (MADI), Leningradsky Prospect 64, 125319 Moscow, Russia; s.sundukov@madi.ru

*   Correspondence: a.filatova@stankin.ru (A.A.F.); a.okunkova@stankin.ru (A.A.O.); Tel.: +7-909-913-1207 (A.A.O.)

Received: 23 October 2020; Accepted: 16 November 2020; Published: 19 November 2020

Abstract: The paper is devoted to researching various post-processing methods that affect surface quality, physical properties, and mechanical properties of laser additively manufactured steel parts. The samples made of two types of anticorrosion steels—20kH13 (DIN 1.4021, X20Cr13, AISI 420) and 12kH18N9T (DIN 1.4541, X10CrNiTi18-10, AISI 321) steels—of martensitic and austenitic class were subjected to cavitation abrasive finishing and vibration tumbling. The roughness parameter R_a was reduced by 4.2 times for the 20kH13 (X20Cr13) sample by cavitation-abrasive finishing when the roughness parameter R_a for 12kH18N9T (X10CrNiTi18-10) sample was reduced by 2.8 times by vibratory tumbling. The factors of cavitation-abrasive finishing were quantitatively evaluated and mathematically supported. The samples after low tempering at 240 °C in air, at 680 °C in oil, and annealing at 760 °C in air were compared with cast samples after quenching at 1030 °C and tempering at 240 °C in air, 680 °C in oil. It was shown that the strength characteristics increased by ~15% for 20kH13 (X20Cr13) steel and ~20% for 12kH18N9T (X10CrNiTi18-10) steel than for traditionally heat-treated cast samples. The wear resistance of 20kH13 (X20Cr13) steel during abrasive wear correlated with measured hardness and decreased with an increase in tempering temperatures.

Keywords: anticorrosion steel; hardness; laser powder bed fusion; microroughness; tensile test; wear resistance

1. Introduction

Analysis of modern aircraft designs shows that about 50%–60% of the parts that form the outer contour of the product and parts of the inner set can be manufactured in monolithic structures by various methods of additive manufacturing [1–5]. The labor intensity of manufacturing parts from metal powders under conditions of mass production will not exceed 35%–60% of the total labor intensity of their production from deformable materials in the form of prefabricated structures, which creates a significant economic effect [6].

Such laser powder bed fusion factors like laser power, beam spot size, laser beam profile, scanning speed, strategy and hatch distance, powder particle shape, and morphology characteristics (bulk density of the powder in a layer that also depends on the way of leveling), in combination, affect not only surface quality, but also the main physical, mechanical, and exploitation properties [7–11]. Regardless of all the research attempts optimizing the factors, the surface quality of the produced parts and physical, mechanical, and exploitation properties stayed under the required level for the real industrial applications and were reported multiple times [12–16].

The main problem of surface quality is the natural waviness of the produced surface and unmelted granules trapped in the molten pool [17–21] that are especially important for inner cavities of complex geometrical products since it can reduce the functionality of the surfaces—their wear resistance in friction pairs that strongly depend on submicron roughness [22–25]. The impossibility of polishing inner and complex-profile areas by most known and widespread post-processing methods hampers additive manufacturing's widespread and consequent transfer to the sixth technological paradigm [26–29] that determines the relevance of developing scientific and technological principles of finishing of the parts obtained with the laser powder bed fusion method.

One of the most popular is mechanical polishing, which strongly depends on the size of the used abrasive, retains quite typical traces of abrasive wear, and provides one of the best polishing effects, but makes it impossible for the application to the inner cavities of the complex shaped parts of real production [30–32]. Creation of the specified polishing tools to achieve inner cavities handsomely hampered by the typical irregular topology of obtained surfaces, has difficulties in control of even geometry, and remains a labor-intensive task [14–16,33]. The method of laser-plasma polishing occurs in a metal vapor that prevents oxidation, has the local impact of the laser beam of a relatively small laser spot [34–36], has a severe problem similar to the mechanical polishing methods related to the linear nature of coherent light propagation that is blocked for inner cavities by the geometry of the complex part. The same problem can be detected during high-current pulsed electron beams polishing, allowing almost the eliminating of porosity and reducing the roughness parameter R_a from tens to several micrometers [37–39]. Besides, it does not reduce the roughness parameter R_a of less than 1 μm when actual mechanical polishing allows the roughness parameter R_a reduction up to approximately 0.04 μm that corresponds to the highest class of surface cleanliness [40,41].

Electrochemical etching is of the disadvantages of beam methods, allowing the finishing of complex parts up to the roughness parameter R_a of ~0.04 μm. However, using a specified electrolyte for each material has a potential threat to the environment and hampers its widespread use for additive manufactured parts, which remains one of the most meaningful methods for a large field of applications [42,43].

One of the most promising post-processing methods for the surface treatment of complex-shaped parts remains underestimated—processing in a gas discharge plasma [44–46]. It is free of the inability to process inner cavities and channels of the part with the most sophisticated geometry. Explosive ablation of surface protrusions, polishing with a concentrated beam of fast neutral argon atoms at a large angle of incidence, surface coating deposition upon sputtering with argon ions of solid magnetron targets, and/or evaporation of a liquid metal magnetron target heated by ions allows reduction of the roughness parameter R_a to 0.1–0.2 μm with a decrease in pulse width to 1.5 ns. It cannot be considered an example of protrusion removal on the part of the surface immersed in a plasma due to explosive ablation when high-voltage pulses are applied. However, there are still a few works searching for the method's full potential for additive manufacturing.

Ultrasonic and vibratory finishing is known as mechanical surface treatment methods based on the complex mechanical nature of the action and is considered as a traditional alternative for post-processing methods requiring the setup of a sophisticated unit [47–52], that can be a strong preference in the conditions of aircraft part production. Another advantage is processing more large-scale functional parts with an overall size of more than 100 mm. An important feature of ultrasonic liquid finishing is that the working bodies are cavitation cavities that arise in a liquid under the action of ultrasonic vibrations, which makes it possible to process surfaces of any complexity. At present, there are no actual results of the experiments on the successful use of ultrasonic treatment to reduce the roughness of laser additively manufactured parts, developed recommendations, and strong mathematical support, especially for the parts made of structural anti-corrosion steels of austenitic and martensitic class that stay traditionally under the demand of the real production.

The same problem is related to the research of heat treatment effect on the physical and mechanical properties of the laser additively manufactured parts produced from anti-corrosion steel and their effect on the samples' wear resistance since most of the work is devoted to the quite well-known cast parts [53,54]. However, laser additively manufactured steel parts have other problems. Since they were already heat-treated and remelted multiple times with a laser beam, they should not be additionally quenched to improve their hardness, but they remain with the strong anisotropy of the properties that can be reduced with a developed complex of the post-treatment based on traditional approaches (low tempering at 240 °C in air, at 680 °C in oil, annealing at 760 °C in air) that needs an experimental approval.

In this regard, the work investigated the prospects for using ultrasound post-processing methods to improve the surface quality parameters, topology and the effect of various heat-treatment methods on the physical and mechanical properties of produced samples compared to the cast parts' wear resistance.

The scientific novelty of the work is in researching post-processing methods, including heat treatment (tempering and annealing) and polishing methods based on mechanical nature (ultrasonic cavitation finishing and vibratory tumbling) and their modes, including mathematical support, for additively manufactured parts from corrosion-resistance steels for the aircraft industry and their influence on mechanical properties and surface roughness of the complex-shaped parts.

The purpose of the work is to determine the effect of post-processing modes on the hardness, resistance to abrasive wear, surface roughness parameters (arithmetic mean deviation (R_a), ten-point height (R_z), and maximum peak-to-valley height (R_{tm})) of additively manufactured parts produced by the laser powder bed fusion method to ensure the required properties of aircraft parts made of corrosion-resistant steels of austenitic and martensitic classes.

The results obtained for 20kH13 (DIN 1.4021, X20Cr13, AISI 420) steel are required for a quarter-turn lock mechanism of the aircraft that includes a pin, washer, and sleeve. Since the parts of the lock mechanism are with a diameter of 11 mm, a height of 7 mm, and are complex shaped, the traditional production route is rather laborious and complicated when the application of the laser powder bed fusion method for its production simplifies the operational way without loss of the part exploitation properties. The material of the washer should be wear-resistant since there are friction surfaces between the two parts. The material should differ in strength from the pin material by 20%, which should provide no sticking effect between the parts; the strength of the lock pin is not less than 1300 MPa. The required hardness is not less than 42 HRC, the density is not less than 7.7 g·cm^{-3}, and roughness parameter R_a is less than 3.2 μm.

Another airplane part made of 12kH18N9T (DIN 1.4541, X10CrNiTi18-10, AISI 321) that requires experimental data is an air intake grille module steel, which is an responsible element for protecting the air intake duct from the objects entering it and is an obstacle to the air intake to the engine with overall dimensions of 180 mm × 100 mm × 30 mm with the minimal thickness of the inclined wall of 0.3 mm. The part should be produced following quality requirements: tensile strength is not less than for a standard semi-finished product with a density of less than 7.9 g·cm^{-3}, roughness parameter R_a of less than 6.3 μm. The traditionally produced grille modules are characterized by significant labor intensity. Their manufacturing path includes many operational steps—cutting, bending, manual assembly of almost seventy parts, welding, soldering, etc. Its direct laser manufacturing from the powder takes the production to a new level [55–58].

2. Materials and Methods

2.1. Materials

A wide range of aircraft parts are manufactured from corrosion-resistant steels of the martensitic and austenitic classes that provide the required exploitation properties. The researching of post-processing methods and modes were conducted with samples made of corrosion-resistant steel of the martensitic class, grade 20kH13 (analog DIN 1.4021, X20Cr13 according to EN 10088-4), and corrosion-resistant chromium-nickel steel of the austenitic class, grade 12kH18N9T (analog DIN 1.4541, X10CrNiTi18-10 according to EN 10088-1) under the request of the aviation industry enterprise. The chemical composition is presented in Table 1. PR 20kH13 powder with a fraction of 40 μm and PR 12kH18N9T powder with 20 to 63 μm (JSC POLEMA, Tula, Russia) were manufactured by dispersing molten metal with a jet of compressed gas [59,60].

Table 1. Chemical composition of the steel powder grades PR 20kH13 and PR 12kH18N9T [%].

Material	C	S	P	Mn	Cr	W	V	Si	Ni	Mo	Cu
PR 20kH13	0.16–0.25	≤0.025	≤0.03	≤0.6	12–14	-	-	≤0.6	≤0.6	-	-
PR 12kH18N9T	≤0.12	≤0.02	≤0.035	≤2.0	17–19	≤0.2	≤0.2	≤0.8	8–9.5	≤0.5	≤0.4

2.2. Equipment

Laser powder bed fusion was carried out on an EOS M280 industrial unit (EOS GmbH, Krailling, Germany) and an ALAM experimental laser powder bed fusion setup (MSTU Stankin, Moscow, Russia) [61,62] equipped with the laser source of continuous radiation LK-200 (IPG LASER GMBH, Fryazino, Russia), equipped with two attenuators, B-Cube and C-Varm (Coherent, Santa Clara, CA, USA), and a Focal-piShaper (πShaper, Berlin, Germany). The experimental unit had the following technical parameters: a wavelength of 1070 nm, a beam divergence of 0.2°, maximum power $P_{l.max}$ of 200 W (Figure 1).

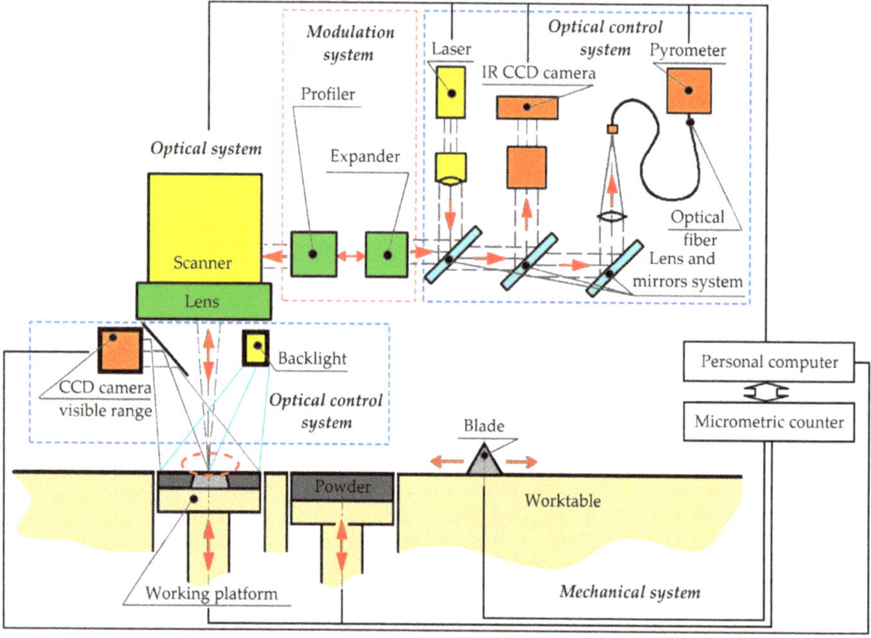

Figure 1. Experimental setup, where *CCD* is a charge-coupled device, *IR* is infrared.

The powders were sifted using an analytical sieving machine AS200 basic (Retsch, Dusseldorf, Germany) with test sieves of 40 μm by ISO 3310-1 and dried by a vacuum oven VO400 (Memmert GmbH + Co. KG, Schwabach, Germany) before processing. The powder drying removed the excess air and contributed to a better powder density in the layers.

A parametric analysis of experimental work was carried out on the ALAM and EOS M 280 setup to select the optimal modes' manufacturing samples.

The chosen laser powder bed fusion factors at the EOS M280 and ALAM setups for growing solids of 20 mm × 20 mm × 20 mm samples for five pieces for each of experimental set are presented in Table 2. For tensile tests, 15 flat specimens were used in accordance with GOST 11,701 (Figure 2). The specimens were grown in the direction of the Z axis, the angle of inclination of the longitudinal axis of the specimen to the printing plane was 0°.

Table 2. Laser powder bed fusion factors for growing solids. 10 ± 0.1.

Material	Layer Thickness (μm)	Laser Radiation Power P_l (W)	Scanning Speed vs. (mm·s⁻¹)
PR 20kH13 (AISI 420)	20	80	390
PR 12kH18N9T (AISI 321)	20	100	100

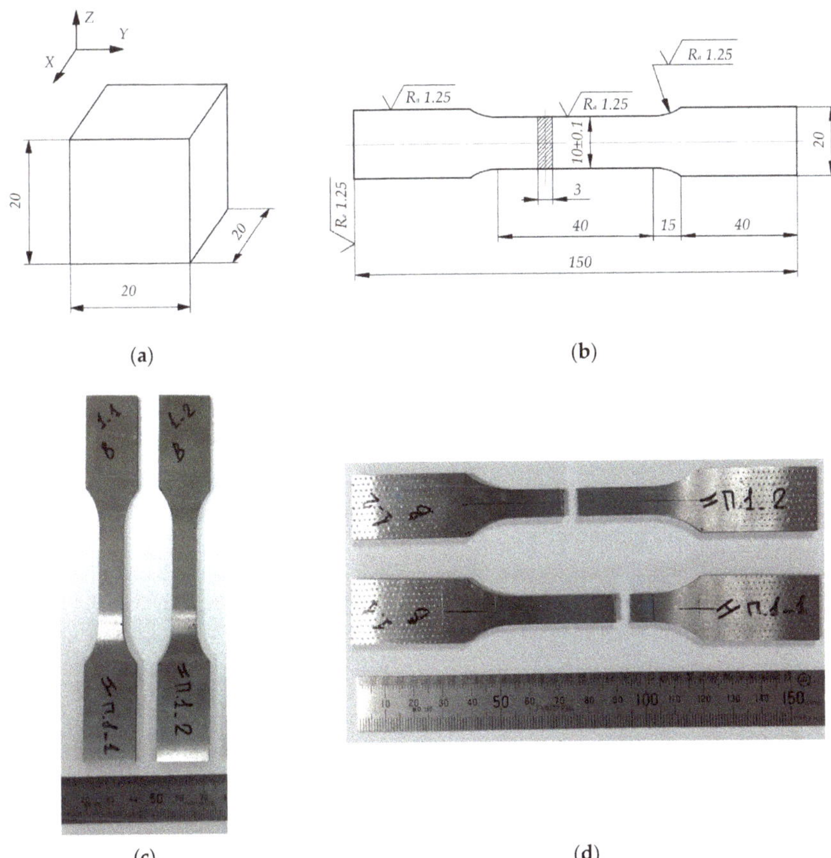

Figure 2. Specimens for tests: (**a**) schemes of specimens for hardness tests; (**b**) schemes of specimens for tensile tests; (**c**) general view of tensile test specimens; (**d**) destroyed specimens.

The quenching and tempering of the samples were produced in a chamber furnace CWF 12/23 with maximal temperature of 1200 °C and capacity of 23 L manufactured by Carbolite Gero (Hope Valley, UK).

The comparison of the hardness and wear resistance test results were done on the casted samples in the state of delivery according to TU 14-1-377-72 produced by LLC "Chelyabinsk Forging—Mechanical Plant" (Chelyabinsk, Russia).

2.3. Surface Morphology

The reduction of the roughness of the finished products was carried out by two methods of ultrasonic processing—cavitation-abrasive finishing and vibratory tumbling in water. It should be noted that dry vibratory tumbling is forbidden in many countries by sanitary norms and rules of production due to destructive action of the ceramic dust on human health (silicosis) [63].

Effects of a mechanical nature exert the most significant influence on the processing in liquid during ultrasonic cavitation abrasive finishing: cavitation, variable sound pressure, radiation pressure, acoustic streams of various scales, sound capillary effect. The introduction of ultrasonic vibrations into a liquid is an effective way of complex-shaped part processing with the internal cavities [44–46,64–66]. The method's effectiveness is due to many specific effects arising in a liquid technological medium under the influence of vibrations [67]. An acoustic pressure arises when ultrasound passes through a liquid as follows:

$$P_a = P_A \sin 2\pi f t, \tag{1}$$

where P_A is a maximal amplitude of acoustic pressure (Pa); f is oscillation frequency (Hz); t refers to the propagating (collapse) time (s) [68,69]:

$$t = 0.915 R_{max} \left(\frac{\rho}{P_m} \right)^{\frac{1}{2}} \left(1 + \frac{P_{vg}}{P_m} \right), \tag{2}$$

where R_{max} is the radius of the cavity at the start of the collapse (m); ρ is the medium density (kg·m^3); P_m is the medium pressure at the time of collapse (Pa); or [70]

$$t = 0.915 R_{max} \left(\frac{\rho}{P_h} \right)^{\frac{1}{2}}, \tag{3}$$

where P_h is the hydrostatic pressure surrounding the cavity (Pa). The maximum pressure developed in bubble P_{max}:

$$P_{max} = P_{vg} \left(\frac{P_m(\gamma - 1)}{P_{vg}} \right)^{\frac{\gamma}{(\gamma-1)}}, \tag{4}$$

where P_{vg} is pressure inside the cavity (of vapor-gas mixture) at maximum radius R_{max} (in the bubble at its maximum size, pressure at the initial stage) (Pa); γ is the polytropic index (exponent) for the gas mixture that is equal to the adiabatic exponent for the adiabatic process following Poisson's equation:

$$P \cdot V^\kappa = \text{const} \tag{5}$$

where V is the volume (m^3), and κ:

$$\kappa = \frac{C_p}{C_v} \tag{6}$$

where C_p and C_v are heat capacity of gas at constant pressure and at constant volume, correspondingly. If:

$$\gamma = \frac{C - C_p}{C - C_v} \tag{7}$$

and C is heat capacity of gas in the given process, then, for adiabatic process:

$$\gamma = \kappa \tag{8}$$

The polytropic exponent γ determines the gas state in the cavity and is in the range of 1–1.33. The amplitude can be presented by acoustic force F_A and system stiffness k_a:

$$A_m = \frac{F_A}{k_a} \tag{9}$$

At the same time, the stiffness of the system is determined by its mass:

$$K_a = 4\pi^2 \frac{m_a}{T^2}, \tag{10}$$

where T is a period of natural oscillations [s]; or by cross-sectional area perpendicular to the line of the force application, Young's modulus and length of an element:

$$K_a = \frac{S_A \cdot E}{L}. \tag{11}$$

Because of the periodic action of tensile and compressive forces in the liquid, cavitation occurs. It consists of discontinuities in the continuity of the liquid with the subsequent collapse of these cavities. A feature of cavitation in the processing of solids is transforming a relatively low energy density of the sound field into a high density of local impulse action when cavitation bubbles collapse. Thus, if we consider the process of bubble collapse to be adiabatic, then the pressure inside is determined by expression:

$$P_m = P_{vg}\left(\frac{R_{max}}{R_{min}}\right)^{3\gamma} \tag{12}$$

or

$$P_m = P_{vg}\left(\frac{R_{max}}{R_{min}}\right)^{3-4} \tag{13}$$

where R_{max} is a maximum bubble radius at the initial stage of collapse (m); R_{min} is a minimum bubble radius at the end of the collapse (mm). The maximum pressure is determined by the ratio $\frac{R_{max}}{R_{min}}$. The results of classical studies [71–73] show that during the stretching period of the liquid R_{max} exceeds the radius of the cavitation nucleus by 100–300 times, and the pressures P_{max} can reach up to 10^7–10^{11} (Pa) at the stage of the collapse of the cavitation bubble, which causes plastic deformation of the solid surface and local destruction of the surface (erosion) when specific pressures are exceeded.

The second effect that determines the efficiency of ultrasonic liquid treatment is acoustic flows, the role of which is in the transfer and distribution of cavitation bubbles over the sound volume, which is especially important for the treatment of complex-profile surfaces. The nature of acoustic flows primarily depends on the mode of ultrasonic treatment, determined by the amplitude of oscillations of the end of the radiator S_m. Thus, large-scale acoustic flows are virtually absent at low-amplitude processing mode ($S_m < 10$–12 µm for water), and random sections of the sound volume are involved in cavitation. The transition to a high-amplitude mode ($S_m > 10$–12 µm) is abrupt and is explained by the strong absorption of acoustic energy during the development of the cavitation region at the end surface of the radiator [74], as a result of which directional hydrodynamic flows are formed, which carry out an active transfer of bubbles from the radiation surface to the treated surface. Formed flows lead to the formation of a stable cavitation area. The height of this area characterizes the depth of penetration of the cavitating liquid flow into the treated volume and depends on the amplitude of vibrations and the absorbing capacity of the process medium:

$$a = \frac{2\pi^2 f^2}{\rho c^3}\left(\frac{4}{3}\eta + \frac{(\gamma-1)\theta}{\gamma C_v}\right), \tag{14}$$

where ρ is medium density (kg·cm^{-3}); c is the speed of sound in a medium (m·s^{-1}); η is the viscosity of a fluid (Pa·s); θ is a coefficient of thermal conductivity of a material (W·(m·K)$^{-1}$); C_v is the molar heat capacity at constant volume (J·(K·mol)$^{-1}$).

One of the methods for intensifying the solids' ultrasonic treatment is adding the abrasive powder to the working fluid—cavitation-abrasive finishing. The addition of insoluble abrasive particles to the sonicated liquid leads to a significant change in the processing. The presence of inhomogeneities in the technological liquid medium leads to a decrease in the liquid's cavitation strength and an increase in the number of cavitation centers, which increases the volume of the effective cavitation region. The mechanism of the effect of cavitation-abrasive finishing on the surface of the product is based as well on the micro-cutting action of abrasive particles, which acquire acceleration due to impulse transmission from shock waves' large-scale acoustic currents.

The ultrasonic cavitation-abrasive finishing was carried out on a half-wave magnetostrictive oscillatory system powered by a UZG 2.0/22 generator (JSK "Ultra-resonance", Yekaterinburg, Russia) (Figure 3a). An ultrasonic emitter of a rod three-half-wave magnetostrictive oscillatory system was immersed in water at a distance of 20 mm from the end face to the workpiece. The processing was carried out with the following parameters: the vibration frequency f = 21,000 Hz, the vibration amplitude of the end face of the emitter S_m = 20 µm, and the processing time t = 120 s. During processing, the ultrasonic generator was operated in the automatic frequency control mode to maintain resonance conditions.

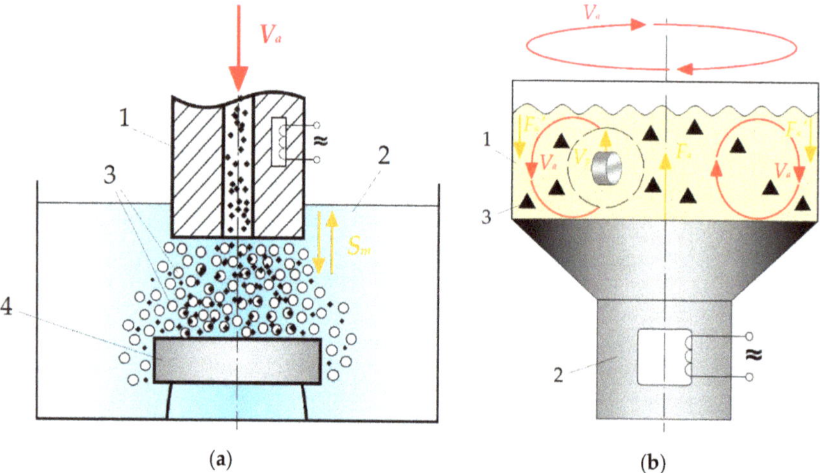

Figure 3. Schemes of ultrasonic treatment: (**a**) cavitation-abrasive finishing, where (**1**) is a transducer, (**2**) is medium, (**3**) is a cavitation bubble, (**4**) is a part to be processed, V_a is a feed of abrasive, S_a is a direction of induced oscillations; (**b**) vibration tumbling, where (**1**) is a tank with the parts, (**2**) is a transducer, (**3**) is an abrasive particle, V_a is an abrasive movement speed, V_p is a part movement speed, F_a is an induced acoustic force, F_a' is a medium force response.

The sample was placed in a radiator, on the bottom of which a layer of Elbor-R abrasive cubic boron nitride (β-BN) powder (JSC SPC "Abrasives and Grinding", Saint-Petersburg, Russia) of 2 mm thick was poured. An axial channel was made in the radiator to ensure the supply of abrasive powder (Elbor-R) to the cavitation zone. The optimal processing parameters were determined based on preliminary studies.

The vibratory tumbling was carried out while moving products and abrasive grains relative to each other in a vibrating container of an 80 L ZHM-80A vibratory tumbler finishing machine planetary drum type (Shengxiang, Zhejiang, China); a filler SCT VFC 10 mm × 10 mm ceramics prism gray (tumbling body) (CFT, Moscow, Russia) made of ceramics with 30% of silicon abrasive with a smooth surface (Figure 3b). The processing was carried out in the following modes: a vibration frequency of 50 Hz, a filler weight of 50 kg, an operating time of 20.5 h, and an engine speed of 1440 rpm. The harmonic vibration amplitude was 6–7 mm.

2.4. Characterization

The dispersed (granulo- and morphometric) composition of the powders was determined on an optical particle shape analyzer 500NANO (Occhio, Liege, Belgium) using static image analysis according to ISO 13322-1: 2014. A scanning electron microscope VEGA 3 LMH 1,000,000× (Tescan, Brno, Czech Republic) determined particle morphology, real-time chemical analysis, and topology of the samples (with a surface inclination up to 30°).

An Empyrean diffractometer Series 2 (PANalytical, Almelo, The Netherlands) ranging from 25° to 70° carried out XRD measurements using monochromatic CuKα radiation (λ = 1.5405981 Å) working at 60 kV and 30 mA in a step-scanning mode from 25° to 70° with a step size of 0.05° and a scan speed of 0.06°/min. The phase composition was analyzed using the PANalytical High Score Plus software and ICCD PDF-2 database. To conduct a quantitative phase analysis, the spectrum fitting method (Rietveld method) was used.

The microstructure and microrelief of the steel surface were studied using an Axio Observer D1m optical microscope (Carl Zeiss Microscopy GmbH, Kelsterbach, Germany) and a Phenom ProX scanning electron microscope following GOST 5639-82, GOST 1583-93 developed by the Euro-Asian Council for Standardization, Metrology, and Certification (EASC).

The measurements of the geometric parameters of the samples were carried out on a ScopeCheck MB multi-sensor coordinate measuring machine for high-precision measurements of large workpieces (Werth Messtechnik GmbH, Giessen, Germany).

The density of the samples ρ was determined by hydrostatic weighing in distilled water using Archimedes' principle and compared with the theoretical values by an XP504 laboratory balance (Mettler Toledo, Columbus, OH, USA) with an accuracy of 0.001 g·cm^{-3} [75,76].

Gas porosity was estimated on panoramic images with an area of 4 mm^2 (GOST 1583-93) using the Axio Observer D1m optical microscope; image processing including evaluation of the volume fraction of pores, the size distribution, and total porosity score was performed by a Thixomet Pro software (Thixomet, Saint Petersburg, Russia).

The following roughness parameters were evaluated during experiments according to EN ISO 4287:1997—arithmetic mean deviation (R_a), ten-point height (R_z), maximum peak-to-valley height (R_{tm}). The roughness was controlled by a high-precision profilometer, Hommel Tester T8000 (Jenoptik GmbH, Villingen-Schwenningen, Germany), by a Dektak XT stylus profilometer (Bruker Nano, Inc., Billerica, MA, USA) with a vertical accuracy of 5 Å (0.5 nm) and a tip radius of 12.5 µm, and by atomic force microscopy using a SMM-2000 multimicroscope (JSC "Plant PROTON" (MIET), Moscow, Zelenograd, Russia). Determination of submicro-roughness parameters was carried out by scanning by the constant height method, based on maintaining a constant distance between the cantilever and the sample.

All cross-sections of the samples were polished down to 1 µm using an ATM Machine Tools sampling equipment (ATM Machine Tools Ltd., Wokingham, UK).

Tensile tests under the application of static loads were carried out by a 5989 Universal Testing System (Instron ITW Company, Norwood, MA, USA) with a force of up to 600 kN at room temperature following GOST 1497-87. Three millimeter thick and ten millimeter width proportional flat specimens were produced for tests following GOST 11701-84.

A Wilson Tukon 2500 Automated Knoop/Vickers Hardness Tester (Instron ITW Company, Norwood, MA, USA) determined Vickers microhardness following GOST R ISO 6507-1-2007. A 574T Series Wilson Rockwell Hardness Tester (Instron ITW Company, Norwood, MA, USA) determined Rockwell's hardness following GOST 9013-59.

The wear resistance was determined by the method of water-jet wear by measuring the depth and width of the formed groove using a Calowear Instrument abrasion tester, which characterizes the resistance to abrasion of a surface (CSM Instruments, Peseux, Switzerland). The tests were carried out under the following conditions: a friction rate of 594 rpm, a static load of 0.25 N, a ball diameter of 25.4 mm, test time of 3, 6, 9 min, sample surface roughness parameter R_a of 1.25 µm. An ODSCAD 6.2 measurement program (GFM LMI Technologies GmbH, Teltow, Germany) investigated the resulting wells' diameter and depth.

3. Results

3.1. Granulomorphometry and Samples' Production

The results of the granulo- and morphometric analysis of the powder showed that spherical particle shape, high fluidity, a microcrystalline structure, equiaxial morphology, and a small number of satellites characterize both of the steels (Figure 4, Table 3).

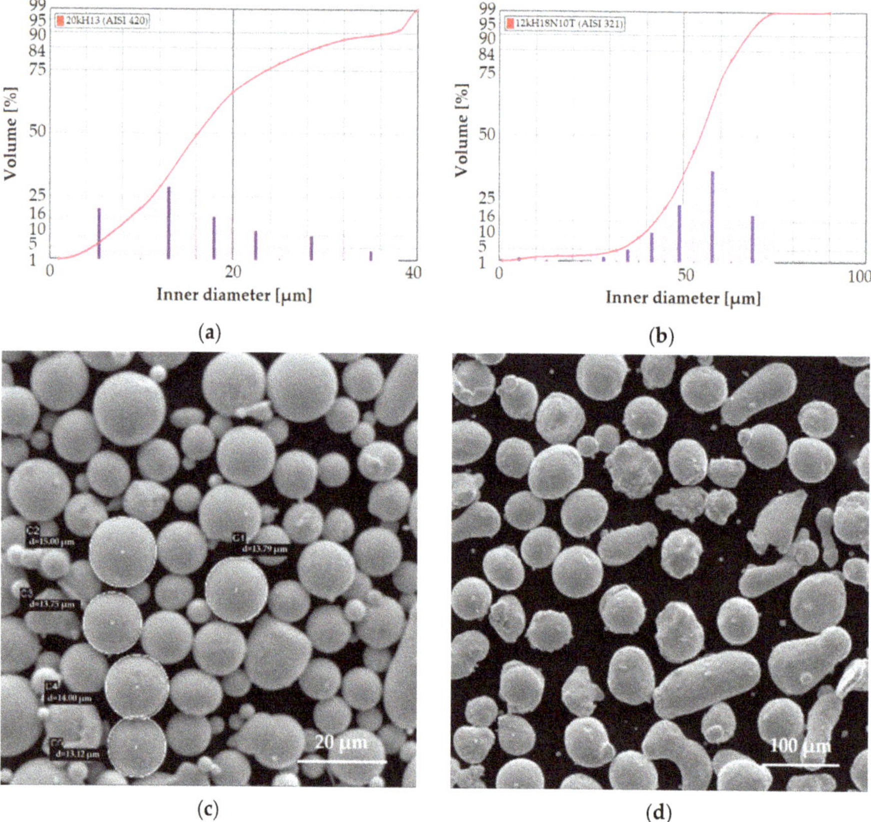

Figure 4. Granulometry and morphology (scanning electron microscopy) of the steel powders: (**a**) measurement results of PR 20kH13 (AISI 420); (**b**) measurement results of PR 12kH18N9T (AISI 321); (**c**) SEM-image of PR 20kH13 (AISI 420) particles, 2.98k×; (**d**) SEM-image of PR 12kH18N9T (AISI 321) particles, 510×.

Table 3. Shape parameters of the powders according to EN ISO 9276-6; 7; 8.

Powder	Shape Factor φ_i (Dimensionless)	Bluntness (%)	Roughness (mm)	Elongation (%)	Circularity (%)	Solidity (%)
PR 20kH13 (AISI 420)	0–1	91.3	0.003	23.0	72.5	97.8
PR 12kH18N9T (AISI 321)		73.9	0.031	22.0	65.2	90.6

Tracks were made from PR 20kH13 (AISI 420) and PR 12kH18N9T (AISI 321) steels at laser power in the range from 40 to 120 W, and scanning speeds from 50 to 150 mm·s^{-1}; the thickness of the powder layer applied to the substrate varied from 20 to 50 μm; the length of the track (melt pool) was 15 mm for all experiments. The height of the track above the substrate, the depth of the penetration zone of the underlying layer, the width of a single track, the width of the penetration zone, and the substrate's wetting angle in the zone of exposure laser radiation were analyzed for each experiment. The following laser powder bed fusion factors were selected for the manufacture of samples with the aim of further research because of the conducted analysis (Table 2):

- Layer thickness of 20 μm, laser power P_l of 80 W, scanning speed vs. of 390 mm·s^{-1} for PR 20kH13 (AISI 420) steel;
- Layer thickness of 20 μm, laser power P_l of 100 W, scanning speed vs. of 100 mm·s^{-1} for PR 12kH18N9T (AISI 321) steel.

3.2. Microstructure

The optical microscopy of the samples showed no visible cracks and pores. The microstructure of samples is shown in Figure 5. The grain structure and size are discerned at 1000× magnification (Figure 5b). The obtained microphotographs demonstrate paths of powder fusion by a laser beam. The average grain diameter of the observed area does not exceed 2 μm for 20kH13 (AISI 420) steel that corresponds to the 15th point (fine-grained group). However, it is not possible to identify individual phases in micrographs since they are very different from equilibrium conditions for the formation of the structure. The detected cells with an oriented structure characteristic of the laser powder bed fusion method are revealed at high magnifications (Figure 5a). The dendrites have no second-order axes in the microstructures of the produced samples. No non-metallic inclusions or traces of intergranular corrosion were found.

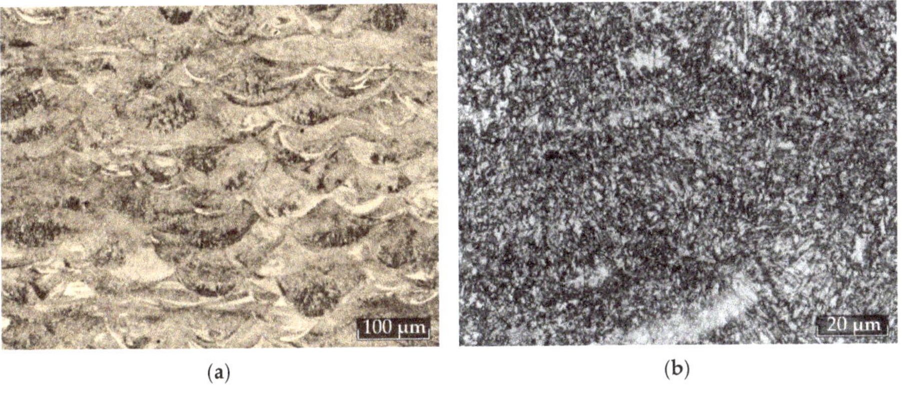

(a) (b)

Figure 5. Optical microscopy of 20kH13 (AISI 420) steel sample after laser powder bed fusion: (**a**) 200×; (**b**) 1000×.

X-ray structural analysis showed the presence of a supersaturated solid solution of the α-phase in the structure of the 20kH13 (AISI 420) sample and γ-phase in the 12kH18N9T (AISI 321) sample. The pore size does not exceed 0.01 mm that corresponds to point 3 by porosity point scale; the volume fraction of pores is in the range of 0.05%–0.1%.

3.3. Physical and Mechanical Properites

The obtained physical and mechanical characteristics of specimens made of 20kH13 (AISI 420) and 12kH18N9T (AISI 321) steels produced by the laser powder bed fusion method in comparison with the traditional manufacturing methods presented in Table 4.

Table 4. Physical and mechanical properties of 20kH13 (AISI 420) and 12kH18N9T (AISI 321) steel samples produced by laser powder bed fusion in comparison with the data obtained for traditionally produced samples.

Material	Producing Method	Density (ρ) (g·cm^{-3})	Young's Modulus (E) (GPa)	Yield Strength ($\sigma_{0.2}$) (MPa)	Tensile Strength (σ_B) (MPa)	Elastic Recovery (δ) (%)
20kH13 (AISI 420)	Laser powder bed fusion	7.709 ± 0.004	208.918 ± 1.025	863.98 ± 7.38	1584.09 ± 3.58	7.31 ± 0.22
	Cast + Quenching + low tempering	7.7 [1,2]	-	987 [1]	1485 [1]	9 [1]
	Cast + Quenching + high tempering	7.7 [1,2]	190–218 [2,3]	635 [1]	830 [1]	10 [1]
12kH18N9T (AISI 321)	Laser powder bed fusion	7.905 ± 0.004	193.012 ± 1.018	576.02 ± 7.31	657.03 ± 3.32	48 ± 0.56
	Cast + Quenching at 1050–1100 °C with cooling in water	7.9 [1,2]	175–195 [2,3]	196 [1]	540 [1]	40 [1]

[1] Provided according to GOST 5949-2018; [2] Provided according to GOST 5632-72; [3] the largest value corresponds to +20 °C and the smallest value corresponds to the +300 °C of working temperatures.

The results show that the mechanical characteristics of steels after laser powder bed fusion are higher or at the same level than those mentioned that are produced by traditional processing methods, which is explained by the features of the structure of steels obtained after the laser powder bed fusion method (supersaturation of the solid solution and mecodispersity of the structure) with almost equal density values.

3.4. Rougness

3.4.1. 20kH13 (AISI 420) Anticorrosion Steel

The roughness parameter R_a of the samples produced by laser powder bed fusion method from 20kH13 (AISI 420) steel was 7.24 ± 0.19 µm. The topology of the surface is shown in Figure 6. The measured results of roughness parameters (R_a, R_z, R_{tm}) are presented in Table 5; the obtained topology is shown in Figure 7 (visible area is 3.8 µm × 3.8 µm). The obtained profiles are presented in Figure 8. The untreated surface has a non-uniform topography with many inclusions formed due to metal splashing in the molten pool. The surface acquires a relatively regular structure and is characterized by the absence of inclusions after cavitation-abrasive finishing.

Table 5. The roughness parameters (R_a, R_z, R_{tm}) of the 20kH13 (AISI 420) steel sample surfaces before and after cavitation-abrasive finishing (after Gaussian regression).

Method	Arithmetic Mean Deviation (R_a) (µm)	Ten-Point Height (R_z) (µm)	Maximum Peak-to-Valley Height (R_{tm}) (µm)
After laser powder bed fusion manufacturing	7.24	72.9	83.7
After ultrasonic cavitation abrasive finishing	3.04	50.1	58.8

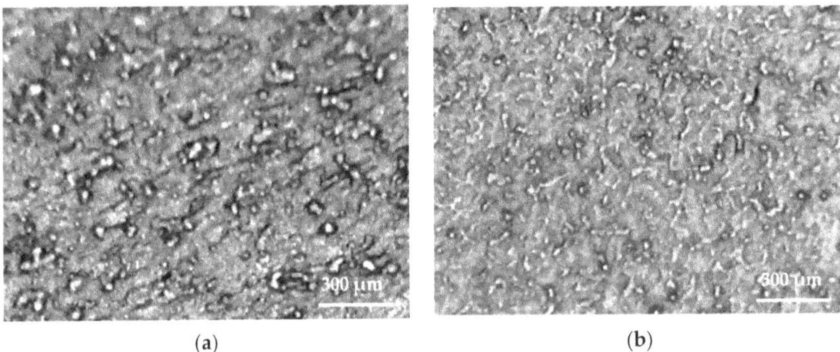

(a)   (b)

Figure 6. Surface topology of the samples (optical microscopy, 300×): (**a**) after production; (**b**) after cavitation-abrasive finishing.

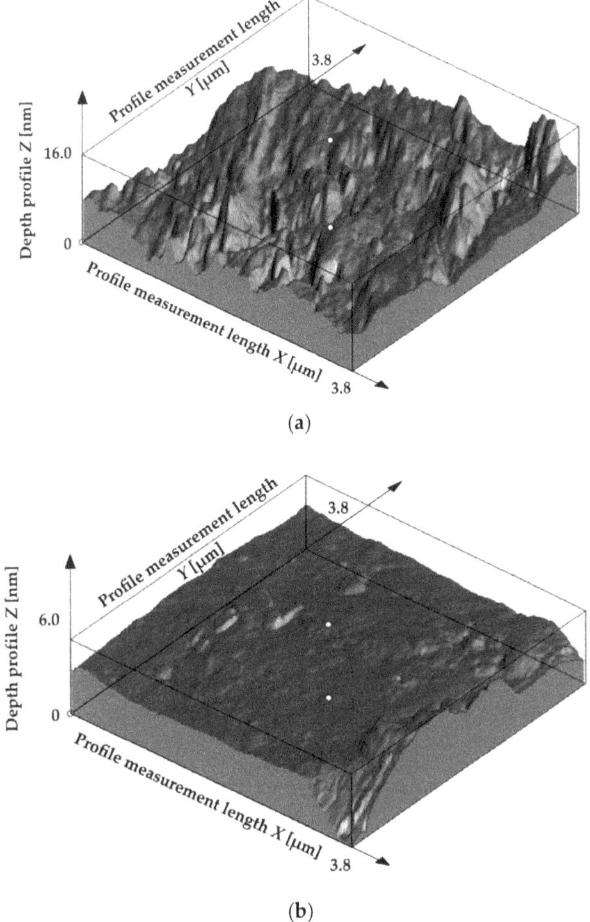

(a)

(b)

Figure 7. Surface topology of the samples (atomic force microscopy): (**a**) after production; (**b**) after cavitation-abrasive finishing.

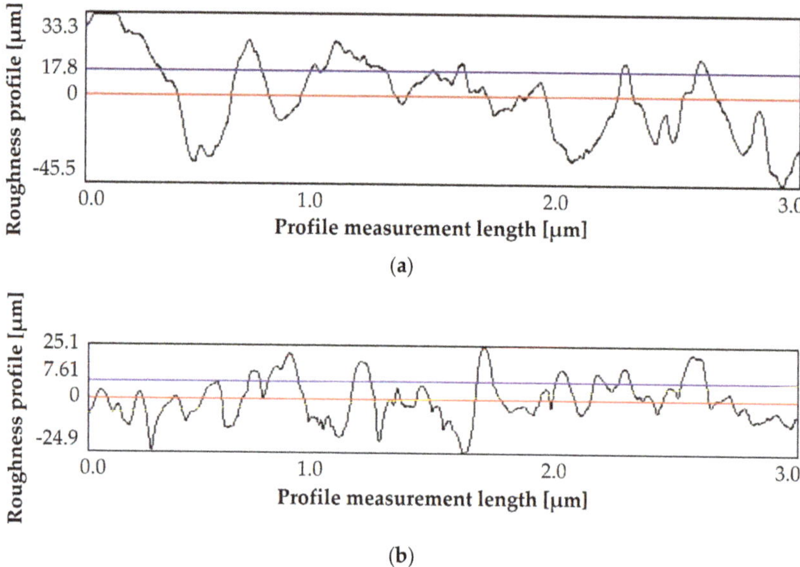

Figure 8. Surface profile of the samples: (**a**) after production; (**b**) after cavitation-abrasive finishing.

It is correlated to the distribution of cavitation activity, which depends on the treated surface's microgeometry parameters. Working fluids, which are cavitation bubbles and abrasive particles, have the most significant effect in places of most significant irregularities. In this case, the abrasive particles have damping functions, i.e., take on the energy arising from the collapse of cavitation bubbles, and the deformation of the surface is carried out due to the impact of the abrasive and not due to the effect of cumulative streams, which can lead to erosion. Thus, the applied cavitation-abrasive finishing is erosion-free and leads to surface smoothing.

It is quantitatively expressed in a decrease in the roughness parameters (R_a, R_z, R_{tm}) by 30%–60%, decreasing the average pitch of irregularities by more than two times. Increasing the processing time (over 120 s) does not increase the effect further.

3.4.2. 12kH18N9T (AISI 321) Anticorrosion Steel

The roughness parameter R_a of the samples produced by laser powder bed fusion method from 12kH18N9T (AISI 321) steel varied in the range of 8.5–14.1 µm. The obtained topology is shown in Figure 9; measuring results are presented in Table 6. 3D-profiles of the surfaces are presented in Figure 10. The roughness parameter R_a for the walls of after the vibratory tumbling was reduced by more than two times—from 14.1 to 5.0 µm.

Table 6. The roughness parameters (R_a, R_z) of the 12kH18N9T (AISI 321) steel surfaces before and after vibratory tumbling (before Gaussian regression).

Method	Wall of the Sample		Top of the Sample	
	Arithmetic Mean Deviation (R_a) (µm)	Ten-Point Height (R_z) (µm)	Arithmetic Mean Deviation (R_a) (µm)	Ten-Point Height (R_z) (µm)
After laser powder bed fusion manufacturing	14.1	53.0	8.5	36.5
After vibratory tumbling	5.0	17.4	2.5	11.3

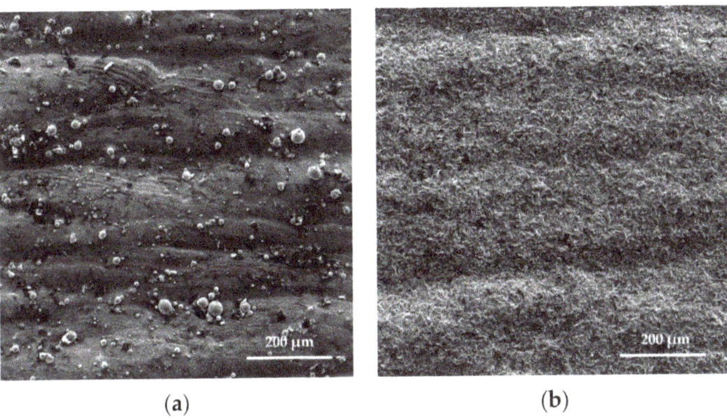

Figure 9. SEM-image in secondary electrons of the surface produced from 12kH18N9T (AISI 321) anticorrosion steel powder: (**a**) after laser powder bed fusion method, 500×; (**b**) after vibratory tumbling, 502×.

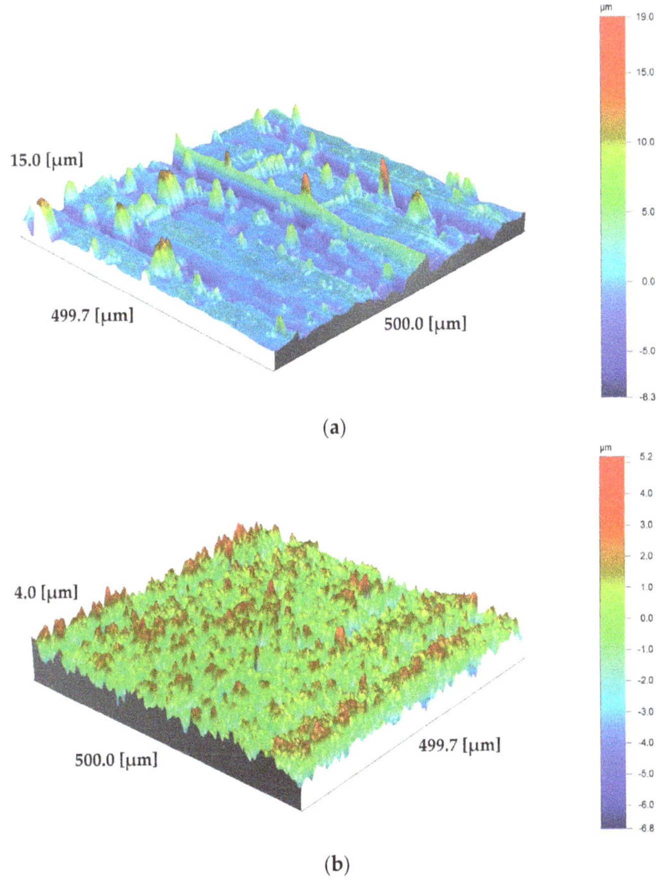

Figure 10. Surface topology of the samples (profilometry): (**a**) after production; (**b**) after vibration tumbling.

The surface longitudinal and traverse profiles on the area of 500 μm × 500 μm after Gaussian regression are presented in Figures 11 and 12. The measured roughness parameters R_a and R_z after regression are:

- 1.45 and 8.55 μm for raw sample after production;
- 1.25 and 7.05 μm for treated sample.

In the first case, the peaks of roughness profile (Figures 11b and 12b a peak of 12 μm) are associated with the presence of the unmelted granules on top of the surface that has metallurgical contact with the built sample. For the samples after vibratory tumbling, this type of morphology is absent; the surface retains a less pronounced wave character, without peaks, but with wells from the used abrasive, which are evenly distributed and regularized (Figure 11d).

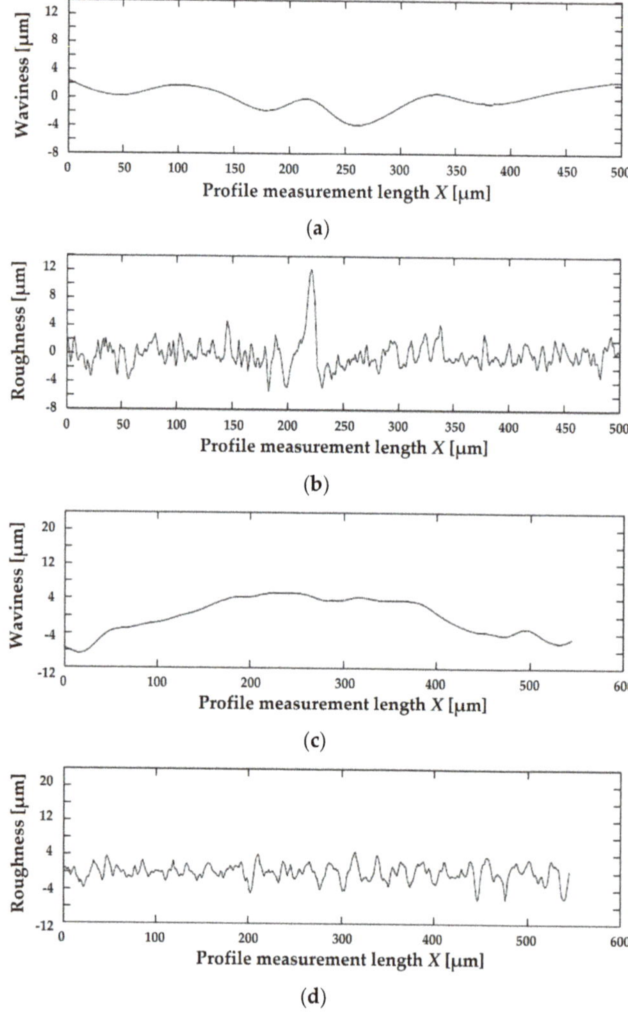

Figure 11. Longitudinal waviness and roughness profile after Gaussian regression (profilometry): (**a**) waviness after laser powder bed fusion method; (**b**) roughness profile after laser powder bed fusion method; (**c**) waviness after vibratory tumbling; (**d**) roughness profile after vibratory tumbling.

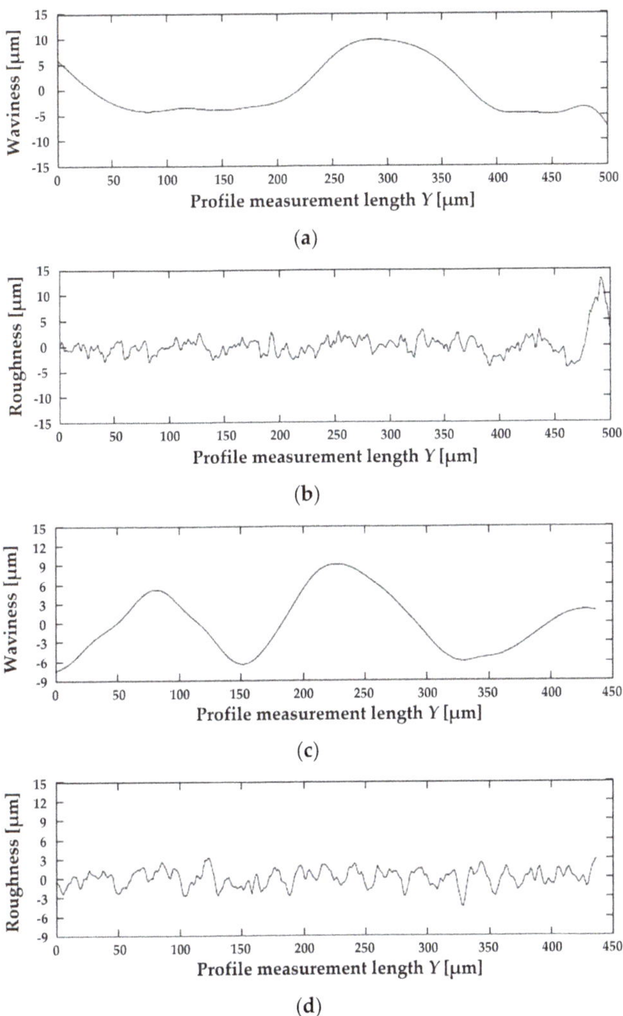

Figure 12. Transverse waviness and roughness profile after Gaussian regression (profilometry): (**a**) waviness after laser powder bed fusion method; (**b**) roughness profile after laser powder bed fusion method; (**c**) waviness after vibratory tumbling; (**d**) roughness profile after vibratory tumbling.

3.5. Heat Treatment, Hardness and Wear Resistance

The effect of subsequent heat treatment on the hardness of the 20kH13 (AISI 420) steel samples produced by laser powder bed fusion and traditional casting are presented in Table 7. The measured microhardness of the samples correlated to the data obtained for Rockwell hardness. The results of studies on water-jet wear of the laser powder bed fused samples and cast samples after combined quenching and low tempering are presented in Figures 13–15.

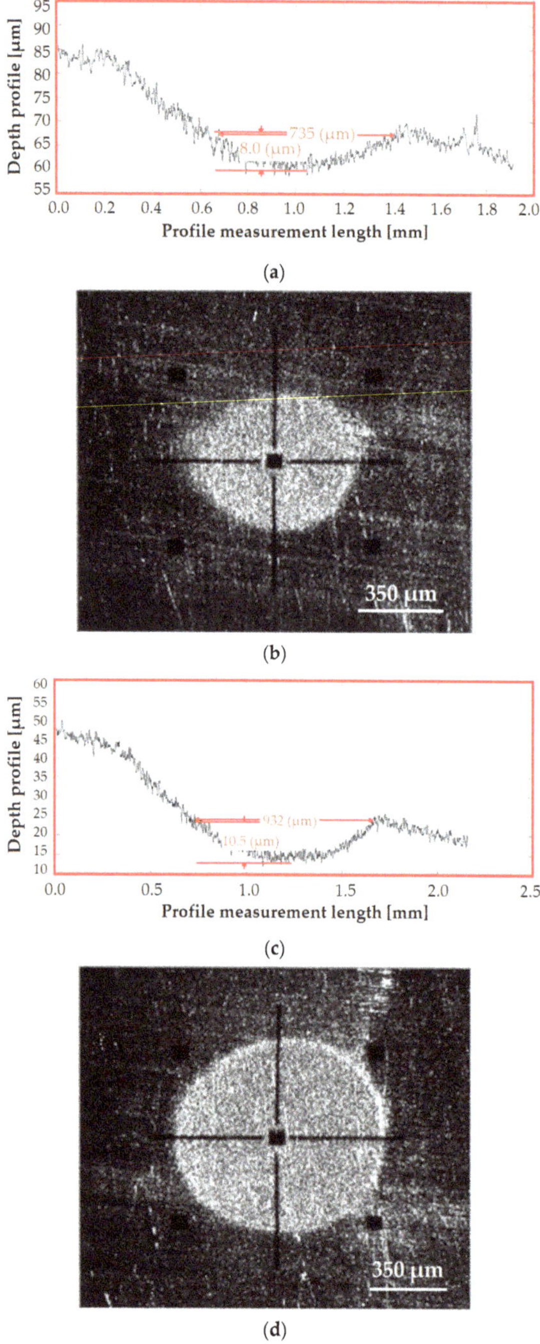

(a)

(b)

(c)

(d)

Figure 13. Profiles and microphotographs of the wells of the 20kH13 (AISI 420) steel samples after 3 min of test time (abrasion wear analyzer): (**a**) profile of the LPBF produced sample; (**b**) well image of the LPBF-produced sample; (**c**) profile of the cast sample treated by quenching with low tempering; (**d**) well image of the cast sample treated by quenching with low tempering.

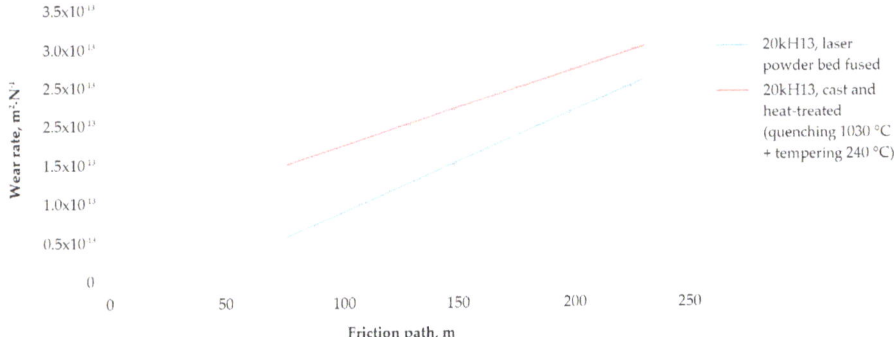

Figure 14. Dependence of the wear rate on the friction path of the 20kH13 (AISI 420) steel samples.

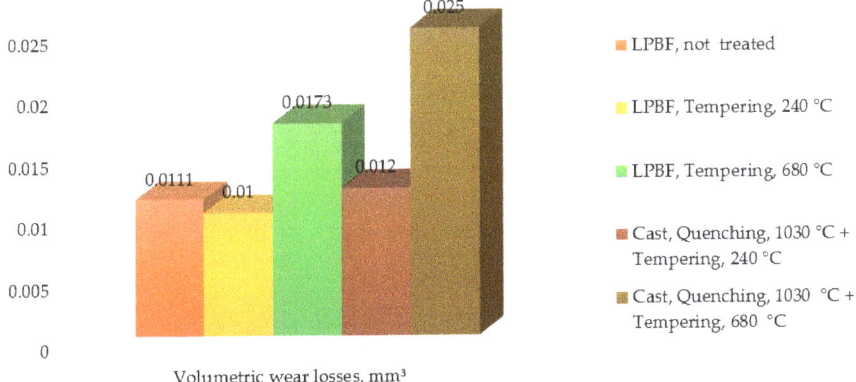

Figure 15. Histogram of volumetric losses during abrasive wear of steel 20kH13 (AISI 420), 9 min of wear time.

Table 7. Rockwell hardness of the samples made of the 20kH13 (AISI 420) steel in raw state after various types of post-processing.

Producing Method	Hardening (°C)	Medium	Tempering (°C)	Medium	Rockwell Hardness HRC_z [1]	Rockwell Hardness HRC_{xy} [1]
Laser powder bed fusion	-	-	240	Air	44.2	46.2
	-	-	680	Oil	38.7	39.1
	-	-	Annealing, 760	Air	33.4	35.8
	-	-	-	-	-	46.25
Cast	1030	Oil	240	Air	43.0	
	1030	Oil	680	Oil	22.3	

[1] Provided for two different directions—along Z-axis and X- and Y-axes for the samples produced by laser powder bed fusion due to known anisotropy (orthotropy) of the properties and one value HRC for cast samples due to known isotropy of the properties.

4. Discussion

4.1. Estimation of the Obtained Effects

The provided study allows researching the possible steps of post-processing for the parts produced from structural anticorrosion chrome-nickel steels—20kH13 (AISI 420) and 12kH18N9T (AISI 321).

The samples' detected fine structure was obtained due to high cooling rates [18,77–79]—the liquid quenching mechanism is implemented at growing solids by the laser powder bed fusion method.

The presence of inhomogeneities in the technological liquid medium during ultrasonic cavitation abrasive finishing led to a decrease in the liquid's cavitation strength and an increase in a cavitation centers number [80]. The abrasive particles acquired acceleration because of the transfer of impulse energy from shock waves of acoustic micro- and macroflows [81]. It was expressed quantitatively in a decrease in the roughness parameters (R_a, R_z, R_{tm}) by 50%–60%, a two-fold reduction in the average pitch of irregularities, and an increase in the profile reference length by 10%. Practice showed that extended processing time, which exceeded 120 s, did not increase the obtained effect.

Besides, abrasive particles have damping functions and protected the surface from cavitation destruction. The surface had no traces of cavitation erosion; there were no grinding marks, which indicated a uniform effect over the entire processing area and intensive removal of the material layer. The layer change occured due to the destruction of the protrusions of the grain surface. The most significant effect was achieved by ultrasonic cavitation abrasive finishing (Table 5, Figure 8).

The analysis of the results (Table 7) shows that the hardness of 20kH13 (AISI 420) samples produced by the laser powder bed fusion method was higher than that of the cast and heat-treated 20kH13 (AISI 420) samples. The low tempering of the laser powder bed-fused samples led to a slight decrease in hardness. Given that low tempering reduces quenching stresses, this kind of heat treatment can be recommended for parts that work for wear, particularly for those chosen in this study part as a locking washer. The hardness after high tempering was significantly lower than after a low one, which was explained by the decomposition of martensite and its transition to sorbitol tempering.

Investigation of the effect of heat treatment on the wear resistance of 20kH13 (AISI 420) steel specimens produced by the laser powder bed fusion method showed that the wear resistance without subsequent heat treatment was higher than that of samples obtained by the traditional casting with special heat treatment (Figures 10 and 11). The abrasion resistance correlates with the measured hardness. The wear resistance decreases when the tempering temperature increases for all the samples under study. The values of the wear resistance of specimens obtained by the laser powder bed fusion method and by the laser powder bed fusion method with low tempering differ slightly. Simultaneously, they are slightly higher than the wear resistance of cast specimens after traditional quenching and low tempering.

4.2. Quantituve Evaluation of Cavitation-Abrasive Finishing Factors

Let us quantitatively evaluate the achieved results. The stiffness of the system for two types of steel samples produced by additive manufacturing will be:

$$K_{420} = 4.18 \cdot 10^9 \left[\frac{N}{m}\right], \tag{15}$$

$$K_{321} = 3.86 \cdot 10^9 \left[\frac{N}{m}\right]. \tag{16}$$

Then the period of the natural oscillations will be

$$T_{420} = \sqrt{4\pi^2 \frac{m_{420}}{k_{420}}} = \sqrt{4\pi^2 \frac{0.0616 [\text{kg}]}{4.18 \cdot 10^9 \left[\frac{N}{m}\right]}} = 2.412 \cdot 10^{-5} [\text{s}], \tag{17}$$

$$T_{321} = \sqrt{4\pi^2 \frac{0.0632 [\text{kg}]}{3.86 \cdot 10^9 \left[\frac{N}{m}\right]}} = 2.542 \cdot 10^{-5} [\text{s}], \tag{18}$$

when the period of forced oscillations is:

$$\tau = \frac{1}{f_{frc}} = 4.762 \cdot 10^{-5} [\text{s}]. \tag{19}$$

The amplitude of the vibration can expressed as follows [82,83]:

$$\overline{A_m} = \overline{A_0} \cdot e^{\overline{\beta} \cdot \tau}, \tag{20}$$

where $\overline{\beta}$ is the damping coefficient expressed as a complex number in the form $a + bi$:

$$\overline{\beta} = \overline{q} - \overline{\mu}, \tag{21}$$

where $\overline{q}$ is the index of excited oscillations, and $\overline{\mu}$ is the coefficient of medium resistance, let us allow that:

$$\overline{\mu} \to 0 \tag{22}$$

and $\overline{q}$ can be expressed as follows:

$$\overline{q} = \sqrt{\frac{\sqrt{h} - k}{2 \cdot m_n}}, \tag{23}$$

where:

$$h = \frac{H_m}{H_0} \tag{24}$$

and

$$k = \frac{K_m}{K_0} = 1 \tag{25}$$

where the values indexed m are for the actual conditions and the values indexed 0 for conditions at the initial stage of processing. If the amplitude of the vibration S_m is 20 µm transit to the sample in the ideal conditions, then:

$$\sqrt{h} = \sqrt{\frac{H_m}{H_0}} = 0.9995 \tag{26}$$

and

$$\overline{\beta_{420}} = \overline{q_{420}} = |0.064|, \tag{27}$$

$$\overline{\beta_{321}} = \overline{q_{321}} = |0.063|. \tag{28}$$

The amplitude of the forced oscillations will be:

$$\overline{A_{420}} = \overline{A_0} \cdot e^{(|0.064| \cdot 4.76 \cdot 10^{-5})} = \overline{A_0} \cdot e^{3.05 \cdot 10^{-6}} = 1.000003 \cdot \overline{A_0} = 2.0 \cdot 10^{-6} [\text{m}], \tag{29}$$

$$\overline{A_{321}} = \overline{A_0} \cdot e^{(|0.063| \cdot 4.76 \cdot 10^{-5})} = \overline{A_0} \cdot e^{3.00 \cdot 10^{-6}} = 1.000003 \cdot \overline{A_0} = 2.0 \cdot 10^{-6} [\text{m}]. \tag{30}$$

The applied force will be:

$$F_{A420} = K_{420} \cdot \delta = 4.18 \cdot 10^9 \left[\frac{\text{N}}{\text{m}}\right] \cdot 6 \cdot 10^{-12} [\text{m}] = 25.08 \cdot 10^{-3} [\text{N}], \tag{31}$$

$$F_{A321} = K_{321} \cdot \delta = 3.86 \cdot 10^9 \left[\frac{\text{N}}{\text{m}}\right] \cdot 6 \cdot 10^{-12} [\text{m}] = 23.16 \cdot 10^{-3} [\text{N}]. \tag{32}$$

and acoustic pressure on 0.0004 m^2 of the sample:

$$P_{A420} = \frac{F_{A420}}{S_A} = \frac{25.08 \cdot 10^{-3} [\text{N}]}{0.0004 [\text{m}^2]} = 62.7 [\text{Pa}], \tag{33}$$

$$P_{A321} = \frac{F_{A321}}{S_A} = \frac{23.16 \cdot 10^{-3} [\text{N}]}{0.0004 [\text{m}^2]} = 57.9 [\text{Pa}].$$ (34)

Thus, it should be noted that the period of the natural oscillations that are determined by the nature of the system was less than the chosen period (frequency) of forced oscillations. Self-oscillations resulting from the action of the internal energy of the system with a fixed frequency f_{slf}, close to the natural frequency f_{nat}, and a fixed amplitude; the reason, in this case, is associated with the low rigidity of the system and fluctuations in the acting force F_A; vibration frequency increases with increasing system rigidity (stiffness K) and decreases with decreasing workpiece thickness. It is recommended that the cavitation finishing be performed outside the free (natural) oscillation area to exclude the resonance phenomenon [84]:

$$0.7 < \frac{f_{frc}}{f_{slf}} < 1.3$$ (35)

or

$$0.7 < \frac{T}{\tau} < 1.3$$ (36)

then the process can be characterized as effective and stable. In the given case:

$$\frac{T_{420}}{\tau} = \frac{2.412 \cdot 10^{-5}}{4.762 \cdot 10^{-5}} = 0.51,$$ (37)

$$\frac{T_{321}}{\tau} = \frac{2.542 \cdot 10^{-5}}{4.762 \cdot 10^{-5}} = 0.53.$$ (38)

As it can be seen, the chosen frequency should be enlarged for further experiments to achieve a more effective and stable processing mode in cavitation finishing; however, the introduction of ceramic abrasive in the working zone changed the conditions in the working tank and allowed reduced frequency in combination with the effectiveness of abrasive granule deformation. For improving parameters of processing up to the ratio of stable processing without adding additives that influence medium resistance, the frequency of the forced oscillations should be in the interval of 29–54 kHz for 20kH13 (AISI 420) steel and 28–52 kHz for 12kH18N9T (AISI 321) steel. If the amplitude of the forced vibrations in the working zone is small and measures in micrometers, then the difference δ between $\overline{A_0}$ and $\overline{A_m}$ is extremely small and is measured in picometers, which 0.01 of an angstrom (Å). With the known acoustic pressure, it is possible to evaluate the acoustic pressure's amplitude in each moment of the cycle. At the same time, it should be noted that frequency higher than 30 kHz and up to 1 MHz could be harmful to the biological process in the human body since arising cavitation with bubble formation with a diameter less than 1 μm (ultrasound surgery) [85], when ultrasound in the range 2–29 MHz is used in echography; the works should be conducted according to the sanitary norms and rules of production.

5. Conclusions

5.1. Structure and Surface Quality

In the process of the laser powder bed fusion method in the steels under study, as a result of high cooling rates, a finely dispersed structure with supersaturated solid solutions is formed, which provides higher strength characteristics of steels after additive manufacturing than after traditional processing methods.

The maximum effect on the change in surface quality was achieved by ultrasonic cavitation-abrasive finishing based on the combination of cavitation, acoustic flows, and abrasive particles that were evaluated quantitatively. A twofold decrease in the parameters of micro- and submicroroughness was achieved (roughness parameter R_a was reduced from 7.24 to 3.04 μm), and no erosion caverns were detected on the researched surfaces. The use of a vibratory grinder reduces the roughness parameter

R_a of the sample walls made of 12kH18N9T (AISI 321) powder from 14.1 to 5 μm. However, it should be borne in mind that using a vibratory grinder usually reflects the service life of parts negatively.

The quantitative evaluation showed that the difference in the chosen material's mechanical and physical properties did not reflect the maximal amplitude of $2.0 \cdot 10^{-6}$ m when δ of the samples was about $6 \cdot 10^{-12}$ m. The chosen frequency ratio to the natural frequency was about 0.5 and can be improved to improve processing effectiveness and stability without adding abrasive. The frequency of the forced oscillations should be in the interval of 29–54 kHz for 20kH13 (AISI 420) steel and 28–52 kHz for 12kH18N9T (AISI 321) steel.

5.2. Resistance to Abrasion Wear

The obtained abrasion resistance correlated with the measured hardness values. The wear resistance decreased with an increase in the tempering temperature for all the samples—the volume of worn material increased by 1.5–2 times for samples with a tempering temperature of 680 °C] in comparison with the raw part or after tempering at 240 °C. The values of wear resistance and hardness of additively manufactured specimens and the specimens that were additionally subjected to low tempering differ insignificantly (HRC of 46.25 and 46.2, correspondingly). However, they are slightly higher than the wear resistance of samples after traditional hardening and low tempering (HRC of 45.9). Considering that low tempering reduces residual stresses, heat treatment can be recommended for additively manufactured parts working for wear.

Analysis of the research results shows that, for complex-shaped parts made of corrosion-resistant steels by the laser powder bed fusion, it is promising to use heat treatment to improve mechanical characteristics and ultrasonic treatment in order to improve the surface quality in post-processing.

Author Contributions: Conceptualization, S.N.G. and M.A.V.; methodology, T.V.T. and S.K.S.; software, Y.A.M. and P.A.P.; validation, A.S.M. and S.K.S.; formal analysis, A.S.M. and A.A.F.; investigation, A.A.F., P.A.P. and Y.A.M.; resources, M.A.V., T.V.T.; data curation, A.A.O. and Y.A.M.; writing—original draft preparation, A.A.F. and S.K.S.; writing—review and editing, T.V.T. and A.A.O.; visualization, A.A.O. and P.A.P.; supervision, S.N.G.; project administration, S.N.G. and M.A.V.; funding acquisition, A.S.M. All authors have read and agreed to the published version of the manuscript.

Funding: This research was funded by the Russian Science Foundation, grant number No. 20-19-00620.

Acknowledgments: The research was done at the Department of High-Efficiency Processing Technologies of MSTU Stankin.

Conflicts of Interest: The authors declare no conflict of interest. The funders had no role in the design of the study; in the collection, analyses, or interpretation of data; in the writing of the manuscript, or in the decision to publish the results.

References

1. Yadroitsev, I.; Bertrand, P.; Antonenkova, G.; Grigoriev, S.; Smurov, I. Use of track/layer morphology to develop functional parts by selective laser melting. *J. Laser Appl.* **2013**, *25*, 052003. [CrossRef]
2. Saprykin, A.A.; Sharkeev, Y.P.; Saprykina, N.A.; Ibragimov, E.A. Selective Laser Melting of Magnesium. *Key Eng. Mater.* **2020**, *839*, 144–149. [CrossRef]
3. Polozov, I.A.; Borisov, E.V.; Sufiiarov, V.S.; Popovich, A.A. Selective Laser Melting of Copper Alloy. *Mater. Phys. Mech.* **2020**, *43*, 65–71.
4. Willner, R.; Lender, S.; Ihl, A.; Wilsnack, C.; Gruber, S.; Brandao, A.; Pambaguian, L.; Riede, M.; Lopez, E.; Brueckner, F.; et al. Potential and challenges of additive manufacturing for topology optimized spacecraft structures. *J. Laser Appl.* **2020**, *32*, 032012. [CrossRef]
5. Verna, E.; Genta, G.; Galetto, M.; Franceschini, F. Planning offline inspection strategies in low-volume manufacturing processes. *Qual. Eng.* **2020**, 1–16. [CrossRef]
6. Malkina, M.; Balakin, R. Sectoral determinants of sub-federal budget tax revenues: Russian case study. *Equilib. Q. J. Econ. Econ. Policy* **2019**, *14*, 233–249. [CrossRef]
7. Nath, S.D.; Irrinki, H.; Gupta, G.; Kearns, M.; Gulsoy, O.; Atre, S. Microstructure-property relationships of 420 stainless steel fabricated for by laser-powder bed fusion. *Powder Technol.* **2019**, *343*, 738–746. [CrossRef]

8. Jamshidinia, M.; Sadek, A.; Wang, W.; Kelly, S. Additive Manufacturing of Steel Alloys Using Laser Powder-Bed Fusion. *Adv. Mater. Process.* **2015**, *173*, 20–24.

9. Zhu, H.; Li, Y.; Li, B.; Zhang, Z.; Qiu, C. Effects of Low-Temperature Tempering on Microstructure and Properties of the Laser-Cladded AISI 420 Martensitic Stainless Steel Coating. *Coatings* **2018**, *8*, 451. [CrossRef]

10. Tarasova, T.V.; Gusarov, A.V.; Protasov, K.E.; Filatova, A.A. Effect of Thermal Fields on the Structure of Corrosion-Resistant Steels Under Different Modes of Laser Treatment. *Met. Sci. Heat Treat.* **2017**, *59*, 433–440. [CrossRef]

11. Nowotny, S.; Tarasova, T.V.; Filatova, A.A.; Dolzhikova, E.Y. Methods for Characterizing Properties of Corrosion-Resistant Steel Powders Used for Powder Bed Fusion Processes. *Mater. Sci. Forum* **2016**, *876*, 1–7. [CrossRef]

12. Doubenskaia, M.; Pavlov, M.; Grigoriev, S.; Tikhonova, E.; Smurov, I. Comprehensive Optical Monitoring of Selective Laser Melting. *J. Laser Micro/Nanoeng.* **2012**, *7*, 236–243. [CrossRef]

13. Smurov, I.; Doubenskaia, M.; Grigoriev, S.; Nazarov, A. Optical Monitoring in Laser Cladding of Ti6Al4V. *J. Therm. Spray Technol.* **2012**, *21*, 1357–1362. [CrossRef]

14. Breidenstein, B.; Brenne, F.; Wu, L.; Niendorf, T.; Denkena, B. Effect of Post-Process Machining on Surface Properties of Additively Manufactured H13 Tool Steel. *J. Heat Treat. Mater.* **2018**, *73*, 173–186. [CrossRef]

15. Crayford, A.P.; Lacan, F.; Runyon, J.; Bowen, P.J.; Balwadkar, S.; Harper, J.; Pugh, D.G. Manufacture, Characterization and Stability Limits of an Am Prefilming Air-Blast Atomizer. In Proceedings of the ASME Turbo Expo: Turbomachinery Technical Conference and Exposition, Phoenix, AZ, USA, 17–21 June 2019; American Society of Mechanical Engineers: New York, NY, USA, 2019; Volume 4B.

16. Hunter, L.W.; Brackett, D.; Brierley, N.; Yang, J.; Attallah, M.M. Assessment of trapped powder removal and inspection strategies for powder bed fusion techniques. *Int. J. Adv. Manuf. Technol.* **2020**, *106*, 4521–4532. [CrossRef]

17. Aleshin, N.P.; Grigor'ev, M.V.; Shchipakov, N.A.; Krys'ko, N.V.; Krasnov, I.S.; Prilutskii, M.A.; Smorodinskii, Y.G. On the Possibility of Using Ultrasonic Surface and Head Waves in Nondestructive Quality Checks of Additive Manufactured Products. *Russ. J. Nondestruct. Test.* **2017**, *53*, 830–838. [CrossRef]

18. Tarasova, T.; Gvozdeva, G.; Ableyeva, R. Innovation in additive manufacturing of parts from aluminium matrix composites. In *MATEC Web of Conferences, Proceedings of the International Conference on Modern Trends in Manufacturing Technologies and Equipment (ICMTMTE)—Materials Science, Sevastopol, Russia, 10–14 September 2018*; Bratan, S., Gorbatyuk, S., Leonov, S., Roshchupkin, S., Eds.; E D P Sciences: Les Ulis, France, 2018; p. 01073.

19. Manzhirov, A.V.; Murashkin, E.V.; Parshin, D.A. Modeling of Additive Manufacturing and Surface Growth Processes. In Proceedings of the International Conference on Numerical Analysis and Applied Mathematics (ICNAAM), Rhodes, Greece, 13–18 September 2018; Simos, T., Tsitouras, C., Eds.; Am. Inst. Physics: Melville, NY, USA, 2019; p. 380011.

20. Gorunov, A.I. Directional Crystallization of 316L Stainless Steel Specimens by Direct Laser Deposition. *Inorg. Mater.* **2019**, *55*, 1439–1444. [CrossRef]

21. Kalashnikov, K.N.; Kalashnikova, A. Surface Morphology of Ti-alloy Samples Obtained by Electron Beam 3D-printing. In Proceedings of the International Conference on Advanced Materials with Hierarchical Structure for New Technologies and Reliable Structures, Tomsk, Russia, 1–5 October 2019; Panin, V.E., Psakhie, S.G., Fomin, V.M., Eds.; Am. Inst. Physics: Melville, NY, USA, 2019; Volume 2167, p. 020145.

22. Upadhyay, R.K.; Kumar, A. Scratch and wear resistance of additive manufactured 316L stainless steel sample fabricated by laser powder bed fusion technique. *Wear* **2020**, *458*, 203437. [CrossRef]

23. Grigoriev, S.N.; Volosova, M.A.; Vereschaka, A.A.; Sitnikov, N.N.; Milovich, F.; Bublikov, J.I.; Fyodorov, S.V.; Seleznev, A.E. Properties of (Cr,Al,Si)N-(DLC-Si) composite coatings deposited on a cutting ceramic substrate. *Ceram. Int.* **2020**, *46*, 18241–18255. [CrossRef]

24. Vereschaka, A.S.; Grigoriev, S.N.; Tabakov, V.P.; Sotova, E.S.; Vereschaka, A.A.; Kulikov, M.Y. Improving the efficiency of the cutting tool made of ceramic when machining hardened steel by applying nano-dispersed multi-layered coatings. *Key Eng. Mater.* **2014**, *581*, 68–73. [CrossRef]

25. Grigoriev, S.N.; Sobol, O.V.; Beresnev, V.M.; Serdyuk, I.V.; Pogrebnyak, A.D.; Kolesnikov, D.A.; Nemchenko, U.S. Tribological characteristics of (TiZrHfVNbTa)N coatings applied using the vacuum arc deposition method. *J. Frict. Wear* **2014**, *35*, 359–364. [CrossRef]

26. Glaziev, S.Y. The discovery of regularities of change of technological orders in the central economics and mathematics institute of the soviet academy of sciences. *Econ. Math. Methods* **2018**, *54*, 17–30. [CrossRef]

27. Korotayev, A.V.; Tsirel, S.V. A spectral analysis of world GDP dynamics: Kondratiev waves, Kuznets swings, Juglar and Kitchin cycles in global economic development, and the 2008–2009 economic crisis. *Struct. Dyn.* **2010**, *4*, 3–57.

28. Perez, C. Technological revolutions and techno-economic paradigms. *Camb. J. Econ.* **2010**, *34*, 185–202. [CrossRef]

29. Wonglimpiyarat, J. Towards the sixth Kondratieff cycle of nano revolution. *Int. J. Nanotechnol. Mol. Comput.* **2011**, *3*, 87–100. [CrossRef]

30. Chen, Y.; Sun, H.; Li, Z.; Wu, Y.; Xiao, Y.; Chen, Z.; Zhong, S.; Wang, H. Strategy of Residual Stress Determination on Selective Laser Melted Al Alloy Using XRD. *Materials* **2020**, *13*, 451. [CrossRef]

31. Wan, H.Y.; Luo, Y.W.; Zhang, B.; Song, Z.M.; Wang, L.Y.; Zhou, Z.J.; Li, C.P.; Chen, G.F.; Zhang, G.P. Effects of surface roughness and build thickness on fatigue properties of selective laser melted Inconel 718 at 650 degrees C. *Int. J. Fatigue* **2020**, *137*, 105654. [CrossRef]

32. Jamshidi, P.; Aristizabal, M.; Kong, W.; Villapun, V.; Cox, S.C.; Grover, L.M.; Attallah, M.M. Selective Laser Melting of Ti-6Al-4V: The Impact of Post-processing on the Tensile, Fatigue and Biological Properties for Medical Implant Applications. *Materials* **2020**, *13*, 2813. [CrossRef]

33. Sova, A.; Okunkova, A.; Grigoriev, S.; Smurov, I. Velocity of the Particles Accelerated by a Cold Spray Micronozzle: Experimental Measurements and Numerical Simulation. *J. Therm. Spray Technol.* **2013**, *22*, 75–80. [CrossRef]

34. Khmyrov, R.; Grigoriev, S.; Okunkova, A.; Gusarov, A. On the possibility of selective laser melting of quartz glass. *Phys. Procedia* **2014**, *56*, 345–356. [CrossRef]

35. Lober, L.; Flache, C.; Petters, R.; Kuhn, U.; Eckert, J. Comparison of different post processing technologies for SLM generated 3161 steel parts. *Rapid Prototyp. J.* **2013**, *19*, 173–179. [CrossRef]

36. Gatto, A.; Bassoli, E.; Denti, L.; Sola, A.; Tognoli, E.; Comin, A.; Porro, J.A.; Cordovilla, F.; Angulo, I.; Ocana, J.L. Effect of Three Different Finishing Processes on the Surface Morphology and Fatigue Life of A357.0 Parts Produced by Laser-Based Powder Bed Fusion. *Adv. Eng. Mater.* **2019**, *21*, 1801357. [CrossRef]

37. Jung, S.; Kang, E.; Park, J.; Kim, T.; Baek, S. A Study on Proposal of E-Beam Drilling Machining Criterion by Using on Vaporized Amplification Sheets in a High-Power Density Electron Beam. *J. Nanosci. Nanotechnol.* **2020**, *20*, 4231–4234. [CrossRef] [PubMed]

38. Gavrilov, N.V.; Men'shakov, A.I. Effect of the electron beam and ion flux parameters on the rate of plasma nitriding of an austenitic stainless steel. *Tech. Phys.* **2012**, *57*, 399–404. [CrossRef]

39. Fedorov, S.V.; Pavlov, M.D.; Okunkova, A.A. Effect of structural and phase transformations in alloyed subsurface layer of hard-alloy tools on their wear resistance during cutting of high-temperature alloys. *J. Frict. Wear* **2013**, *34*, 190–198. [CrossRef]

40. Grigoriev, S.N.; Metel, A.S.; Fedorov, S.V. Modification of the structure and properties of high-speed steel by combined vacuum-plasma treatment. *Met. Sci. Heat Treat.* **2012**, *54*, 8–12. [CrossRef]

41. Volosova, M.A.; Grigor'ev, S.N.; Kuzin, V.V. Effect of titanium nitride coating on stress structural inhomogeneity in oxide-carbide ceramic. Part 4. Action of heat flow. *Refract. Ind. Ceram.* **2015**, *56*, 91–96. [CrossRef]

42. Dong, G.; Marleau-Finley, J.; Zhao, Y.F. Investigation of electrochemical post-processing procedure for Ti-6Al-4V lattice structure manufactured by direct metal laser sintering (DMLS). *Int. J. Adv. Manuf. Technol.* **2019**, *104*, 3401–3417. [CrossRef]

43. Rotty, C.; Mandroyan, A.; Doche, M.-L.; Monney, S.; Hihn, J.Y.; Rouge, N. Electrochemical Superfinishing of Cast and ALM 316L Stainless Steels in Deep Eutectic Solvents: Surface Microroughness Evolution and Corrosion Resistance. *J. Electrochem. Soc.* **2019**, *166*, C468–C478. [CrossRef]

44. Metel, A.; Bolbukov, V.; Volosova, M.; Grigoriev, S.; Melnik, Y. Equipment for deposition of thin metallic films bombarded by fast argon atoms. *Instrum. Exp. Tech.* **2014**, *57*, 345–351. [CrossRef]

45. Volosova, M.A.; Gurin, V.D. Influence of vacuum-plasma nitride coatings on contact processes and a mechanism of wear of working surfaces of high-speed steel cutting tool at interrupted cutting. *J. Frict. Wear* **2013**, *34*, 183–189. [CrossRef]

46. Semenov, A.P.; Baldanov, B.B.; Ranzhurov, T.V. A Source of Nonequilibrium Argon Plasma Based on a Volume Gas Flow Discharge at Atmospheric Pressure. *Instrum. Exp. Tech.* **2020**, *63*, 284–287. [CrossRef]

47. Tan, K.L.; Yeo, S.H. Surface finishing on IN625 additively manufactured surfaces by combined ultrasonic cavitation and abrasion. *Addit. Manuf.* **2020**, *31*, 100938. [CrossRef]
48. Wang, J.; Zhu, J.; Liew, P.J. Material Removal in Ultrasonic Abrasive Polishing of Additive Manufactured Components. *Appl. Sci.* **2019**, *9*, 5359. [CrossRef]
49. Tan, K.L.; Yeo, S.H. Surface modification of additive manufactured components by ultrasonic cavitation abrasive finishing. *Wear* **2017**, *378–379*, 90–95. [CrossRef]
50. Gou, J.; Wang, Z.; Hu, S.; Shen, J.; Tian, Y.; Zhao, G.; Chen, Y. Effects of ultrasonic peening treatment in three directions on grain refinement and anisotropy of cold metal transfer additive manufactured Ti-6Al-4V thin wall structure. *J. Manuf. Process.* **2020**, *54*, 148–157. [CrossRef]
51. Zhou, C.; Jiang, F.; Xu, D.; Guo, C.H.; Zhao, C.Z.; Wang, Z.Q.; Wang, J.D. A calculation model to predict the impact stress field and depth of plastic deformation zone of additive manufactured parts in the process of ultrasonic impact treatment. *J. Mater. Process. Technol.* **2020**, *280*, 116599. [CrossRef]
52. Bankowski, D.; Spadlo, S. Vibratory Machining Effect on the Properties of the Aaluminum Alloys Surface. *Arch. Foundry Eng.* **2017**, *17*, 19–24. [CrossRef]
53. Blinn, B.; Krebs, F.; Ley, M.; Teutsch, R.; Beck, T. Determination of the influence of a stress-relief heat treatment and additively manufactured surface on the fatigue behavior of selectively laser melted AISI 316L by using efficient short-time procedures. *Int. J. Fatigue* **2020**, *131*, 105301. [CrossRef]
54. Teixeira, Ó.; Silva, F.J.G.; Ferreira, L.P.; Atzeni, E. A Review of Heat Treatments on Improving the Quality and Residual Stresses of the Ti–6Al–4V Parts Produced by Additive Manufacturing. *Metals* **2020**, *10*, 1006. [CrossRef]
55. Filatova, A.; Tarasova, T.; Peretyagin, P. Developing processes for manufacturing metal aviation technology components using powder bed fusion methods. *MATEC Web Conf.* **2019**, *298*, 00116. [CrossRef]
56. Cortina, M.; Arrizubieta, J.I.; Calleja, A.; Ukar, E.; Alberdi, A. Case Study to Illustrate the Potential of Conformal Cooling Channels for Hot Stamping Dies Manufactured Using Hybrid Process of Laser Metal Deposition (LMD) and Milling. *Metals* **2018**, *8*, 102. [CrossRef]
57. Zhao, Y.; Li, F.; Chen, S.; Lu, Z.Y. Unit block-based process planning strategy of WAAM for complex shell-shaped component. *Int. J. Adv. Manuf. Technol.* **2019**, *104*, 3915–3927. [CrossRef]
58. Han, P. Additive Design and Manufacturing of Jet Engine Parts. *Engineering* **2017**, *3*, 648–652. [CrossRef]
59. Liu, Z.; Huang, C.; Gao, C.; Liu, R.; Chen, J.; Xiao, Z. Characterization of Ti6Al4V Powders Produced by Different Methods for Selective Electron Beam Melting. *J. Min. Metall. Sect. B-Metall.* **2019**, *55*, 121–128. [CrossRef]
60. Gao, C.F.; Xiao, Z.Y.; Zou, H.P.; Liu, Z.Q.; Chen, J.; Li, S.K.; Zhang, D.T. Characterization of spherical AlSi10Mg powder produced by double-nozzle gas atomization using different parameters. *Trans. Nonferrous Met. Soc. China* **2019**, *29*, 374–384. [CrossRef]
61. Metel, A.S.; Stebulyanin, M.M.; Fedorov, S.V.; Okunkova, A.A. Power Density Distribution for Laser Additive Manufacturing (SLM): Potential, Fundamentals and Advanced Applications. *Technologies* **2019**, *7*, 5. [CrossRef]
62. Gusarov, A.V.; Grigoriev, S.N.; Volosova, M.A.; Melnik, Y.A.; Laskin, A.; Kotoban, D.V.; Okunkova, A.A. On productivity of laser additive manufacturing. *J. Mater. Process. Technol.* **2018**, *261*, 213–232. [CrossRef]
63. Rando, R.J.; Vacek, P.M.; Glenn, R.E.; Kwon, C.W.; Parker, J.E. Retrospective Assessment of Respirable Quartz Exposure for a Silicosis Study of the Industrial Sand Industry. *Ann. Work Expo. Health* **2018**, *62*, 1021–1032. [CrossRef]
64. Aleksandrov, V.A.; Sundukov, S.K.; Fatyukhin, D.S.; Filatova, A.A. Ultrasonic Methods for Improving Object Surface Quality Prepared by Corrosion-Resistant Steel Powder Selective Laser Melting. *Met. Sci. Heat Treat.* **2018**, *60*, 381–386. [CrossRef]
65. Kropotkina, E.; Zykova, M.; Shein, A.; Kapustina, N. Application of roller burnishing technologies to improve the wear resistance of submerged pump parts made of powder alloys. *Mech. Ind.* **2018**, *19*, 705. [CrossRef]
66. Naydenkin, E.; Mishin, I.; Khrustalyov, A.; Vorozhtsov, S.; Vorozhtsov, A. Influence of Combined Helical and Pass Rolling on Structure and Residual Porosity of an AA6082-0.2 wt % Al_2O_3 Composite Produced by Casting with Ultrasonic Processing. *Metals* **2017**, *7*, 544. [CrossRef]
67. Livanskiy, A.N.; Prikhodko, V.M.; Sundukov, S.K.; Fatyukhin, D.S. Research on the Influence of Ultrasonic Vibrations on Paint Coating Properties. *Trans. Famena* **2016**, *40*, 129–138.

68. Salmi, T.O.; Mikkola, J.P.; Warna, J.P. *Chemical Reaction Engineering and Reactor Technology*, 1st ed.; CRC Press: Boca Raton, FL, USA, 2011; p. 644.

69. High-Intensity Ultrasonic Fields (Ultrasonic Technology). In *Ultrasonic Technology. A Series of Monographs*, 1st ed.; Rozenberg, L. (Ed.) Springer: New York, NY, USA, 1971; p. 443.

70. Khodabandeloo, B.; Landro, M.; Hanssen, A. Acoustic generation of underwater cavities-comparing modeled and measured acoustic signals generated by seismic air gun arrays. *J. Acoust. Soc. Am.* **2017**, *141*, 2661–2672. [CrossRef]

71. Neppiras, E.A. Mechanics and Physics of Bubbles in Liquids—*Vanwijngaarden*, L. Ultrasonics **1983**, *21*, 238–239. [CrossRef]

72. Sirotyuk, M.G. Resonant Electrodynamic Sound Generator for Acoustic Levitation. *Sov. Phys. Acoust. USSR* **1986**, *32*, 399–403.

73. Agranat, B.A.; Khavskii, N.N. Experience in Incorporation of High-Intensity Ultrasonic Lines in Continuous Annealing Plants. *Sov. Phys. Acoust. USSR* **1976**, *22*, 80.

74. Badalyan, V.G.; Vorontsova, N.N.; Kazantsev, V.F.; Nazarov, A.V. Change of Copper Dislocational Structure after Action of Statical and Ultrasound Stresses. *Fiz. Met. Metalloved.* **1982**, *54*, 1191–1193.

75. Hayu, R.; Sutanto, H.; Ismail, Z. Accurate density measurement of stainless steel weights by hydrostatic weighing system. *Measurement* **2019**, *131*, 120–124. [CrossRef]

76. Firsov, K.N. Influence of Partial Wetting on the Results of Verification of Glass Hydrometers for Ethanol. *Meas. Tech.* **2017**, *60*, 92–95. [CrossRef]

77. Banoth, S.; Li, C.-W.; Hiratsuka, Y.; Kakehi, K. The Effect of Recrystallization on Creep Properties of Alloy IN939 Fabricated by Selective Laser Melting Process. *Metals* **2020**, *10*, 1016. [CrossRef]

78. Rafieazad, M.; Ghaffari, M.; Nemani, A.V.; Nasiri, A. Microstructural evolution and mechanical properties of a low-carbon low-alloy steel produced by wire arc additive manufacturing. *Int. J. Adv. Manuf. Technol.* **2019**, *105*, 2121–2134. [CrossRef]

79. Hung, C.H.; Sutton, A.; Li, Y.; Shen, Y.Y.; Tsai, H.L.; Leu, M.C. Enhanced mechanical properties for 304L stainless steel parts fabricated by laser-foil-printing additive manufacturing. *J. Manuf. Process.* **2019**, *45*, 438–446. [CrossRef]

80. Endo, H. Thermodynamic Consideration of the Cavitation Mechanism in Homogeneous Liquids. *J. Acoust. Soc. Am.* **1994**, *95*, 2409–2415. [CrossRef]

81. Wang, Q. Local energy of a bubble system and its loss due to acoustic radiation. *J. Fluid Mech.* **2016**, *797*, 201–230. [CrossRef]

82. Grigoriev, S.N.; Kozochkin, M.P.; Porvatov, A.N.; Volosova, M.A.; Okunkova, A.A. Electrical discharge machining of ceramic nanocomposites: Sublimation phenomena and adaptive control. *Heliyon* **2019**, *5*, e02629. [CrossRef] [PubMed]

83. Melnik, Y.A.; Kozochkin, M.P.; Porvatov, A.N.; Okunkova, A.A. On adaptive control for electrical discharge machining using vibroacoustic emission. *Technologies* **2018**, *6*, 96. [CrossRef]

84. Chen, Y.; Yang, M.; Long, J.; Hu, K.; Xu, D.; Blaabjerg, F. Analysis of oscillation frequency deviation in elastic coupling digital drive system and robust notch filter strategy. *IEEE Trans. Ind. Electron.* **2019**, *66*, 90–101. [CrossRef]

85. Andreeva, T.A.; Berkovich, A.E.; Bykov, N.Y.; Kozyrev, S.V.; Lukin, A.Y. High-Intensity Focused Ultrasound: Heating and Destruction of Biological Tissue. *Tech. Phys.* **2020**, *65*, 1455–1466. [CrossRef]

Publisher's Note: MDPI stays neutral with regard to jurisdictional claims in published maps and institutional affiliations.

MDPI

St. Alban-Anlage 66

4052 Basel

Switzerland

Tel. +41 61 683 77 34

Fax +41 61 302 89 18

www.mdpi.com

Metals Editorial Office

E-mail: metals@mdpi.com

www.mdpi.com/journal/metals